Advances in Behavior Genetics

Series editor

Yong-Kyu Kim

More information about this series at http://www.springer.com/series/10458

Briana N. Horwitz · Jenae M. Neiderhiser
Editors

Gene-Environment Interplay in Interpersonal Relationships across the Lifespan

 Springer

Editors
Briana N. Horwitz
Department of Psychology
California State University
Fullerton, CA
USA

Jenae M. Neiderhiser
Department of Psychology
The Pennsylvania State University
University Park, PA
USA

Advances in Behavior Genetics
ISBN 978-1-4939-2922-1   ISBN 978-1-4939-2923-8   (eBook)
DOI 10.1007/978-1-4939-2923-8

Library of Congress Control Number: 2015941110

Springer New York Heidelberg Dordrecht London

Printed on acid-free paper

Springer Science+Business Media LLC New York is part of Springer Science+Business Media
(www.springer.com)

Preface

A rapidly accumulating literature in recent years has focused on delineating how genes and environments work together to influence interpersonal relationships. These relationships include those within the family, like parent-child, marital, and sibling relationships and those outside the family, like peer and work relationships. Until fairly recently interpersonal relationships were considered purely environmental without consideration of the role of the individual in influencing his or her relationships via a variety of mechanisms including genes. Many investigators in the areas of interpersonal relationships and development now recognize the important role of the individual in shaping his or her interpersonal relationships and in helping to explain how those relationships impact development. This shift coincides with a rapidly growing body of work that has documented that interpersonal relationships are influenced at least in part by genetically-influenced characteristics of the individual.

In addition, these questions have been embedded within a lifespan developmental framework. Thus, great strides have been made in understanding pathways linking interpersonal relationships to individual adjustment across the lifespan. This volume will provide an overview of studies examining genetic and environmental influences on interpersonal relationships across the lifespan. We have identified experts within a variety of fields to describe behavioral genetic research focused on key developmental transitions and interpersonal relationships with a focus on gaps in the literature in order to identify important future directions of work in this area.

We thank the senior editor, Dr. Yong-Kyu Kim and each contributing author of this volume. Funding for this volume was provided by the National Institute on Aging (F32 AG039165).

Contents

1 Gene-Environment Interplay, Interpersonal Relationships,
and Development: A Volume Introduction 1
Briana N. Horwitz and Jenae M. Neiderhiser

2 Gene-Environment Interplay in Parenting Young Children 13
Jeffrey Henry, Michel Boivin and George Tarabulsy

3 Parenting in Childhood. 57
Alison Pike and Bonamy R. Oliver

4 The Sibling Relationship as a Source of Shared Environment 83
Shirley McGuire, Meenakshi Palaniappan and Taryn Larribas

5 Gene-Environment Transactions in Childhood
and Adolescence: Problematic Peer Relationships. 97
Mara Brendgen and Michel Boivin

6 Gene-Environment Processes in Adolescent Family Relationships . . . 131
Nan Chen and Kirby Deater-Deckard

7 Toward a Developmentally Sensitive
and Genetically Informed Perspective on Popularity 151
S. Alexandra Burt and M. Brent Donnellan

8 Spouse, Parent, and Co-workers: Relationships
and Roles During Adulthood 171
Erica L. Spotts and Jody M. Ganiban

9   Interpersonal Relationships in Late Adulthood   203
Carol E. Franz, Ruth Murray McKenzie, Ana Ramundo,
Eric Landrum and Afrand Shahroudi

10   The Family System as a Unit of Clinical Care:
The Role of Genetic Systems   241
David Reiss

Index   275

Editors and Contributors

About the Editors

Briana N. Horwitz Ph.D. is an assistant professor in the Department of Psychology at California State University, Fullerton. She completed her undergraduate studies at The University of California, Los Angeles and her doctoral studies at the University of California, Irvine. She also served as a postdoctoral fellow under the advisement of Dr. Neiderhiser at the Pennsylvania State University. Her research interests are in understanding the sources and directions of the associations among interpersonal relationships and functioning across the lifespan and more specifically to investigate how genetic and environmental factors explain these associations.

Jenae M. Neiderhiser Ph.D. is Liberal Arts Research Professor in the Department of Psychology and Professor of Human Development and Family Studies at The Pennsylvania State University. She completed her undergraduate studies in Psychology at The University of Pittsburgh and her doctoral studies in Human Development and Family Studies at The Pennsylvania State University. She is currently serving as Associate Editor for the *Journal of Research on Adolescence* and is on the editorial board of *Development and Psychopathology* and *Child Development*. Her research interests are in understanding the interplay between genes and environment throughout the lifespan. The environmental influences that she has examined most closely are interpersonal relationships sibling, and peer relationships-including parent-child, spouse, sibling, and peer relationships, and most recently she has included the study of prenatal enviornmental influences in her research. Examining how individuals influence their environments, in part because of their genetically-influenced characteristics, has long been a focus on her work.

Contributors

Michel Boivin École de psychologie, Pavillon Félix-Antoine-Savard, Université Laval, Quebec, Canada

Mara Brendgen Department de Psychologie, Université du Québec à Montréal, Montreal, Canada

S. Alexandra Burt Department of Psychology, Michigan State University, East Lansing, USA

Nan Chen Department of Psychology, Virginia Polytechnic Institute and State University, Blacksburg, VA, USA

Kirby Deater-Deckard Department of Psychology, Virginia Polytechnic Institute and State University, Blacksburg, VA, USA

M. Brent Donnellan Department of Psychology, Michigan State University, East Lansing, USA

Carol E. Franz Department of Psychiatry, University of California, San Diego, San Diego, USA; Center for Behavioral Genomics, Twin Research Laboratory, UCSD School of Medicine, La Jolla, CA, USA

Jody M. Ganiban Department of Psychology, The George Washington University, Washington, USA

Jeffrey Henry École de psychologie, Pavillon Félix-Antoine-Savard, Université Laval, Quebec, Canada

Briana N. Horwitz Department of Psychology, California State University, Fullerton, CA, USA

Eric Landrum Department of Psychiatry, University of California, San Diego, San Diego, USA

Taryn Larribas Los Angeles Unified School District, Los Angeles, USA

Shirley McGuire College of Arts and Sciences, University of San Francisco, San Francisco, CA, USA

Ruth Murray McKenzie Boston University, Boston, USA

Jenae M. Neiderhiser The Pennsylvania State University, Pennsylvania, USA

Bonamy R. Oliver School of Psychology, University of Sussex, Brighton, UK

Meenakshi Palaniappan University of Florida, Gainesville, USA

Alison Pike School of Psychology, University of Sussex, Brighton, UK

Ana Ramundo Department of Psychiatry, University of California, San Diego, San Diego, USA

David Reiss School of Medicine, Yale Child Study Center, Yale University, New Haven, CT, USA

Afrand Shahroudi Department of Psychiatry, University of California, San Diego, San Diego, USA

Erica L. Spotts Office of Behavioral and Social Sciences Research, National Institutes of Health, Bethesda, MD, USA

George Tarabulsy École de psychologie, Pavillon Félix-Antoine-Savard, Université Laval, Quebec, Canada

Chapter 1
Gene-Environment Interplay, Interpersonal Relationships, and Development: A Volume Introduction

Briana N. Horwitz and Jenae M. Neiderhiser

Interpersonal relationships are critical in shaping our development throughout the lifespan. For example, more warmth and support from parents is linked to better adjustment in children and adolescents and harsh or negative parenting to the development of behavioral and emotional problems (e.g., Demo and Cox 2000; Fletcher et al. 2004). As another example, marital quality is associated with spouses' adjustment, with marriages of better quality linked to wellbeing and poor quality marriages to more depressive symptoms (e.g., Beach et al. 1998; Proulx et al. 2007). A family systems perspective is often used to understand how family relationships shape development (Cox and Paley 1997). Within this framework, the family environment is a complex, hierarchically organized system that encompasses many subsystems, such as parent-child and marital relationships. These sub-systems are posited to be interconnected and permeable. As a consequence, the emotional and behavioral dynamics of one subsystem affects other subsystems and the adjustment of individual family members, in turn. For example, conflict arising from the marital relationships has been proposed to spillover over to the parent-child relationship and to cause subsequent adjustment problems in children (Margolin et al. 1996). Thus, family systems theory provides a framework for understanding how family subsystems serve as environmental influences that impact one another as well as individual development.

B.N. Horwitz (✉)
Department of Psychology, California State University, 6848,
Fullerton, CA 92834, USA
e-mail: bhorwitz@fullerton.edu

J.M. Neiderhiser
The Pennsylvania State University, Pennsylvania, USA

© Springer Science+Business Media New York 2015
B.N. Horwitz and J.M. Neiderhiser (eds.), *Gene-Environment Interplay in Interpersonal Relationships across the Lifespan*,
Advances in Behavior Genetics 3, DOI 10.1007/978-1-4939-2923-8_1

1

Consistent with family systems theory, the vast majority of work that has examined interpersonal relationships has considered them as "environmental" factors. For example, links between parenting behaviors and child outcomes have been thought to be due to the direct environmental effects of the parents' on the child. There have been a number of exceptions to this unidirectional thinking (e.g. Russell and Russell 1987) but only in the past two-to-three decades has the role of an individual's genes in their relationships with others been systematically considered. Beginning in the 1980's, however, genetically informed studies have examined the extent to which genetic and environmental factors account for variance in interpersonal relationships (Rowe 1981, 1983). When genetic influences on interpersonal relationships are identified, this is not an indication that individuals' genes directly impact the way that others treat them. Rather, these genetic influences indicate that individuals' heritable characteristics are operating, at least in part, to influence their relationships. For example, parents may parent a child who is temperamentally challenging differently than a child who is more easy-going. Because temperament is partially influenced by genes, the effect of this differential response is also partially due to the child's genes. Numerous studies have consistently shown that genetic influences account for variance in interpersonal relationships, although environmental factors often explain as much or more of the variance (see reviews by Horwitz et al. 2011; Horwitz and Neiderhiser 2011; Kendler and Baker 2007).

1.1 The Current Chapter

In the current chapter, we discuss how behavioral genetic studies help us to understand the role of both genes and environments in interpersonal relationships. We begin with a review of some of the basic definitions of behavioral genetics. Accumulating evidence suggests that the interplay of genes and environments, including gene-environment correlation (rGE) is involved in interpersonal relationships across the lifespan. Thus, we next describe how behavioral genetic studies assist in the interpretation of different types of rGE in interpersonal relationships. We then provide an overview of which, if any, type of rGE contributes to interpersonal relationships and associations between interpersonal relationships and individual development. Next, we describe how another form of gene-environment interplay – gene × environment interaction (G × E)—may be operating in interpersonal relationships and individual adjustment. The function of genes (e.g., gene expression and telomere length) in relation to interpersonal relationships across the lifespan is also addressed. Avenues for future research examining how gene-environment interplay operates to shape interpersonal relationships across the lifespan are then presented. Finally, we address how behavioral genetic studies can inform prevention and intervention research to improve interpersonal relationships and individual adjustment.

1.2 Basic Definitions

Twin studies take advantage of the variation in genetic relatedness of identical (MZ: share 100 % of their genes) and fraternal (DZ: share 50 % of their segregating genes, on average) twins to disentangle genetic and environmental influences on a wide range of constructs. Typically, behavioral genetic studies estimate genetic, shared environmental (non-genetic factors that make family members similar), and nonshared environmental influences (non-genetic factors that make family members different, including measurement error). This approach can be used to understand interpersonal relationships and associations between interpersonal relationships and individual adjustment. In children, shared environmental influences on interpersonal relationships suggest that the way twins are raised creates similarities in their relationships. In adults, the presence of shared environmental influences indicates that the way twins were raised as children or their current contact as adults contribute to their relationships. Nonshared environmental influences on interpersonal relationships represent the role of environmental factors that twins do not share with their co-twins (e.g., exposure to differential parental treatment or having different friends) in their relationships.

As described in the chapters throughout the current volume, a rapidly growing literature has suggested that gene-environment interplay plays an important role in interpersonal relationships and their links to individual adjustment across the lifespan. To date, genetically informed studies have focused on two broad forms of gene-environment interplay, including genotype-environment correlation (rGE) and gene × environment interaction (G × E). rGE refers simply to a correlation between genotype and environment (Plomin et al. 1977; Scarr and McCartney 1983). Gene × environment interaction is the result of a particular environment moderating the influence of genetic factors on a particular phenotype or genes moderating the influence of environments on the phenotype. The current volume will provide an overview of studies examining genetic and environmental influences on interpersonal relationships across the lifespan, with a focus on gaps in the literature in order to identify key future directions of work in this area. We have identified experts within a variety of fields to describe behavioral genetic research focused on interpersonal relationships and key developmental stages ranging from infancy, early childhood, middle childhood, adolescence, early and middle adulthood, and older adulthood.

1.3 Genotype-Environment Correlation

Behavioral genetic studies assist in the interpretation and understanding of genetic and environmental factors on interpersonal relationships (see review by Knafo and Jaffee 2013). When interpersonal relationships are explained, in part, by genetic influences, this suggests the presence of rGE. Three types of rGE have been typically described: passive, evocative, and active (Plomin et al. 1977). Passive *r*GE results when genes and environments are correlated because the parent provides both to the child. In the case of parenting, parents may pass on genes related to "difficult temperament" to their children. In the parents, these "difficult temperament" genes may be exhibited as irritable and negative parenting. Second, evocative *r*GE is the result of individuals evoking behaviors from the environment via their heritable characteristics. To illustrate, emotional reactivity in the child, a heritable characteristic of the child, may elicit more negative responses from his/her parent. As another example, an aggressive spouse may elicit more hostile responses from his/her partner, resulting in more marital conflict. Third, active *r*GE is the result of individuals selecting environments or interpersonal relationships that are correlated with their heritable characteristics. Active *r*GE can more easily explain how genetic influences contribute to interpersonal relationships that are selected. For example, individuals are generally high in positive emotionality may tend to select friends who also have higher levels of positive emotionality. Assortative mating is a specific type of active *r*GE that pertains to spouse selection, in which individuals may select spouses with similar heritable characteristics. The chapters in this volume help to clarify if and when these forms of *r*GE explain interpersonal relationships.

1.3.1 Interpersonal Relationships in Childhood

One of the most studied interpersonal relationships is that between parents and their children. Four of the chapters in this volume address the role of passive and/or evocative *r*GE in this relationship. Boivin et al. (2015) discuss pathways through which parents' behaviors towards their infants and parent-infant attachment patterns arise. The authors review research, which has shown that child-based genetic factors contribute to parents' negative behaviors towards their infants, but that environmental influences primarily explain parent-infant attachment patterns. The authors include in their review reports that have examined associations between specific genes and parent-infant attachment to more directly assess what genes may explain how *r*GE influences the parent-child relationship. As such, this chapter illustrates how parents' behaviors arise both as a reaction to infants' heritable characteristics and from environmental influences.

Pike and Oliver (2015) address the role of *r*GE in parenting across a range of child ages (from <1 year to 12 years), parenting behaviors (e.g., positivity and negativity), informants (parent vs. child vs. observation vs. interviewer), and designs (twin and adoption designs). The authors describe a consistent pattern of findings based on this literature, such that there is evidence of both passive and evocative *r*GE in parenting behaviors. Oliver and Pike further describe evidence that genetic influences also contribute to the covariance between parenting behaviors and child externalizing problems. This may indicate that parenting behaviors are at least partially a response to heritable externalizing problems in children.

Chen and Deater-Deckard (2015) further discuss the importance of genetic influences on adolescents' family subsystems, such as their parents' marital relationship, parenting, and family conflict. In this review, the authors describe evidence suggesting that passive *r*GE, evocative *r*GE, and environmental factors contribute to parenting behaviors towards adolescent children, but that these influences vary based upon the type of parenting and informant (i.e., mother, father, adolescent, or observational report). Together, the literature reviewed in these chapters helps to clarify how parenting behaviors and parent-child attachment are shaped in infancy, childhood, and adolescence.

Although parenting has been one of the most thoroughly studied family relationships in quantitative genetic research, other interpersonal relationships in childhood have also been examined, including sibling relationships. Sibling relationships have been found to be some of the most reciprocal relationships within the family. Specifically, sibling relationships have tended to show large shared environmental influences, especially during adolescence and when siblings report on themselves and their sibling/twin. McGuire et al. (2015) discuss results from studies that have focused on the role of genes and environments in the quality of sibling relationships. Based on consistent evidence across several studies, the authors conclude that sibling relationships are primarily attributable to shared environmental influences, with negligible influences from genetic factors. Thus, environments that siblings share, including their reciprocal relationships with each other, exert important influences on the quality of their relationship with one another.

Studies have also been focused on the extent to which genetic and environmental influences explain how children's relationships with peers are formed and maintained. As discussed within the chapter by Burt and Donnellan (2015), there is evidence of genetic variance in popularity with peers in middle childhood. That is, tendencies to be popular may partially be the result of children evoking positive responses from peers or selecting popular peer groups, via evocative and active *r*GE, respectively. While genetic influences on peer relationships indicate that children's heritable characteristics play a role in these relationships, they do not necessarily indicate what heritable characteristics, exactly, are involved or whether these processes vary by child sex. Thus, Burt and Donnellan (2015) further present novel data, which suggests that personality correlates of popularity, including positive emotionality and constraint, are moderated by sex of the individuals in emerging adulthood. Brendgen and Boivin (2015) review literature that has focused on the extent to which genetic and environmental influences account for variance in peer deviancy in childhood and adolescence. A main conclusion from this chapter is that peer deviancy is explained by environmental variance in middle childhood and genetic variance in adolescence. This may indicate that as children become more autonomous in adolescence, they tend to affiliate and select deviant peer groups based upon their own heritable characteristics through evocative and active *r*GE processes, whereas in middle childhood, environmental factors may play a stronger role.

In sum, several of the chapters included in this volume are concerned with understanding how both genetic and environmental factors influence children's interpersonal relationships. A number of conclusions may be drawn from these chapters. First, passive rGE, evocative rGE, and environmental influences contribute to the parent-child relationship, but the extent of contribution from these processes may vary depending on the particular facet of the parent-child relationship and the developmental stage of the child. Second, research has shown that the shared environment, which potentially suggests reciprocal influences of siblings on their relationship with one another, primarily influences children's sibling relationships. Third, genetic variance, suggesting evocative and/or active rGE contributes to children's popularity, but the association between personality traits (i.e., positive emotionality and constraint) and popularity is moderated by sex of the child in emerging adulthood. Fourth, contributions from genes and environments on peer deviancy may vary depending on child age, such that environmental factors influence peer deviancy in childhood, whereas genetic influences explain peer deviancy in adolescence. Thus, these chapters underscore that the importance of both genes and environments in interpersonal relationships across child development.

1.3.2 Interpersonal relationships in adulthood

Behavioral genetic studies have also helped us to understand how adults' interpersonal relationships are shaped. As discussed by Spotts and Ganiban (2015), many of these studies have focused on general aspects of marriage, including the propensity to marry, risk of divorce, remarriage, and marital instability, as well marital relationship quality. Spotts and Ganiban (2015) discuss that results typically indicate moderate genetic and substantial nonshared environmental influences on these marital constructs, with little evidence of shared environmental influences. Evidence of shared environmental influences on adults' interpersonal relationships indicate that environmental factors that make twins similar (e.g., the early rearing environment or current contact as adults) contribute to their relationships. In the case of adults' interpersonal relationships, shared environmental influences indicate that their relationships are shaped by the authors further present evidence which suggests that genetic influences also explain common variance among family relationships, including marital and parent-child relationship. This suggests that the same set of genes which explain the ways that parents behave towards or evoke behaviors from the spouses also influence their parenting behaviors. Finally, Spotts and Ganiban (2015) review work that suggests that genetic influences account for covariance between marital quality and adult mental health (e.g., depressive symptoms and positive mental health), though nonshared environmental influences account for a majority of this covariance. Thus, the body of work reviewed in this chapter illustrates that adults' interpersonal relationships are influenced by the way they evoke behaviors from others in their relationships, via rGE and by environments that are specific to them.

Franz and McKenzie (2015) describes research that has focused on interpersonal relationships in older adulthood. According to Franz, one of the focuses of genetically informed studies of interpersonal relationships in older adulthood has been to understand genetic and environmental variance in social support. Genetic influences have been found to play a key role in adults' social support; thus, individuals' heritable characteristics are involved in the degree of social support they perceive or experience. Additional literature, as reviewed in this chapter, has concentrated on the extent to which dyadic interactions and assortative mating explains spouse similarity for a range of phenotypes, including health, education, and cognitive abilities in older adults. Importantly, Franz and McKenzie (2015) points out that, as a whole, behavioral genetics research that is focused on interpersonal relationships in samples of older adults remains sparse. She also highlights key areas where more attention is needed. For example, she points out that genetically informed studies of bereavement often focus on younger adults and conclusions from these studies may not generalize to older adults for whom death of loved ones may be particularly salient. Thus, the chapter by Franz and McKenzie (2015) calls attention to the necessity for more genetically informed research that focuses on interpersonal relationships in older adulthood.

In sum, the chapters in this volume suggest that there is now a sizable literature that has examined genetic and environmental influences on interpersonal relationships at different key developmental stages across the lifespan. A clear conclusion across several of these chapters is that rGE plays an important role in shaping interpersonal relationships ranging from childhood to adulthood. Yet, it is also evident from the work reviewed in these chapters that environments also play key roles in interpersonal relationships across the lifespan. For example, shared environments primarily explain variance in children's sibling relationships and nonshared environments often account for a majority of the variance in adults' interpersonal relationships. Collectively, the research presented in this volume underscores the role of both genes and environments in interpersonal relationships at different stages in development are shaped.

1.4 Gene × Environment Interaction

Behavioral genetic studies have also examined the extent to which interpersonal relationships can influence individual adjustment through genotype × environment interaction (G × E). For example, research has addressed the extent to which differences in genetic, shared environmental, and nonshared environmental influences on child adjustment are moderated by varying levels of interpersonal relationship quality, typically described as G × E (e.g., Purcell 2002). Another strategy for examining G × E is to use an adoption design in which children are adopted at birth by genetically unrelated adoptive parents. If both adoptive families and birth parents are assessed, the direct environmental influences of the rearing environment (indexed by adoptive parents' behaviors and characteristics), genetic factors (indexed by biological parents' behaviors and characteristics), and their interactions on child adjustment can be examined. This design can be used to examine environmental moderation of genetic factors or genetic moderation of the environment.

Several of the chapters in this volume have addressed how G × E may operate in interpersonal relationships and individual development. For example, Henry et al. (2015) discuss the role of G × E in parenting and infant adjustment. Specifically, they discuss evidence from both cross-sectional and longitudinal adoption studies which suggest that adopted children at genetic risk for several developmental problems might inherit specific vulnerabilities that make them more sensitive to environmental risk factors (e.g., adoptive mothers' harshness or depression). Brendgen and Boivin further review literature which has examined contributions from G × E in adolescent adjustment, including the extent to which parenting behaviors and deviant peers moderate genetic influences on adolescent externalizing problems. The presence of G × E has also been investigated with respect to adults' interpersonal relationships and adjustment. Spotts and Ganiban (2015) describe research in which the magnitude of child-based genetic influences (i.e., evocative rGE) on parents' behaviors is moderated by varying levels of adolescent temperament. Thus, it is clear from the evidence presented in these chapters that G × E plays an important role in interpersonal relationships and adjustment across development.

1.5 Gene Function and Interpersonal Relationships

In the final chapter in this volume, Reiss (2015) discusses several contemporary approaches to increase our understanding of the function of genes (e.g., the role of specific genes, gene expression, and telomere biology) that are related to interpersonal relationships (Walum et al. 2008, 2012). For example, Reiss (2015) describes research to suggest that variation of the oxytocin receptor gene in adolescent girls is associated with social competence and the quality of bonding with romantic partners. Reiss (2015) also discusses that there has been an increased understanding of the influence of family relationships on mechanisms by which genes are expressed. Reiss (2015) highlights research that may suggest a change in what genes are expressed across toddlerhood (e.g., Forget-Dubois et al. 2007). For example, child-based genetic effects that evoked hostile and negative parenting at 5 months were uncorrelated with genetic factors that evoked the same parental behavior at 18 months (Forget-Dubois et al. 2007). Reiss (2015) further discusses research that has drawn links between interpersonal processes and telomere length. Telomeres – extended strands of non-replicating DNA at both ends of the chromosome that serve to protect genes during cell division. If telomeres shorten too much, chromosomal replication is impaired and leads to cell death. In one study reviewed by Reiss, caretakers, who as children, were exposed to adverse rearing environments show greater telomere shortening when as older adults they become caretakers of a disabled spouse (Kiecolt-Glaser et al. 2011). As a whole, Reiss' chapter suggests that gene function, including the role of specific genes, gene expression, and telomere length have important associations with interpersonal relationships.

1.6 Conclusion

In conclusion, the focus of this volume is to clarify the role of both genetic and environmental factors in interpersonal relationships and associations between interpersonal relationships and individual adjustment across the lifespan. A clear pattern is evident from the studies reviewed in this volume, such that rGE is operating to shape interpersonal relationships across development as well as the associations between interpersonal relationships and individual adjustment. Another conclusion from the work presented in this volume is that environmental influences (shared and nonshared) are critical for explaining variance in interpersonal relationships. Specifically, shared environmental influences are indicated during childhood and nonshared environmental influences are found during adulthood. Finally, this volume shows that G × E is involved in interpersonal relationships and adjustment in childhood and adulthood.

Yet another common theme throughout this volume is that there are several important areas where additional attention is needed to further clarify how interpersonal relationships are shaped across the lifespan. Here we suggest several broad avenues for future investigation and more future directions for specific developmental stages are presented in the subsequent chapters. Specifically, future research should continue to employ longitudinal investigations to delineate the reciprocal influences interpersonal relationships and individual adjustment over time. More studies are also needed to delineate how different types of rGE and direct environmental influences operate longitudinally. More studies are also needed to identify the specific characteristics of family members contributing to the development of interpersonal relationships of individuals contributing to the development of interpersonal processes and adjustment.

Additional research is needed to clarify how G × E operates in interpersonal relationships and individual adjustment. For example, an important future direction of G × E-focused studies is to understand how interpersonal relationships influence individual adjustment at different developmental stages. Another future direction is to explore different mechanisms through which environmental factors moderate genetic influences on development, such as the environmental context triggering negative genetic predispositions, compensating for negative genetic predispositions, preventing genetically influenced behaviors, and enhancing adaptation (Shanahan and Hofer 2005). Further, researchers have only begun to explore how rGE and G × E operate together to influence interpersonal relationships (e.g., Ulbricht et al. 2013), and more studies are needed to delineate how these complex processes influence different relationships and development. Finally, as raised in the chapter by Reiss (2015), a critical avenue for future research is to conduct investigations that seek to understand the function of genes (e.g., gene expression and telomere length) in interpersonal relationships and development.

Until fairly recently, research in behavioral genetics and in prevention science has been conducted independently, with little effort to use findings across areas to inform interpretation or design. There have been recent reviews and calls to action (e.g., Jaffee and Price 2007; Leve et al. 2010; Reiss and Leve 2007), as well as at least one empirical report (Brody et al. 2009), that have attempted to combine findings from genetic studies with prevention science as well as more general developmental research to advance our understanding of mechanisms. Research focused on understanding the interplay of genes and environments has the potential to inform prevention and intervention research by helping to better specify where to target interventions (e.g., Leve et al. 2010). Studies that have focused on how *r*GE operates to shape interpersonal relationships can provide directions for intervention research. In the case of parenting, for example, if passive *r*GE is found to explain the impact of a mother's positive parenting on her child's internalizing behavior, a strategy focused more on changing the child's behavior than on mother's parenting may be more effective. On the other hand, if evocative *r*GE is found to explain negative parenting, parent training, where a parent is taught to respond differently to her child, could be a viable approach to changing the child's adjustment problems. Similarly, if parents evoke negative responses from their spouses based upon the parents' heritable characteristics, marital training where spouses are taught to respond differently to one another should be considered. Studies that evaluate whether interpersonal relationships moderate genetic and environmental influences on adjustment via G × E can further specify where to target intervention efforts. For example, when genetic influences on child adjustment problems are greater in the context of higher levels of negative parenting, this would suggest that more intensive intervention is required for children who have both a genetic risk and an environmental risk (e.g., negative parenting), compared with children with who are only at genetic risk or only at environmental risk (Feinberg et al. 2007). At this point, using the findings of genetic research to direct intervention on an individual level is not feasible. We suggest, however, that using findings from genetically informed research can help to guide intervention strategies. Most importantly, genetic research, especially behavioral genetic research, underscores the need to consider individual differences and highlights how approaches that allow for differences among individuals are likely to be most effective.

References

Beach, R. H., Fincham, F. D., & Katz, J. (1998). Marital therapy in the treatment of depression: Toward a third generation of therapy and research. *Clinical Psychology Review, 18*, 635–661.

Boivin, J. H., Boivin, M., & Tarabulsy, G. (2015). Gene-environment interplay in parenting young children. In B. N. Horwitz & J. M. Neiderhiser (Eds.), *The behavioral genetics of interpersonal relationships in childhood and adulthood*. Thousand Oaks: Sage Publications.

Brendgen, M., & Boivin, M. (2015). Gene-environment transactions in childhood and adolescence: Problematic peer relationships. In B. N. Horwitz & J. M. Neiderhiser (Eds.), *The behavioral genetics of interpersonal relationships in childhood and adulthood*. Thousand Oaks: Sage Publications.

Brody, G. H., Beach, S. R. H., Philibert, R. A., Chen, Y., & Murry, V. M. (2009). Prevention effects moderate the association of 5-HTTLPR and youth risk behavior initiation: Gene × environment hypotheses tested via a randomized prevention design. *Child Development, 80*, 645–661.

Burt, A. S., & Donnellan, B. M. (2015). Toward a developmentally sensitive and genetically informed perspective on popularity. In B. N. Horwitz & J. M. Neiderhiser (Eds.), *The behavioral genetics of interpersonal relationships in childhood and adulthood*. Thousand Oaks: Sage Publications.

Chen, N., & Deater-Deckard, K. (2015). Gene-environment processes in adolescent family relationships. In B. N. Horwitz & J. M. Neiderhiser (Eds.), *The behavioral genetics of interpersonal relationships in childhood and adulthood*. Thousand Oaks: Sage Publications.

Cox, M. J., & Paley, B. (1997). Families as systems. *Annual Review of Psychology, 48*, 243–267.

Demo, D. H., & Cox, M. J. (2000). Families with young children: A review of research in the 1990's. *Journal of Marriage and the Family, 62*, 876–895.

Feinberg, M. E., Button, T. M., Neiderhiser, J. M., Reiss, D., & Hetherington, E. M. (2007). Parenting and adolescent antisocial behavior and depression: evidence of genotype × parenting environment interaction. *Archives of General Psychiatry, 64*, 457–465.

Fletcher, A. C., Steinberg, L., & Williams-Wheeler, M. (2004). Parental influences on adolescent problem behavior: Revisiting Stattin and Kerr. *Child Development, 75*, 781–796.

Forget-Dubois, N., Boivin, M., Dionne, G., Pierce, T., Tremblay, R. E., & Pérusse, D. (2007). A longitudinal twin study of the genetic and environmental etiology of maternal hostile-reactive behavior during infancy and toddlerhood. *Infant Behavior and Development, 30*, 453–465.

Franz, C., & McKenzie, R. M. (2015). Interpersonal relationships in late adulthood. In B. N. Horwitz & J. M. Neiderhiser (Eds.), *The behavioral genetics of interpersonal relationships in childhood and adulthood*. Thousand Oaks: Sage Publications.

Henry, J., Boivin, M., & Tarabulsy, G. (2015). Gene-environment interplay in parenting young children. In B. N. Horwitz & J. M. Neiderhiser (Eds.), *The behavioral genetics of interpersonal relationships in childhood and adulthood*. Thousand Oaks: Sage Publications.

Horwitz, B. N., & Neiderhiser, J. M. (2011). Gene-environment interplay, family relationships, and child adjustment. *Journal of Marriage and Family, 73*, 804–816.

Horwitz, B. N., Marceau, K., & Neiderhiser, J. M. (2011). Family relationship influences on development: What can we learn from genetic research? In K. Kendler, S. Jaffee, & D. Romer (Eds.), *The dynamic genome and mental health: The role of genes and environments in development* (pp. 128–144). New York: Oxford University Press.

Jaffee, S. R., & Price, T. S. (2007). Gene—environment correlations: A review of the evidence and implications for prevention of mental illness. *Molecular Psychiatry, 12*, 432–442.

Kendler, K. S., & Baker, J. H. (2007). Genetic influences on measures of the environment: A systematic review. *Psychological Medicine, 37*, 615–626.

Kiecolt-Glaser, J. K., Gouin, J. P., Weng, N. P., Malarkey, W. B., Beversdorf, D. Q., & Glaser, R. (2011). Childhood adversity heightens the impact of later-life caregiving stress on telomere length and inflammation. *Psychosomatic Medicine, 73*, 16–22.

Knafo, A., & Jaffee, S. R. (2013). Gene-environment correlation in developmental psychology. *Development and Psychopathology, 25*, 1–6.

Leve, L. D., Harold, G. T., Ge, X., Neiderhiser, J. M., & Patterson, G. (2010). Refining intervention targets in family-based research: Lessons from quantitative behavioral genetics. *Perspectives on Psychological Science, 5*, 516–526.

Margolin, G., Christensen, A., & John, R. (1996). The continuance and spillover of everyday tensions in distressed and nondistressed couples. *Journal of Family Psychology, 10*, 304–321.

McGuire, S., Palaniappan, M., & Larribas, T. (2015). The sibling relationship as a source of shared environment. In B. N. Horwitz & J. M. Neiderhiser (Eds.), *The behavioral genetics of interpersonal relationships in childhood and adulthood*. Thousand Oaks: Sage Publications.

Pike, A., & Oliver, B. R. (2015). Parenting in childhood. In B. N. Horwitz & J. M. Neiderhiser (Eds.), *The behavioral genetics of interpersonal relationships in childhood and adulthood*. Thousand Oaks: Sage Publications.

Plomin, R., DeFries, J. C., & Loehlin, J. C. (1977). Genotype—environment interactionand correlation in the analysis of human behavior. *Psychological Bulletin, 84*, 309–322.

Proulx, C. M., Helms, H. M., & Buehler, C. (2007). Marital quality and personal well-being: A meta-analysis. *Journal of Marriage and Family, 69*, 576–593.

Purcell, S. (2002). Variance components models for gene—environment interaction in twin analysis. *Twin Research, 5*, 554–571.

Reiss, D. (2015). The family system as a unit of clinical care: The role of genetic systems. In B. N. Horwitz & J. M. Neiderhiser (Eds.), *The behavioral genetics of interpersonal relationships in childhood and adulthood*. Thousand Oaks: Sage Publications.

Reiss, D., & Leve, L. D. (2007). Genetic expression outside the skin: Clues to mechanisms of genotype × environment interaction. *Development and Psychopathology, 19*, 1005–1027.

Rowe, D. C. (1981). Environmental and genetic influences on dimensions of perceived parenting: A twin study. *Developmental Psychology, 17*, 203–208.

Rowe, D. C. (1983). A biometrical analysis of perceptions of family environment: A study of twin and singleton sibling kinships. *Child Development, 54*, 416–423.

Russell, G., & Russell, A. (1987). Mother–child and father–child relationships in middle childhood. *Child Development, 58*, 1573–1585.

Scarr, S., & McCartney, K. (1983). How people make their own environments: A theory of genotype→environment effects. *Child development*, 424–435.

Shanahan, M. J., & Hofer, S. M. (2005). Social context in gene—environment interactions: Retrospect and prospect. *Journals of Gerontology: Series B, 60B*, 65–76.

Spotts, E. L., & Ganiban, J. (2015). Spouse, parenting, and co-workers: Relationships and roles during adulthood. In B. N. Horwitz & J. M. Neiderhiser (Eds.), *The behavioral genetics of interpersonal relationships in childhood and adulthood*. Thousand Oaks: Sage Publications.

Ulbricht, J. A., Ganiban, J. A., Button, T. M. M., Feinberg, M., Reiss, D., & Neiderhiser, J. M. (2013). Marital adjustment as a moderator for genetic and environmental influences on parenting. *Journal of Family Psychology, 27*, 43–52.

Walum, H., Lichtenstein, P., Neiderhiser, J. M., Reiss, D., Ganiban, J. M., Spotts, E. L., et al. (2012). Variation in the oxytocin receptor gene in associated with pair-bonding and social behavior. *Biological Psychiatry, 71*, 419–426.

Walum, H., Westberg, L., Henningsson, S., Neiderhiser, J. M., Reiss, D., Igl, W., et al. (2008). Genetic variation in the vasopressin receptor 1a gene (AVPR1A) associates with pair-bonding behavior in humans. *Proceedings of the National Academy of Sciences of the United States of America, 105*, 14153–14156.

Chapter 2
Gene-Environment Interplay in Parenting Young Children

Jeffrey Henry, Michel Boivin and George Tarabulsy

Parental behaviors are generally perceived as a cornerstone of socio-emotional development in infancy and early childhood (Boivin et al. 2005). For instance, parental sensitivity or sensitive responsiveness—the caregiver's ability to detect the infant's needs and respond to them fittingly—has long been posited to contribute to a secure parent-child attachment relationship, thus creating a positive context for the child's later socio-emotional adjustment (Bowlby 1982; Bretherton and Waters 1985; De Wolff and van IJzendoorn 1997). Conversely, insensitive interactive behavior marked by inconsistencies in parental responses and a tendency to adopt hostile, restrictive and punitive behaviors toward the child, has been associated with the development of insecure attachment and later externalizing problems (Boivin et al. 2005; Bradley and Corwyn 2007; Campbell et al. 2007; Garai et al. 2009). Over-soliciting parental practices, including insensitive care, over-responsiveness, overprotection, and intrusion also predict concurrent or future internalizing problems in the child, such as social phobia, depression, and agoraphobia (Becker et al. 2010; Gray et al. 2011; Kim 2011; Lieb et al. 2000; Nishikawa et al. 2010; Spokas and Heimberg 2009).

Thus, many aspects of early parental behavior appear to be involved in the infant's socio-emotional development, and specific practices, especially those involving punishment, overprotection and lack of sensitivity, have been associated with attachment disorganization, as well as externalizing and internalizing psychopathology. It has been suggested, however, that rather than being a reflection of

J. Henry · M. Boivin (✉) · G. Tarabulsy
École de psychologie, Pavillon Félix-Antoine-Savard, Université Laval,
2325, rue des Bibliothèques, Quebec G1V 0A6, Canada
e-mail: Michel.Boivin@psy.ulaval.ca

© Springer Science+Business Media New York 2015
B.N. Horwitz and J.M. Neiderhiser (eds.), *Gene-Environment Interplay
in Interpersonal Relationships across the Lifespan,*
Advances in Behavior Genetics 3, DOI 10.1007/978-1-4939-2923-8_2

environmental contributions on socioemotional development, these associations could result from shared genetic vulnerabilities between parent and infant (Collins et al. 2000). In such a scenario, genetic contributions would be confounding factors and the parenting behaviors would mediate the relation between parental genotype and infant development. For instance, short-tempered parents that tend to display insensitive, punitive parenting behaviors may have short-tempered children that tend to show irritable and disorganized behaviors. While it is possible that the parenting behavior led to infant developmental characteristics, it is also possible that both behavior patterns are attributable to genetic similarities.

Most models of early development have argued that parenting is rooted in a complex social system where parental characteristics, contextual factors and child characteristics interact over time (Belsky 1984; Belsky and Jaffee 2006; Vondra et al. 2005). Parental personality and life experience may influence their early interaction with their infant, often in complex ways through their interface with immediate environmental conditions (Bornstein 2002). For example, teen parenthood (Morley et al. 2011) and parent mental health (Atkinson et al. 1999; Tarabulsy et al. 2008) have been linked with adverse parenting behaviors. Contextual stressors, such as economic hardship and stressful life experiences (e.g., domestic violence) may also add to the burden of parenting (Lee et al. 2011; McConnell et al. 2011; McLoyd 1998). There is also empirical evidence to suggest that a child's early behavioral characteristics predict parental behaviors and perceptions (e.g., Caspi et al. 2004; Jaffee et al. 2004a; Kiff et al. 2011; Putnam et al. 2002), although there is disagreement about the importance and meaning of these potential "child effects" (Collins et al. 2000; Dodge 1990).

It is clear that a comprehensive study of early parenting should consider a variety of factors related to the infant, the parent and the family context as they interact over time, as well as possible bidirectional associations between them. However, the understanding of such a complex developmental system is limited by an overreliance on studies based on simple correlational designs, even when the latter are longitudinal (i.e., predictive). Given that experimental manipulations other than randomized interventions (which imply important logistic and financial challenges) are often precluded for ethical and legal considerations, behavior-genetic designs are useful for testing hypotheses regarding environmental-developmental processes. Although behavior-genetics designs, also being correlational by nature, cannot provide clear-cut demonstrations of causality, they can, by statistically disentangling genetic from non-genetic sources of inter-individual variance, test hypotheses regarding the environmental nature and direction of the association between child behavior and parenting. For instance, they may help evaluate the extent to which a child's genetic predispositions account for caregiver involvement in specific parental practices. They may also test theoretical models regarding the complex gene-environment processes underlying the predictive association between parental behavior and subsequent child outcomes.

Behavior-genetics studies on early parenting have occupied a growing portion of the recent child development literature. In the past decade, studies using twin, adoption, step-family and linkage (i.e., molecular) designs have provided important new evidence regarding early parenting and the nature of its association with child socio-emotional development. The goal of this chapter is to review this emerging evidence. In this chapter, we first introduce the reader to attachment theory, a dominant figure of the theoretical landscape regarding early parenting. This theoretical framework will serve as a starting point to posit specific empirical questions relating to the developmental role of early parenting in child development. We then provide a review of extant empirical evidence from behavioral-genetics studies on early parenting and infant socio-emotional development, and discuss its significance for our understanding of the developmental role of early parenting. We conclude with a discussion of methodological concerns and future directions regarding behavior-genetics literature on early parenting.

2.1 Attachment Theory: A Theoretical Framework

Attachment theory, as conceived within an evolutionary framework, posits that the human infant is born with a set of innate neurological and behavioral systems selected for increasing the chances of survival (Bowlby 1982). These mechanisms are deployed within a dyadic regulatory system where associated bonding behaviors are normally activated by impending or perceived external danger and by states of internal distress (i.e., illness, fatigue); and there is a predetermined selectivity of caregivers (i.e., personal relationships) that may provide protection and soothing to the infant. An important theoretical feature is that attachment relationships are formed during the course of interactions with caregivers. Infants gather information on the reliability of caregiver responses in different interactive circumstances, especially in situations where they are alarmed, and by the end of the first year of life, specific representations are formed regarding the caregivers, the self and the nature of interpersonal relationships. These representations influence children's attachment patterns concurrently, and serve as blueprints for the manner in which they will initiate subsequent social relationships (Ainsworth 1985).

2.2 Attachment and Environmental Causation

Bowlby's attachment theory aims to explain individual differences in attachment patterns by individual variations in caregiver behavior. The assumption of an environmental mediation of attachment is grounded in early empirical evidence linking observed sensitive and responsive caregiving behavior at home and characteristic secure attachment behavior patterns in the laboratory-based Strange Situation (SS; Ainsworth et al. 1978). Although many subsequent studies confirmed a significant link between early care and attachment, they varied greatly in the estimated strength of the relationship. De Wolff and van IJzendoorn (1997) reviewed 66 studies to evaluate the mean association between caregiver sensitivity and attachment. They showed that sensitivity, however measured, was far from being an exclusive determinant of the quality of attachment (effect sizes ranged from 0.24 to 0.32). In fact, several other characteristics of parental interactive behavior were identified as playing an equally important role, including mutuality, synchrony, positive attitude, emotional support, and stimulation (De Wolff and van IJzendoorn 1997). Thus, while the theoretical importance attributed to sensitivity as a precursor of attachment is warranted, empirical evidence suggests that other factors may well be involved in the elaboration of this first relationship.

In addition, demographic risk factors, especially if accumulated (e.g., Cummings and Davies 2002; Nair and Murray 2005), and the psychological health of parents, especially mothers (e.g., Martins and Gaffan 2000; Murray 1992), may be involved the development of attachment, presumably through their proximal or distal influence on early parenting. A recent review of attachment-related findings found that family SES and maternal depression, along with maternal sensitivity, were both independent predictors of specific patterns of attachment (Campbell et al. 2004; McElwain and Booth-LaForce 2006). Again, these results underline that sensitive parenting may not be the only parental factor involved in child development.

Much of the above-cited empirical evidence points toward determinants other than Bowlby's (1982) sensitivity-focused, proximal environmental causation hypothesis. However, it is not clear whether these determinants operate through environmentally mediated processes. Behavior-genetics designs may help to more thoroughly put to test the central assumptions of attachment theory regarding the developmental role of early parental practices, including but not limited to sensitivity. A first assumption is that although attachment relationships are formed in the course of interactions with significant caregivers, children's behavior are to a significant extent influenced by sensitive responsiveness of caregivers, over and above initial "child effects". From a broader, behavior-genetic standpoint, this claim can be verified, for instance, by determining the extent to which genetic and environmental factors respectively account (1) for parental behaviors, and then, (2) for the association between early child characteristics and later parental behaviors, thereby isolating "pure" environmental variance from genetically mediated "child effects" on parenting. A second assumption is that the quality of the relationship with the caregiver influences children's behavior contemporaneously and subsequently. Again, this can be verified by determining the extent to which parental behavior is associated with various child developmental outcomes, and whether this association is mediated by genetic factors or not. This question can also be extended to examining the possible moderating effect of child or parent genetic vulnerability (i.e., gene-environment interaction) qualifying that association. These forms of G-E analyses are made possible when genetic factors are assessed directly (i.e., measured genes) or indirectly (through twin or adoption designs). They help delineate the genetic and environmental architecture of caregiving contribution to child behavior. Specifically, three empirical questions may be analyzed regarding the theoretical claims of attachment theory on the developmental role of early parenting: (1) To what extent is parent-infant attachment environmentally mediated? (2) Is early parenting a "pure" environmental factor? (3) What type of gene-environment interplay is involved in early parenting?

2.3 To What Extent Is Parent-Infant Attachment Environmentally Mediated?

Most univariate genetically sensitive studies of parent-infant attachment have focused on observed parent-infant interactions. Six twin studies have estimated the genetic and environmental contributions to parent-infant interaction quality, as assessed through child-, parent- or child-parent-focused observational procedures. These six studies typically used a univariate ACE approach where genetic (A), shared environmental (C) and non-shared environmental (E) sources of interindividual differences on a given measure of parent-infant interaction quality were assessed.

Adoption studies, which typically examine the correspondence in attachment patterns in biologically related and unrelated infant-parent dyads, also provide insight on the possible contribution of shared genes on infant attachment. As adoptive parents and adopted children do not, in principle, share genes by common descent, significant associations between parent and infant attachment-related experience in adoption contexts may arise as a consequence of dyadic interaction histories (Verissimo and Salvaterra 2006). Three such adoption studies examined similarities in attachment among biologically unrelated mother–infant dyads.

Finally, another method used to assess the genetic and environmental contributions to infant attachment patterns is to examine potential independent contributions of functional variants in genotype. Seven such linkage studies have estimated the association between specific genotype variations and infant attachment disorganization.

2.4 Twin Studies of Infant Attachment

The earliest twin studies to examine potential genetic contributions to early attachment patterns used a modified version of the SS, the Louisville Twin Study Procedure (LTS; Matheny et al. 1984), as an index of parent-infant attachment. Similar to the SS, the LTS procedure implies having each twin experience two separations and two reunions with the mother. In a first study, videotapes of 34 MZ pairs and 26 DZ 18–24-month pairs at ages 18 and 24 months were rated using the LTS procedure (Finkel et al. 1998). MZ concordance for attachment was 67.6 %, significantly greater than the DZ concordance of 38.5 %, thus suggesting significant heritability (Finkel et al. 1998). Such findings are in contrast to those from an early quantitative review of parent-infant attachment data on twins

(Ricciuti 1993), which found high concordance in both MZ and DZ pairs, suggesting little genetic mediation. Another report using archival data from a sample of 99 MZ pairs and 108 DZ pairs (mean age = 24 months) from the Finkel et al. (1998) study found an MZ concordance for attachment of 62.6 %, significantly greater than the 44.4 % DZ concordance, with 25 % of the variability in attachment attributable to genetic factors, and the remaining 75 % attributable to non-shared environment (or measurement error; Finkel and Matheny 2000).

Two twin studies assessed parent-infant attachment using the SS (Ainsworth et al. 1978), a well-known seven-episode procedure designed to assess how a child uses the parent as a secure base for exploration. In a first study, 110 42–45-month twin pairs were coded using conventional four-way classifications and a continuous measure of attachment security (O'Connor and Croft 2000). Intraclass correlations were equally high in MZ and DZ twin pairs, suggesting little genetic contribution. Intraclass correlations on the continuous measure of attachment security were 0.48 and 0.38 for MZ and DZ pairs respectively, also suggesting modest heritability. However, contrary to Finkel and Matheny (2000), significant contributions of shared and non-shared environment were found (O'Connor and Croft 2000).

A second study assessed 57 MZ and 81 DZ 12–14-month twin pairs using the SS (Bokhorst et al. 2003). For secure/non-secure attachment classification, 52 % of the variance was accounted for by shared environment, leaving 48 % to non-shared environment. Non-shared environment, which includes measurement error, mainly explained the variance in organized/disorganized classification. Differences in temperamental reactivity, which were heritable at 77 %, were not associated with attachment concordance (Bokhorst et al. 2003).

In a first twin study on *infant-father* attachment, mothers of 14–16-month MZ ($N = 21$) and DZ ($N = 91$) twin pairs sorted the Attachment Q-Sort (AQS; Vaughn and Waters 1990) with a focus on the infant's behaviors in the presence of the father (Bakermans-Kranenburg et al. 2004). Attachment security was explained by shared (59 %) and non-shared environment (41 %), and it was uncorrelated with infant dependency (a contrasting construct assessed with the AQS), which was heritable at 66 % (Bakermans-Kranenburg et al. 2004).

Taken together, findings from twin studies of parent-infant interactive behavior suggest that early secure base behavior towards the mother *and* the father is mainly a function of environments that differ between families (i.e., parenting, family environment). However, two studies using a modified version of the SS found genetic (Finkel et al. 1998) and non-shared environmental mediation (Finkel and Matheny 2000). Unfortunately, these different results could not be replicated, as no other study employed this specific procedure.

Several twin studies assessed constructs neighboring to parent-infant attachment. Two combined studies assessed observed parent-child dyadic mutuality (i.e., shared positive affect, responsiveness, and cooperation; Deater-Deckard and O'Connor 2000). The first study included 62 MZ twin pairs and 58 DZ twin pairs (mean age = 3 years). Heritability and non-shared environment each accounted for half of the variance in mother–child dyadic mutuality, with an non-significant shared environmental contribution. These findings were replicated in a second observational study of 102 pairs of adoptive and biological siblings in matched comparison families (Deater-Deckard and O'Connor 2000). Using the same observational procedures and measures from videotaped interactions of 3-year children with their parents, full siblings (who share 50 % of their genes as do fraternal twins) were found to correlate at the same level as fraternal twins in the first study, indicating environmental mediation of parent-child mutuality (Deater-Deckard and O'Connor 2000). Another study (Roisman and Fraley 2006) assessed infant-mother relationship quality using the Nursing Child Assessment Teaching Scale (NCATS; Summer and Spietz 1995). In this procedure, the infant's primary caregiver is asked to teach the target child a task just beyond his capacity. A total infant-caregiver score is derived by summing all items within a single indicator characterizing the degree to which parent and child fruitfully employ a "teaching loop", whereby the primary caregiver (a) is observed to properly alert the child, thereby setting up expectations, (b) effectively instructs the child by making suggestions, asking questions, etc., (c) provides time for the child to respond to the instruction and (d) offers adequate and sensitive feedback to the child, and the target infant (e) sends clear cues to the caregiver and (f) is appropriately responsive to caregiver cues. Data was collected on 127 MZ and 333 DZ 9-month twin pairs. Genetic variation was non-significant and the shared and non-shared environmental contributions were substantial in accounting for the infant-caregiver relationship quality (Roisman and Fraley 2006).

Findings from twin studies of parent-infant interactions have been remarkably consistent. In most cases, the estimated genetic contribution to parent-infant attachment security was modest or close to zero. Also, there was evidence in most reports of substantial shared and non-shared environmental contributions to parent-infant attachment security and disorganization, respectively. Inversely, studies using modified versions of well-validated paradigms (Finkel et al. 1998; Finkel and Matheny 2000) found genetic and non-shared environmental (or measurement error) contributions to the parent-infant attachment security.

Despite this notable consistency, several features of the samples of the above-mentioned studies preclude any firm and definitive conclusion regarding genetic-environmental mediation of parent-infant attachment. First, one of the major limitations of this work is its modest-sized samples of unknown representativeness, which may bias the ACE estimates in an unknown way (ACE estimates always depend on the accessible variance of each component in the sample), as well as limit the external validity of results. Second, a huge part of the published reports are based on the same sample or subsamples thereof, albeit with different focus (i.e., mother–child security, father–child security; Roisman and Fraley 2008), which may partly account for the relative uniformity of results across studies. Moreover, the nature of the observational procedures used may have led to a programming effect of context, which might partly explain the shared environmental findings. As context may drive human behavior, especially in stressful settings, a procedure whereby the mother is subjected within a short period of time to the same stressful interactive task is likely to yield a stream of similar behaviors regardless of who this person interacts with. This could lead to an overestimation of shared environment linked to the context of the observational settings, such as in the SS (the AQS, being a home-visit procedure, is considered more naturalistic). Yet, such settings may also provide the opportunity for understanding the diversity of maternal behavior by giving context for interpretation. For instance, observation of micro-processes within mother–infant interactions often leads coders to recognize the considerable variability of mother responses to infant outstretched arms, as well as the variability of infant signals that may trigger the same response in two mothers.

Finally, most recent work in this area moved beyond univariate ACE models to examine the G-E etiology of the covariance between parent-infant attachment and measured parenting. This particular type of bivariate analysis is especially important in that it may—and does, as will be discussed—provide insight on possible child effects involved in the process, while at the same time providing evidence that some mechanisms underlying the predictive significance of parenting for several developmental outcomes are non-genetic in origin.

2.5 Adoption Studies of Infant Attachment

A first adoption study investigated the concordance between foster mothers' attachment state of mind (i.e., quality of mother's processing of thoughts and feelings regarding her own attachment experience with her child, as assessed through a process of discourse analysis; Main and Goldwyn 1998) and foster infants' attachment quality in a sample of 50 mother–infant (12–24 months) dyads (Dozier et al. 2001). The correspondence between maternal state of mind and infant attachment quality was 72 %, similar to the level seen among biologically intact mother–infant dyads, thus pointing toward a non-genetic mechanism in the inter-generational transmission of attachment (Dozier et al. 2001). A second study examined whether the adoptive mother's internal attachment representation predicted infant attachment security in 106 mother–infant dyads at 3 years (Verissimo and Salvaterra 2006). Scores reflecting the presence and quality of maternal secure base scripts concurrently predicted infant security according to the AQS (Verissimo and Salvaterra 2006). A third study tested associations between maternal state of mind regarding attachment upon their adopted infant and emotional themes appearing in doll play narratives obtained from their recently adopted and previously maltreated 4–8-year children (Steele et al. 2003). Significant associations were found between maternal state of mind and infants' story-completions. Specifically, mothers judged insecure (dismissing or preoccupied) tended to have adopted children who, 3 months after placement, provided story-completions with higher levels of aggressiveness as compared to the stories provided by children adopted by secure-autonomous mothers. Moreover, children whose adoptive mothers displayed unresolved mourning regarding past loss or trauma provided story completions with higher scores for emotional themes such as 'parent appearing child-like' and 'throwing out or throwing away' (Steele et al. 2003). Thus, three adoption studies of attachment-related experience in adopted (i.e., biologically unrelated) mother–infant dyads bring additional support to the notion that intergenerational transmission of attachment-related experience involves non-genetic (i.e., most likely environmental) processes.

2.6 Linkage Studies of Infant Attachment

Another exciting avenue for research on genetics of infant attachment is provided by molecular genetics. Once identified, measured genes can be incorporated into regular studies and analyzed together with other measured variables, such as attachment. Evidence so far has linked infant disorganization with the DRD4, 5-HTT and OXTR gene polymorphisms.

The DRD4 gene polymorphism has been linked to novelty seeking, pathological impulsive and compulsive behavior in adults (Benjamin et al. 1996; Comings et al. 1999), as well as ADHD in children (Faraone et al. 2005); most of these phenotypes potentially having in common impairments of regulation of emotional arousal and a possible dysregulation of the reward system. Functional variants of the DRD4 gene polymorphism may thus be involved in infant engagement and activity level during interactions with caregivers, thus eliciting adaptive or maladaptive responses from the caregiving environment (Mills-Koonce et al. 2007). Therefore, significant associations between DRD4 variations and infant attachment disorganization (i.e., the lack of coherent behavioral strategy to cope with social stresses) may be plausible on theoretical grounds.

The earliest linkage study of infant attachment assessed attachment disorganization of a low-social-risk sample of 90 20-month infants with the SS (Lakatos et al. 2000). A significant association was found between the DRD4 7-repeat (i.e., long allele) and attachment disorganization: the DRD4 7-repeat was significantly more frequent in disorganized infants than in non-disorganized infants (Lakatos et al. 2000). The estimated relative risk for disorganized attachment among carriers of the DRD4 7-repeat was fourfold (Lakatos et al. 2000). The authors extended these findings by genotyping the same infants for the functional -521 C/T single nucleotide polymorphism in the upstream regulatory region of the DRD4 gene, in order to test the association with disorganization both alone and in interaction with the DRD4 7-repeat (Lakatos et al. 2002). There was a significant interaction between the DRD4 7-repeat and the 521 C/T promoter polymorphism, the odds ratio for disorganized attachment increasing tenfold in the presence of both risk alleles (Lakatos et al. 2002). Findings from these two reports indicate that the DRD4 7-repeat promotes independent genetic risk for infant disorganization. This risk may be amplified if an individual also possesses the 521 C/T promoter polymorphism risk factor.

In another report using the same sample, observed response to a novel, anxiety-provoking stimulus (i.e., SS) was investigated for 90 12-month infants genotyped for the DRD4 7-repeat and for 5-HTTLPR risk alleles (Lakatos et al. 2003). Combined genotype contributions were found: infants with at least one copy of both the DRD4 7-repeat and the long variant of 5-HTTLPR responded with less anxiety than other infants. Inversely, infants with the DRD4 7-repeat and who were homozygous for the short form of 5-HTTLPR showed more anxiety and resistance to the stranger's initiation of interaction (Lakatos et al. 2003).

A replication of the first two Lakatos studies (2000, 2002) in a larger sample did not confirm the contribution of the DRD4 7-repeat and the -521 C/T promoter gene on disorganized attachment (Bakermans-Kranenburg and van IJzendoorn 2004). Although the sample used was larger ($N = 132$; mean age $= 50$ months), which may have enhanced the power to find significant DRD4-C/T interactions, the association was not found. Even when the authors combined their sample with the Lakatos sample, the interaction of the DRD4 and -521 C/T polymorphisms on disorganization was not confirmed (Bakermans-Kranenburg and van IJzendoorn 2004). These results contradict those obtained by Lakatos et al., which indicated independent contribution of the DRD4 7-repeat and interaction with the -521 C/T polymorphism.

Following up the results of their previous population linkage studies, the Lakatos group performed extended transmission disequilibrium tests (ETDT) on the same sample to determine whether biased transmission of the DRD4 7-repeat occurred to infants displaying disorganized and secure attachment behavior with their mothers (Gervai et al. 2005). A trend for preferential transmission of the DRD4 7-repeat to disorganized infants and a significant non-transmission of this allele to secure infants were observed, suggesting that results from the Lakatos et al. studies were not due to population stratification (Gervai et al. 2005).

Finally, a recent investigation examined the oxytocin receptor OXTR gene, purportedly involved in social stress regulation, as a possible source of variation in infant attachment, as assessed by the SS (Chen et al. 2011). In a sample of 176 12–16-month infants, the A allele of OXTR rs2254298 was significantly associated with attachment security, but only in non-Caucasian infants (Chen et al. 2011).

Findings from linkage studies of infant attachment have been rather inconclusive. In the earliest studies (Lakatos et al. 2000, 2002), significant links were found between infant disorganization and the DRD4 7-repeat. These findings were later extended using ETDT (Gervai et al. 2005). However, one study using a larger sample, even when combining its own sample with the sample from the Lakatos studies (Bakermans-Kranenburg and van IJzendoorn 2004), could not replicate such findings. There may be multiple reasons for non-replication of such association studies (e.g., low statistical power, population stratification). The primary cause of non-replications may lie in the nature of polygenic inheritance (Comings 1998). Complex human behaviors, such as parent-attachment, are likely accounted for by multiple genes (and functional variants) each individually having a small effect. There may be an inverse relation between the odds of finding significant independent genetic contributions to a human trait, and the sophistication of the trait of interest (Turkheimer and Waldron 2000). The possible genetic heterogeneity across populations (i.e., specific combinations in genotype being more prevalent in some populations than in others) may also explain the lack of replication. Negative findings may also reflect the low power of individual studies that, when combined in a meta-analysis, may yield a significant albeit small effect.

In brief, findings from linkage studies remain equivocal regarding the importance of the DRD4 7-repeat and its association with infant disorganization. Yet, DRD4 variations have previously been linked to high reward-dependence, low behavioral inhibition and low self-regulation traits or conditions. These associations may be plausible on theoretical grounds, as infant activity level and capacities of regulation of emotional arousal could have—through possible "child effects"—a proximal or distal effect on early parenting. Subsequently, early adaptive or maladaptive interactions between the parent and the infant may contribute to the development of infant secure base behavior (Bowlby 1982). Conclusive demonstration of such assumptions within behavioral-genetics designs is however still lacking.

With only a few exceptions, findings from univariate twin and adoption studies generally suggest that parent-infant attachment is mainly environmentally mediated, with shared environment prevailing in explaining variance on attachment security (e.g., Roisman and Fraley 2006), and non-shared environment (or measurement error) explaining most of the variance on attachment disorganization (Bokhorst et al. 2003), which raises issue of the validity of these measures to assess disorganization. Moreover, most findings from linkage studies indicate a significant contribution of specific genes on infant disorganization (e.g., Lakatos et al. 2000, 2002), although a more powerful study yielded negative findings (Bakermans-Kranenburg and van IJzendoorn 2004). Therefore, no clear-cut conclusion can be drawn regarding the possible links between infant genotype and early attachment phenotype.

In brief, results from twin and linkage studies generally suggest little genetic influence on infant attachment behavior. The possibility of genes interacting with putative environmental risk factors still remains. As will be discussed in further sections, G-E processes involved in early parenting and infant attachment were investigated in numerous bivariate studies.

2.7 Is Early Parenting a "Pure" Environmental Factor?

Analyzing parenting in a behavior-genetic design allows for the assessment of the G-E etiology of parental behaviors. In univariate ACE twin designs, finding significant heritability for a parental behavior may be indirect evidence of "child effects" on parenting. It is a first step in assessing possible G-E correlation involving parenting behaviors, as will be further discussed. Inversely, finding significant associations between maternal genotype and parental behaviors tend to mitigate the importance of these child effects, as will also be discussed. Four univariate ACE twin studies estimated the G-E etiology of parental perceptions and self-reported styles, one adoption study assessed home environment in biologically related and unrelated siblings, and three linkage studies estimated the association between maternal sensitivity and maternal genotype.

2.8 Twin Studies of Early Parenting

Four reports have documented the genetic and environmental contributions to parents' self-reported perceptions and behaviors toward their infants. In a first report, a sample of *twin parents* of children under 8 completed the Parental Attitudes Toward Childrearing Questionnaire (PATC; Easterbrooks and Goldberg 1984) assessing positive support and negative control (Losoya et al. 1997). Moderate heritability was found for all parenting variables, and shared environment mediation was significant for non-affective control. These results suggest child effects on all parenting variables but non-affective control, which may be a function of parental characteristics. Yet, small sample size made statistical distinctions among models difficult in most cases (Losoya et al. 1997). In a second study, *twin parents* completed a questionnaire assessing four parenting styles: over-protective, rejecting, supportive, authoritarian (Spinath and O'Connor 2003). Genetic and non-shared environmental mediation were found for all parenting styles but the rejecting one, mainly mediated by non-shared environment (Spinath and O'Connor 2003). Thus, most parenting styles may be accounted for by genetic factors in the child and non-shared environmental factors; except for the rejecting style, which may be a function of within-family differences and/or measurement error.

In a third report, parents of 5-month twins completed a questionnaire assessing four parenting dimensions: self-efficacy, perceived parental impact, hostile-reactivity and overprotection (Boivin et al. 2005). Shared environment mainly accounted for each parenting dimension. Maternal hostile-reactive behaviors were also moderately heritable, and this association was mainly mediated by infant difficultness (Boivin et al. 2005). In a follow-up of this study, the same group performed genetic analyses on 292 mothers' self-reported hostile-reactive behaviors toward each of their twins at 5, 18 and 30 months (Forget-Dubois et al. 2007). The heritability of maternal hostile-reactive behavior was modest and longitudinal analyses indicated that genetic factors at five and 30 months, although present, were uncorrelated. Shared environment was the main source of variance at the three ages and were highly correlated through time (Forget-Dubois et al. 2007). It was concluded that children's heritable characteristics may evoke maternal hostility at specific times, but were not responsible for its stability from infancy to toddlerhood (Forget-Dubois et al. 2007).

Findings from univariate studies of self-reported parenting are consistent. First, studies of twin infants *and twin parents* have shown that most parenting perceptions may be a function of characteristics that differ between families, although time-specific dimensions of parenting behavior involving harshness toward the child are partly a function of child effects or shared genes. Thus, while most forms of parental perceptions seem to be driven by characteristics of the early caregiving environment, self-reported harsh parenting may be child-evoked at specific times (e.g., during challenging phases of development).

This pattern of results is similar for parenting styles assessed in *twin parents* of infants. One study found that most parenting styles are heritable among twin parents, whereas a highly maladaptive style (i.e., rejecting) is mainly a function of non-shared environment (i.e., characteristics that differ between twin parents from a same family). As with parental perceptions, a wide range of parenting styles may be a function of—potentially heritable—parental features, while self-reported negative parenting within a normative range (i.e., harshness, hostility towards the child) may be partly a function of child characteristics (see also the work by Jaffee described in the rGE section of the present chapter).

2.9 Adoption Studies of Early Parenting

One adoption study examined the resemblance of 105 non-adoptive and 85 adoptive sibling pairs on the Home Observation for Measurement of the Environment (HOME), at 12 and 24 months of age (Braungart-Rieker et al. 1992). Non-adoptive sibling correlations were found to be greater than those for adoptive sibling pairs at both ages, suggesting genetic contributions on the HOME. Moreover, phenotypic and cross-sibling correlations between family environment and subsequent behavior problems were greater for non-adoptive siblings than for adoptive pairs at 24 months, suggesting genetic mediation of this association (Braungart-Rieker et al. 1992). These results suggest that early home environment quality may be a function of child effects or shared genetic vulnerabilities between parent and infant.

2.10 Linkage Studies of Early Parenting

Three linkage studies examined possible contributions of genotype to sensitive parenting. A first study tested the association between mothers' serotonin transporter (5-HTT) and oxytocin receptor (OXTR) genes, both posited to modulate affiliation responses to offspring during interactions, and observed sensitive parenting in 159 mothers toward their 2-year-old infants at risk for externalizing problems (Bakermans-Kranenburg and van IJzendoorn 2008). Significant contributions of 5-HTTLPR SCL6A4 and OXTR rs53576 to maternal sensitivity were found. Controlling for maternal education, depression and marital discord, parents with less efficient variants of these genes showed lower sensitivity (Bakermans-Kranenburg and van IJzendoorn 2008). Another study examined the links between oxytocin receptor (OXTR), peripheral oxytocin (OT) and sensitive parenting of 272 mothers and fathers toward their 4–6-month infants (Feldman et al. 2012). CD38 risk allele of OXTR, which mediates the release of brain OT, was also assessed. Reduced plasma OT and both OXTR and CD38 risk alleles were related to less observed parental touch. The interaction of high plasma OT and low-risk CD38 alleles predicted longer durations of observed parent-infant gaze synchrony. Parents reporting greater parental care showed higher plasma OT, low-risk CD38 alleles, and more touch toward their infants (Feldman et al. 2012). Finally, a recent study examined the associations between the arginine vasopressin receptor 1A (*AVPR1A*) and observed parenting in a normative sample of mothers of infants (M_{age} = 3.5 years; Avinun et al. 2012). The ABPR1A RS3 allele has been linked with stress hyperreactivity (see Avinun et al. 2012). Multilevel regression analyses revealed that mothers who are carriers of the *AVPR1A* RS3 allele tend to show less structuring and support throughout the interaction independent of the child's sex and RS3 genotype (Avinun et al. 2012). Taken together, findings from three linkage studies suggest independent contributions of genotype to sensitive parenting in high- and low-risk samples. Functional variants of genes involved in affiliation across various interpersonal contexts may also predict the quality of the infant-caregiver relationship, as parental involvement in sensitive care toward the infant was significantly linked with 5-HTT and OXTR variations. Moreover, genes involved in stress hyperreactivity may also predict the quality of parenting, as *AVPR1A* was linked to less structuring and support throughout a parent-child interaction.

Overall, findings from univariate twin, adoption and linkage studies of early parenting have been relatively consistent. Four twin studies indicated that most early self-reported parenting dimensions are a function of parents' own characteristics (i.e., shared environmental variance in twin infants designs, and genetic variance in twin parents designs), although highly maladaptive forms of parenting may partly be mediated by child genetic vulnerability at certain periods of development (i.e., heritability variance in twin infants designs, non-shared environmental variance in twin parents designs). One adoption study also indicated that child genetic risk extends to early environment (Braungart-Rieker et al. 1992). This pattern of results is consistent with several non-genetic studies suggesting that child temperamental features such as frustration or fearfulness may elicit maladaptive parenting (e.g., Martini et al. 2004; Rubin et al. 2003). A recent meta-analytic review of twin studies of parenting (n.b., including, but not limited to, parenting in infancy and early childhood; Klahr and Burt 2013) also supports this idea. This study indicated that 40 % of the individual differences in parental negativity are accounted for by genetic factors, while other dimensions of normative parenting (i.e., control, warmth) included significant, but very small (23–26 %) genetic contributions.

At the same time, recent univariate linkage studies have revealed independent contributions of maternal genes known to be involved in social bond formation (e.g., Feldman et al. 2012), suggesting some maternal genetic underpinning of sensitive parenting, over and above child genetic risk. Thus, it appears that parents' own genetic characteristics predict their parenting perceptions and child-rearing practices. This idea is also supported by the above-mentioned meta-analysis (Klahr and Burt 2013), which concluded to significant contributions of parental genetic makeup to parental negativity and warmth. Such key genetic contribution from the parent to early parenting behaviors could, across generations, take the form of shared environmental variance in the context of child-based twin studies of early parenting.

In other words, there seems to be a dual intergenerational process where genetic factors could play a role in parenting: (1) an elicitation process (i.e., "child effects") driven by child genetic risk associated to early environment and parenting, and (2) a direct parental genetic contribution to sensitive parenting that is independent of child-driven evocations. However, maternal genetic vulnerability may not operate independently of child genetic risk, as shared genes with the child could bolster parental involvement in maladaptive parenting. As will be further discussed, parental genetic risk may also *interact* with child genetic risk in predicting specific parental practices and child outcomes. Notwithstanding these more intricate G-E joint contributions, this review suggests that while child effects or shared genes may contribute to time-specific negative parenting, parental characteristics may independently account for a wider range of self-reported and observed parenting behaviors. This however excludes the prospect that early parenting is a "pure", parent-driven environmental factor, and thus provides indirect support for Belsky's (1984) multi-causal model of early parenting.

As stated earlier, finding evidence of child genetic contribution to purportedly "environmental" features such as early parenting is a first step toward assessing more directly potential G-E correlations (Scarr and McCartney 1983) involving parenting behaviors. Such G-E correlations may take the form of two processes. A first process, referred to as an evocative (or reactive) G-E correlation (Plomin et al. 1977), may be seen as child effects on the relationship with caregivers. A second process, the passive G-E correlation (Plomin et al. 1977), refers to shared genetic vulnerabilities between parent and child. Although genes do not have to be directly measured to test for a potential G-E correlation within a behavioral-genetic framework, child behavior and the putative environmental variable (e.g., harsh parenting) have to be measured directly. When the measured child behavior is significantly associated with the measured environmental feature, a twin design makes it possible to evaluate the extent to which this association is accounted for by the child's genes, thus pointing to a potential G-E correlation; although the exact type of correlation—evocative or passive—is not specified. Evidence for G-E correlation in early parenting comes mainly from studies using twin, adoption and step-family genetic designs.

2.11 Gene-Environment Correlations in the Context of Early Parenting

2.11.1 Twin Studies

Three pioneering studies examined the possibility of G-E correlation using direct behavioral observation of parent-child interactions. In a first study, the question of differential treatment during observed interactions involving parents of twins and of male singletons was investigated (n.b., undetermined age; Lytton 1977). Four conclusions were drawn by the authors from correlational analyses: (1) parents treat

MZ twins more alike than DZ twins in some respects; (2) they do not introduce systematically greater similarity of treatment for MZ twins in actions which they initiate themselves; (3) the greater homogeneity of treatment of MZ twins, where it occurs, is in line with their actual, rather than their perceived, zygosity; (4) parents respond to, rather than create, differences between the twins (Lytton 1977). In a second study, the same group used biometrical genetic analysis to study interactive behavior of 24-month male twins with their parents in home and laboratory (Lytton et al. 1977). A model, which included non-shared and shared environmental mediation best fitted most variables, except for instrumental independence and speech rate which showed significant heritability (Lytton et al. 1977). A third study examined differential observed maternal treatment as a function of 7–9-month twins' zygosity (DiLalla and Bishop 1996). Mothers tended to treat both children similarly, regardless of zygosity, suggesting that maternal traits drove the mother–infant interactions. Therefore, even though identical twins were more similar than fraternal twins, mothers tended to treat both types of twins comparably (DiLalla and Bishop 1996).

Thus, two pioneering studies using correlational (Lytton 1977) and univariate twin designs (Lytton et al. 1977) provided indirect evidence of G-E correlation in the context of infant-parent interactions, though in one study this result was specific to precise components of the interaction (Lytton et al. 1977). However, another study of differential maternal treatment (DiLalla and Bishop 1996) found that mothers tend to treat both twins similarly, regardless of zygosity, which reduces the likelihood of child effects on parental behavior.

In a direct test of attachment theory's main assumption—caregiver's sensitivity directly affects infant attachment security through environmental causation—two studies examined the G-E etiology of the association between sensitive parenting and infant attachment. One study assessed maternal sensitivity in the home at 9–10 months, and infant attachment security was observed in the laboratory at 12 months (Fearon et al. 2006). Shared environment in maternal sensitivity was able to account for some of the similarity in attachment security, and weak non-shared associations appeared to suppress the magnitude of the correlation between attachment and sensitivity; thus suggesting little genetically mediated child effects (though child effects may also come out through a non-shared environmental pathway; Fearon et al. 2006). In a second study, 485 same-sex twin pairs were used to test for G-E associations between observed parenting quality and infant attachment security (Roisman and Fraley 2008). In line with the results of Fearon et al. (2006), both constructs observed at 24 months, as well as their covariation, were accounted for by shared (85 %) and non-shared (15 %) environmental variance (Roisman and Fraley 2008).

Thus, findings from two twin studies suggest that infant security during the first two years of life is mainly a function of characteristics of the early caregiving environment, but also, that considerable within-family differences may be involved. Though mainly pointing to shared environmental contribution, these findings raise a doubt on attachment as due to an absolute, all-embracing shared environmental effects. They underline the continuing challenge posed to attachment theory by within-family differences (which may involve child effects) in socio-emotional processes.

A great deal of bivariate reports examined the possibility of child effects in explaining maladaptive forms of early parenting such as harshness or negativity. Indeed, significant genetic correlations were found between parental negativity and various parent-rated developmental problems in the infant: conduct problems (although shared environment mediation was found for observations; mean age = 43 months; Deater-Deckard 2000; mean age = 4 years; Alemany et al. 2013), low prosocial behavior (ages 3, 4 and 7; Knafo and Plomin 2006), externalizing problems (but only in boys; 7, 9, 14, 24 and 36 months; Boeldt et al. 2012), antisocial behavior (ages 4 and 7; Larsson et al. 2008). However, there is evidence of environmental mediation of the risk for antisocial behavior via parental negative feelings toward the infant (Larsson et al. 2008). Shared environment also contributes to several protective cycles of parenting, namely the association of high parental positivity with prosocial behavior (Knafo and Plomin 2006) and low externalizing behavior (but only in girls; Boeldt et al. 2012).

Moreover, four studies using a MZ twin differences design investigated non-shared environmental associations of parental positivity/negativity with various developmental outcomes. Because MZ twins do not differ genetically, associations between early parenting and their behavior can be directly ascribed to non-shared environment. Thus, although this method does not directly test for G-E correlations, it is still somewhat informative of the etiology of associations of parenting with child behavior. In a first study, parental warmth, control and responsiveness covaried in expected ways with twin differences in temperament, prosocial behavior and behavior problems; mean age = 3, 5 years; Deater-Deckard 2001). The twin who received more supportive/positive and less punitive/negative parenting was also higher in positive mood/prosocial behavior and lower in negative mood/behavior problems when compared to his twin; suggesting non-shared environmental mediation (Deater-Deckard 2001). Another study found non-shared environmental associations between two early parenting measures (harsh discipline, negative feelings) and four infant outcomes (anxiety, prosocial behavior, hyperactivity, conduct problems), especially for the extreme portion of the parenting- and behavior-discordant distributions (mean age = 4 years; Asbury et al. 2003). More non-shared associations were found between overprotection and boys' social reticence, and between hostile parenting and girls' social reticence, only with high levels of depressive symptoms in fathers (mean age = 30 months; Guimond et al. 2012). Non-shared associations were also found between high maternal negativity/low warmth and child antisocial behavior (mean age = 5 years; Caspi et al. 2004).

Thus, findings from twin studies mainly point to "child effects" in the form of disruptive behavior on early parental negativity. Within-family differences are also likely to be involved in explaining links between parental negativity and various behavioral and interpersonal outcomes. This pattern of results excludes shared environmental contributions to maladaptive parenting. Rather, shared environment might account for protective cycles linking parental positivity with child prosociality and low externalizing behavior.

Potential G-E correlations were also investigated in explaining the well-documented association between physical maltreatment and later antisocial behavior.

A study using a sample of 1116 5–7-year twin pairs found that child antisocial behavior at 5 (i.e., inferred "child effects") did not account for the prospective association between mother-reported physical maltreatment at 5 and teacher-reported child antisocial behavior at 7 (Jaffee et al. 2004b). These findings suggest that early maltreatment plays a causal role in the later development of antisocial behavior, over and above "child effects" (Jaffee et al. 2004a). In a second study using data from the same sample, the limits of such "child effects" were tested on parental behavior that ranged from the normative (i.e., corporal punishment) to the non-normative (i.e., physical maltreatment; Jaffee et al. 2004a). Shared environment accounted for most of the variation in corporal punishment, as well as in and physical maltreatment. However, corporal punishment was partly genetically mediated (which was not the case for physical maltreatment), and the genetic factors that accounted for corporal punishment were largely the same as those that accounted for child antisocial behavior, suggesting "child effects" (Jaffee et al. 2004a). Thus, risk factors for maltreatment are unlikely to reside within the child and more likely to reside in features varying across families; though normative discipline in the form of corporal punishment may partly be a function of child effects or shared genes. This is consistent with results from epidemiological studies identifying more extreme adverse early environments, such as inadequate housing (Palusci and Loeb 2011) or public aid as a source of income (Parrish et al. 2011), as risk factors for child maltreatment. As stated earlier, bidirectional links between early parenting and these broader psychosocial risk factors are likely.

In this regard, one MZ twin differences study investigated non-shared associations of two broad psychosocial risk factors, namely birthweight-discordance and early family environment, with behavior problems and academic achievement at 7 (Asbury et al. 2006). MZ differences in anxiety, hyperactivity, conduct and peer problems and academic achievement correlated significantly with MZ differences in birthweight-discordance and family environment (Asbury et al. 2006). Associations increased at the extremes of discordance, even in a longitudinal, cross-rater design, with effect sizes reaching 12 %. Some of these associations operated partly as a function of SES, family chaos and maternal depression. Higher-risk families generally showed stronger negative associations (Asbury et al. 2006). This suggests that broad risk indicators of early adversity may be linked with child behavioral and social development through a non-shared environmental pathway. Another study attempted to identify the factors comprising the shared environmental variance on cognitive performance in early childhood (mean age = 3, 5 years; Petrill and Deater-Deckard 2004). SES and parental warmth, taken together, accounted for most shared environmental covariance between task engagement and cognitive skills, indicating that early indicators of environmental adversity such as SES may also contribute to child cognitive development.

Overall, most bivariate findings point to the shared and non-shared environmental etiology of the associations of characteristics of the early caregiving environment with child behavioral and socio-emotional outcomes. This is especially true in the context of protective cycles. This pattern of results points to the importance of features of the early caregiving environment and of specific parental treatment within mother–infant dyads, over and above initial child effects. Inversely, there is evidence of child effects on more adverse forms of parenting in predicting various developmental problems in the child. Thus, in the normative range, there seems to be some kind of a continuum between adaptive parenting, mainly a function of parental characteristics and family environment, and maladaptive parenting, possibly driven by child characteristics having a proximal or distal effect on the caregiving environment (or shared genes). Nevertheless, there seems to be limits to such child effects on adverse parenting, as infant disruptive behavior cannot account for extreme forms of maladaptive parenting ranging in the non-normative spectrum—such as physical maltreatment—, which may partly be a function of broader social-risk factors and their possible bidirectional relations with early parenting.

2.11.2 Adoption Studies

Using a prospective adoption design, one study investigated possible genetic contributions to associations between early family environment and behavior problems at 7 (Braungart-Rieker et al. 1995). Patterns of correlations for non-adopted and adopted boys indicated that links between quality of family environment (i.e., conflict, cohesion, expressiveness) and externalizing behavior in home and school were genetically mediated; indirectly suggesting a G-E correlation. For girls, these links were associated with shared environmental mediation (Braungart-Rieker et al. 1995).

Another recent study investigated the developmental underpinnings of children's socially disruptive behavior using a genetically sensitive design that allowed examination of parent-on-child and child-on-parent (evocative G-E correlation) effects (Elam et al. 2013). Using an adoption-at-birth design, this controlled for passive G-E correlation and directly examined evocative G-E correlation while examining the associations between family processes and children's peer behavior. In 316 linked dyads of birth mothers, adoptive parents and adopted children, this study examined the evocative effect of genetic influences underlying toddler low social motivation on mother—child and father—child hostility and the subsequent influence of parent hostility on disruptive peer behavior during the preschool period (Elam et al. 2013). Results showed that birth mother low behavioral motivation predicted toddler low social motivation, which predicted both adoptive mother–child and father–child hostility. This suggests the presence of an evocative G-E correlation (Elam et al. 2013).

2.11.3 Step-Family Studies

One study used a step-family quantitative genetic design to estimate G-E etiology of 4-year children's behavior problems and prosocial behavior, as well as negativity in their relationships with their mothers and mothers' partners (mean age = 4 years; Deater-Deckard et al. 2001). Behavior problems and partner-child negativity were mainly heritable, and shared environmental variance accounted for mother- and partner-child negativity. 1/5 to 2/3 of the variance was accounted for by non-shared environment. The link between parental negativity and behavior problems was mediated by genetic covariance, and the link between parental negativity and prosocial behavior was mediated by environmental covariance (Deater-Deckard et al. 2001). This pattern of results suggests that the adverse association of parental negativity with behavior problems may be a function of "child effects". The protective link between low parental negativity and prosocial behavior may be a function of the early caregiving environment (Deater-Deckard et al. 2001).

Five general conclusions can be drawn according to results from twin, adoption and step-family studies that examined G-E correlations in the context of early parenting. First, a wide range of early parenting practices providing a positive context for child development are mainly a function of characteristics of the caregiving environment, and are hardly accounted for by child heritable characteristics. Thus, protective factors for a several child outcomes seem to lie in environmental features that differ between families. However, as will be further discussed, several features of adaptive parenting—although independent of "child effects"—may not be free of genetic factors, as they are partly driven by parental genotype. Part of the prominence of statistically estimated environmental mediation within bivariate twin studies may, in fact, reflect parent-driven genetic processes. Second, adverse parental practices within a normative range may be regarded as developmental incidents partly involving "child effects", whereby parental negativity or corporal punishment, for instance, are mainly explained by child heritable characteristics. Shared genes may also partly account for such associations. This is consistent with results from univariate studies indicating that negative parenting, especially time-specific, is mainly a function of child characteristics or shared genes (Forget-Dubois et al. 2007), but also that a wider range of practices, especially adaptive ones, are driven by parental characteristics (i.e., shared environment or parental genotype). Third, more extreme forms of adverse parenting (e.g., physical maltreatment) may not result from customary developmental incidents, as child evocation does not account for physical maltreatment. Hence, there are limits to child effects on maladaptive parenting, which may be parent-driven in its more extreme forms.

2.12 Which Gene-Environment Interactions Are Involved in Early Parenting?

Several univariate studies suggest that a wide range of early parenting behaviors are environmentally-driven. However, evidence from bivariate twin studies suggests that early parenting does not necessarily operate independently of genes. As suggested earlier, most developmental outcomes may result from joint, rather than additive, contributions of genetic and environmental factors (Rutter and Silberg 2002). Therefore, another G-E process of interest regarding early parenting and its association with child behavior refers to a possible G-E interaction (i.e., interacting processes between a putative environmental risk factor and an infant or parental genetic vulnerability), which would be indicated if, for example, the association of harsh parenting with child externalizing problems were differentially manifested as a function of genetic risk for such problems. This mechanism is consistent with the diathesis-stress model of psychopathology, according to which an environmental stressor is more likely to lead to maladjustment if pre-existing genetic vulnerabilities are present (Zuckerman 1999). Beyond the diathesis-stress model, the broad concept of G-E interaction encompasses differential susceptibility, which refers to variations in the degree to which an individual is affected by specific environments—not only to adverse but also to protective ones—according to his or her genotype (Belsky and Pluess 2009).

To evaluate a potential G-E interaction, genes do not have to be directly measured, but a putative environmental variable has to be. Moreover, finding statistical evidence for a G-E interaction is most likely if the measured behavior is under strong genetic influence and if, in contrast, the measured environmental variable has little relation to genetic factors (Rutter and Silberg 2002). In the absence of a G-E interaction, genes and exposure to a specific environment may independently, thus additively, contribute to child outcomes; indicating general (main) rather than conditional (joint) contributions of environment and genotype.

2.12.1 Twin Studies

To our knowledge, four twin reports have examined potential G-E interactions involving early parenting and infant behavior. A first study tested whether the association of parent-reported physical maltreatment with risk for parent- and teacher-rated conduct problems was strongest among infants who were at high genetic risk for externalizing behavior (i.e., co-twin conduct disorder status and the pair's zygosity), using data from 1116 5-year twin pairs and their families (Jaffee et al. 2005). The experience of maltreatment was associated with an increase of 2 % in the probability of conduct disorder among children at low genetic risk, but an increase of 24 % among children at high genetic risk (Jaffee et al. 2005). Thus, context of adversity in the form of physical maltreatment seemed to interact with child genetic vulnerability to predict conduct problems in early childhood (Jaffee et al. 2005). A second study investigated G-E interactions linking three early environmental indices (i.e., family chaos, instructive vs. informal parent-child communication) with verbal ability in a sample of 4-year twins (Asbury et al. 2005). Heritability for verbal ability was greater in high-risk environments, all in the direction of diathesis-stress models (i.e., high family chaos, high instructive communication, low informal communication), rather than in low-risk environments (i.e., low family chaos, less instructive communication and more informal communication) environments, suggesting increased heritability in high-adversity environments in the case of verbal ability (Asbury et al. 2005).

Evidence for a G-E interaction linking early familial adversity (as assessed via a composite score of seven prenatal and postnatal risk factors) and cortisol reactivity was also found in a sample of 346 19-month twins (Ouellet-Morin et al. 2008). In that initial paper, in low-adversity settings, genetic and non-shared environmental factors accounted for cortisol reactivity, with heritability explaining the similarity observed within twin pairs. Under conditions of high adversity, shared and unique environmental factors accounted for variance in cortisol reactivity. In other words, genetic risk for cortisol reactivity was found in low-adversity settings, but not in high-adversity settings. In a second paper on the same sample, this time focusing on cortisol data at 6 months (i.e., one year earlier), an inverse pattern was found: significant heritability for morning cortisol was found under high family adversity, but not in low adversity (Ouellet-Morin et al. 2009). Taken together, these studies indicate a shift in the G-E etiology of the physiological stress response in the context of adversity; consistent with the diathesis-stress model, gene expression (i.e., heritability) was significant at 6 months, but was shut down and replaced by shared environment at 19 months, where genetic factors now played a role in the context of low adversity. This pattern of results indicates that G-E contributions to the cortisol response vary as a function of family adversity (i.e., thus statistically a G-E), but also as a function of age. This evolving pattern of G-E is consistent with the progressive establishment of an environmental programming process in the course of the second year of life (Ouellet-Morin et al. 2009). As the main proximal environment of the child, early parenting is likely to be involved. However, further research is necessary as these studies used broad indicators of early environment, including but not limited to early parenting practices.

Findings from twin studies suggest that early high-risk environments interact with child genetic risk to predict various behavioral, physiological and cognitive outcomes. Some environments seem to increase heritability for various outcomes; in line with diathesis-stress models. In other cases, adverse environments, such as those in which economic and social hardship promote enduring contextual stress (Ouellet-Morin et al. 2008), may supersede genetic contribution to other outcomes (e.g., physiological stress response), perhaps through epigenetic processes. Yet, it is not clear whether similar G-E processes operate for other phenotypes during infancy. Further longitudinal research is needed at this time, as single assessments may be of little help in determining sequential variations in such G-E processes.

2.12.2 Adoption Studies

Adoption studies documenting G-E interactions typically test moderation of child genetic vulnerability (as assessed through birth parent characteristics and/or child temperament) on the association between early environment (assessed through adoptive parents' behavior, quality of the home environment, etc.) and a child outcome. Significant moderation of child genetic risk suggests a G-E interaction.

One prospective adoption study using adoptive parent(s)-birth parent(s)-adopted infant triads found interactions between infant genetic risk for behavior problems (i.e., birth mother's history of depression) and adoptive mother's structured parenting on infants' behavior problems (mean age = 18 months; Leve et al. 2009). This pattern of results suggests that the contribution of structured parenting to infant behavior problems varies as a function of child genetic risk (Leve et al. 2009). Using a similar method, another study found that infants at genetic risk for externalizing problems (i.e., birth parents' externalizing behavior) showed heightened attention to frustrating events only when the adoptive mother had higher levels of anxious/depressive symptoms (mean age = 9 months; Leve et al. 2010). A similar pattern of results emerged in yet another study testing the interaction of adoptive parents' depression and responsiveness with genetic risk for depression (i.e., birth mother's history of depression) in predicting fussiness in 9–18-month adopted children (Natsuaki et al. 2010). Independent contribution of adoptive mothers' depression to infant fussiness was found, as well as a significant interaction between birth mothers' depression and adoptive mothers' responsiveness. Indeed, children of birth mothers with depression showed higher levels of fussiness at 18 months when adoptive mothers had been less responsive (Natsuaki et al. 2010). Another study found moderation of birth mother anger/frustration on the link between adoptive parental harsh discipline (9 months) and infant anger/frustration (18 months; Rhoades et al. 2009). Yet another study found moderation of externalizing problems at 3 years on the link between adoptive mother's over-reactive parenting (9 months) and infant negative emotionality (18 and 27 months), as well as moderation of birth mother negative emotionality on the association of adoptive mother's over-reactive parenting with the same outcome (Lipscomb et al. 2012).

Taken together, findings from prospective cross-sectional and longitudinal adoption studies suggest that adopted children at genetic risk for several developmental problems might inherit specific vulnerabilities that makes them more sensitive to environmental risk factors (e.g., adoptive mother's harshness, depression), suggesting significant G-E interaction, although independent contributions of infant genetic risk (e.g., birth parents' history of affective or externalizing problems) are occasional.

2.12.3 Linkage Studies

More evidence of G-E interplay in early parenting comes from molecular studies testing for interactions between gene variations and measured risk factors in predicting a developmental outcome. Two approaches allow testing of such G-E interactions in molecular designs: (1) to assess the moderating effect of *child genotype* on the association of a putative environmental risk factor (e.g., parenting) with a child outcome; (2) to assess the moderating effect of *parental genotype* on the association of a putative risk factor—in the child or in the parent—with a parental outcome (e.g., parenting). Research using infant and parent molecular markers has examined

G-E interactions involving genes modulators of self-regulatory capacities during stressful social situations, social bond formation (i.e., DRD4, DRD2, 5-HTT, MR, FKBP5, COMT, DAT1), as well as externalizing problems (i.e., MAOA).

2.12.4 DRD4 and COMT

Molecular genetic studies have found significant interactions between the presence of the DRD4 7-repeat (i.e., long allele) and early maternal care (i.e., sensitive parenting, affective communication) or very close correlates (e.g., parental stress) in predicting various outcomes in infancy: externalizing behavior (mean age = 10 months; Bakermans-Kranenburg and van IJzendoorn 2006; ages = 18–30 months; Propper et al. 2007; mean age = 3–4 years; DiLalla et al. 2009; mean age = 3 months; Zohsel et al. 2014), attachment disorganization (mean age = 12 months; Gervai et al. 2007; mean age = 14 months; Luijk et al. 2011a, b), peer problems (DiLalla et al. 2009) and temperamental sensation-seeking (mean age = 18–21 months; Sheese et al. 2007). Some of these studies were conducted in the context of differential vulnerability hypothesis. For instance, it was shown that infants carriers of the DRD4 7-repeat were differentially susceptible to early care in predicting externalizing behavior (e.g., Bakermans-Kranenburg and van IJzendoorn 2006): carriers of the DRD4 7-repeat had more externalizing problems than the non-carriers when exposed to early adverse conditions, but less externalizing problems in early adaptive conditions.

Two randomized controlled trials brought further support to the differential vulnerability hypothesis. There was a moderating role of the DRD4 7-repeat in the effect of a video-feedback intervention to promote positive parenting and sensitive discipline on 1-to-3-years infants' daily cortisol (Bakermans-Kranenburg et al. 2008a) and externalizing behavior (Bakermans-Kranenburg et al. 2008b). The intervention led to a lesser amount of daily cortisol and less externalizing problems in carriers of the long allele of the DRD4 7-repeat, but no intervention led to more daily cortisol and more externalizing problems in carriers of the same genotype, suggesting differential susceptibility as a function of DRD4 7-repeat polymorphism.

Gene-environment interactions were also investigated in a more extreme form of adverse parenting: child maltreatment. A recent study examined the extent to which variation in DRD4 and 5-HTTLPR genotype were differentially associated with the development of attachment security and disorganization in maltreated and non-maltreated 13-month infants, and the extent to which the effect of preventive interventions aimed at promoting attachment security was moderated by genes. Among maltreated infants, DRD4 and 5-HTTLPR variations had minimal associations with improvement in attachment disorganization (Cicchetti et al. 2011). However, among non-maltreated infants, both polymorphisms accounted for attachment security and disorganization at age 2 and the stability of attachment disorganization over time (Cicchetti et al. 2011). In line with earlier findings on the cortisol response (Ouellet-Morin et al. 2008), this pattern of results suggests that early conditions of high adversity, including maltreatment, could have a programming effect on indicators of emotional regulation in the infant, overriding gene expression. However, it is unclear whether HPA axis activity, the main neurobiological underpinning of physiological regulation of stress response, is involved such associations between early maltreatment and various outcomes understood as behavioral indicators of emotional dysregulation. Still, several studies have found associations between the presence of early physical abuse and heightened cortisol stress response (e.g., Carpenter et al. 2011), which points towards the likelihood of HPA axis involvement in the above-cited G-E processes.

G-E interactions were also investigated in regard to maternal personal history. In one study, maternal unresolved loss or trauma (i.e., the quality of the processing and integration of childhood experiences of loss and/or trauma in mothers, assessed via questionnaire) was associated with infant disorganization, but only in the presence of the long allele of the DRD4 7-repeat in the 10–11-month child, suggesting infant genotype of the inter-generational transmission of maladaptive attachment-related experience (van IJzendoorn and Bakermans-Kranenburg 2006).

Two studies examined interactions between parent dopamine-system genotype and several risk factors in predicting sensitive parenting toward the infant. Specifically, COMT variations are associated with variations in emotional resilience against negative mood states (Smolka et al. 2005; van IJzendoorn et al. 2008) and in cognitive distractibility (Drabant et al. 2006). As parental sensitivity requires constant attention of the infant's signals even in stressful circumstances, low distractibility and efficient self-regulation may help parents to remain focused on their child. One study tested whether COMT and DRD4 polymorphisms moderated the negative association of levels of daily hassles with sensitive parenting (mean age = 23 months; van IJzendoorn et al. 2008). In parents with the combination leading to the least efficient dopaminergic functioning (COMT val/val or val/met, DRD4 7-repeat long allele), more daily hassles were associated with less sensitive parenting, and lower levels of daily hassles were associated with more sensitive parenting. This suggests differential susceptibility to hassles depending on parental genotype (van IJzendoorn et al. 2008). Another study tested the interaction of the DRD4 7-repeat with 6 months infants' fussy-difficult temperament in predicting sensitive parenting (Kaitz et al. 2010). Mothers with the long allele of the DRD4 7-repeat were more sensitive to fussier infants and less sensitive to less fussy infants, suggesting differential susceptibility to infant fussiness according to the DRD4 7-repeat (Kaitz et al. 2010).

Overall, molecular findings suggest that in infants, the long version of the DRD4 7-repeat polymorphism, allegedly involved in the early development of self-regulation, may increase infants' sensitivity (i.e., differential susceptibility) to adaptive or maladaptive early environments in the development of various behavioral, interpersonal and physiological outcomes. Moreover, as early conditions of extreme adversity tend to program attachment behavior over and above genetic contributions during the second year of life, early conditions of low adversity are associated with heightened DRD4 gene expression.

Furthermore, G-E interactions involving parental dopamine-system polymor-phisms may explain why some parents are more and others less impacted by con-textual stressors and infant irritability in responding sensitively to their offspring's signals. Variations in genes involved in regulation of emotional arousal (e.g., DRD4, COMT) may contribute to hostility of parental reactions to infant signals. However, reports in this literature appear equivocal, as one study indicates less sensitivity is linked with daily contexts of fussiness in parents carrying the DRD4 7-repeat, reflecting hostile reactions to contextual infant signals, and another study indicates more sensitivity is linked with high infant fussiness in carriers of this genotype. This apparent contradiction may be explained by the fact that, as stated earlier, enduring parenting behaviors may be a function of parents' own genetically-driven character-istics, over and above initial child effects; while time-specific negative parenting is mainly a function of contextual stressors (e.g., child evocation, temporary economic hardship, etc.; Boivin et al. 2005), regardless of parental characteristics.

2.12.5 DRD2

The dopamine receptor D2 gene (DRD2) Aþ1 polymorphism has been linked with sensitivity to reward (Suhara et al. 2001), novelty seeking (Noble et al. 1998) and substance abuse disorders (Oscar-Berman and Bowirrat 2005) in adults. However, data remains inconclusive in children (see Mills-Koonce et al. 2007). In a recent study, it was hypothesized that, along with DRD4, the DRD2 gene polymorphism could be involved in infant difficultness and activity level, as assessed in the course of interactions with caregivers at 6, 12 and 36 months; both potentially evoking spe-cific classes of parental behavior (Mills-Koonce et al. 2007). This study examined a possible interaction between the DRD2 Aþ1 risk allele in mothers and children, and maternal sensitivity in predicting subsequent child affective problems (Mills-Koonce et al. 2007). Evidence was found for a moderating role of child genotype in the association between maternal sensitivity and later child affective problems (Mills-Koonce et al. 2007). A second study from the same group revealed that the DRD2 Aþ1 allele interacted with insensitive parenting to predict low vagal withdrawal in response to maternal separation during the first year of life (Propper et al. 2008). These findings suggest that DRD2 genotype moderates associations of parental sen-sitivity with early behavioral and physiological indicators of emotional regulation, thus making the infant vulnerable to specific environments in the development of affective regulation impairments (see above).

2.12.6 DAT1

Activation of the mesolimbic dopamine tract is necessary for maternal behavior in rats (see Numan 2007), and DAT1 binding in the nucleus accumbens is correlated with maternal behavior in rats (Champagne et al. 2004). Inversely, DAT1 knock-out in mice is associated with disrupted maternal behavior (Spielewoy et al. 2000). Robust data is however still lacking in adult humans. One study tested the association of the 40-bp variable number tandem repeat polymorphism of the dopamine transporter (DAT1) gene with three dimensions of observed parenting: positive parenting, negative parenting, total maternal commands (Lee et al. 2010). A significant interaction was found between maternal DAT1 and 5-year children disruptive behavior, as the association between DAT1 and negative parenting was stronger among mothers whose children were highly disruptive during a mother–child interaction task (Lee et al. 2010). Significant non-additive associations were also found between maternal DAT1 and both negative parenting and total commands during the same task, even after controlling for demographic factors, maternal psychopathology and disruptive child behavior (Lee et al. 2010). This pattern of results suggests that maternal genetic vulnerability to maladaptive parenting is partly moderated by infant difficultness. Thus, more difficult infants (i.e., fussy, difficult) may trigger parental genetic vulnerabilities to predict the use of adverse (i.e., negative, harsh) parenting practices during early parent-child interactions.

2.12.7 5-HTT

Individuals who are either homozygous for the short allele (ss) or heterozygous (sl) of the 5-HTTLPR have been found to be at risk for a range of emotional and behavioral maladaptive conditions such as under-regulated, impulsive, aggressive and risk-taking behavior, executive function deficits, alcohol use, as well as depressive/anxious symptoms (e.g., Brown and Hariri 2006; Lesch and Bengel 1996; Lucki 1998; Posner et al. 2007). Again, in accordance to a self-regulation framework, 5-HTT variations could modulate infant self-regulatory capacities. Such early capacities are susceptible to elicit reactions from the caregiving environment, thus shaping G-E correlations between early parenting and infant behavior. Moreover, human and macaque evidence suggests complex associations of 5-HTT genotypes with activation of brain regions involved in imitation, social cognition and communication (e.g., Canli and Lesch 2007; Watson et al. 2009). In light of this evidence, 5-HTT variations may have adaptive or maladaptive consequences in several interpersonal contexts, such as during mother–infant interactions (Mileva-Seitz et al. 2011).

So far, molecular studies have found significant interactions between variations in infant 5-HTTLPR polymorphism and early maternal care (i.e., responsiveness, quality of parenting behavior) in predicting attachment security (mean age = 7 months; Barry et al. 2008), attachment disorganization (mean age = 12 months; Spangler et al. 2009) and later negative emotionality and fear (mean age = 18 months; Pauli-Pott et al. 2009). Significant interactions were also found with child attachment-related experiences (i.e., child-mother attachment relationship, attachment representation, etc.) in predicting alpha amylase response to separation (ages = 12 to 18 months; Frigerio et al. 2009), electrodermal reactivity (mean age = 7 years; Gilissen et al. 2008) and self-regulation (mean age = 15 months; Kochanska et al. 2009). These findings indicate that the quality of early care and of infant-caregiver attachment relationship both serve to amplify or offset the risk conferred by the 5-HTT genotype in the development of interpersonal (i.e., attachment disorganization), behavioral (i.e., self-regulation, negative emotionality) and physiological (i.e., electrodermal reactivity, alpha amylase reactivity) child outcomes which possibly have common impairments in regulation of emotional arousal.

One study tested if mothers' early life experiences and 5-HTT genotype interact in predicting self-reported sensitive parenting (mean age = 6 months; Mileva-Seitz et al. 2011). Main contributions and significant G-E interactions were found: mothers with no S or L (G) alleles oriented away more frequently from their babies if they also reported more negative early care, and mothers with the S allele and with positive early care scored higher on ratings of perceived attachment to their infant. Regression results also showed that with increasing care quality, mothers with the L(A)L(A) genotype (no S or L(G) allele) oriented away less frequently, while S or L(G) allele carriers showed no significant change. In contrast, with increasing early care quality, L(A)L(A) (no S or L(G) allele) mothers scored lower on perceived attachment to their infants, whereas S or L(G) allele carrying mothers scored higher (Mileva-Seitz et al. 2011).

2.12.8 MAOA

The MAOA gene encodes the MAOA enzyme, which helps metabolizing neurotransmitters such as norepinephrine, serotonin and dopamine (Shih et al. 1999). In adult humans, genetic deficiencies in MAOA activity have been linked with male antisocial behavior (e.g., Brunner and Nelen 1993). As with the DRD2 gene polymorphism, data remains inconclusive in children. Two studies have tested G-E interactions involving the MAOA gene during infancy. In their seminal study, Caspi and colleagues found that MAOA gene polymorphism moderated the association of maltreatment at 3 years with adult antisocial behavior, pointing that maltreatment predicted antisocial behavior only when there was an infant genetic vulnerability in metabolizing neurotransmitters involved in emotion regulation (Caspi et al. 2002). Another study found that, in girls, MAOA-LPR interacted with stressful life events and family adversity (early parenting being likely involved in both risk factors) from 6 months to 3½ years to predict hyperactivity at ages 4 and 7 (Enoch et al. 2012). In boys, the interaction between MAOA-LPR and stressful life events between 1½ and 2½ years predicted hyperactivity at age 7 (Enoch et al. 2012). This is in line with the well-documented role of MAOA in overactive and impulsive behavior (Brunner and Nelen 1993).

2.12.9 MR

The glucocorticoid receptor (GR) and the mineralocorticoid receptor (MR) have been implicated in the variability of HPA axis responses to social stressors

(DeRijk and De Kloet 2008). Their role in infants' behavioral regulation during their very first interpersonal relationships is thus theoretically plausible. One study found a significant interaction between the minor MR allele and sensitive responsiveness during infancy in predicting attachment security (mean age = 14 months; Luijk et al. 2011a, b). Carriers of the minor MR allele (vs. carriers of the major allele) had a more secure attachment if their mothers showed more sensitive responsiveness and a less secure attachment if their mothers showed more insensitivity, suggesting differential susceptibility to sensitive parenting according to MR genotype (Luijk et al. 2011a, b).

2.12.10 FKBP5

As differences in physiological stress response during the SS have been predominantly attributed to the quality of attachment (Oosterman and Schuengel 2007) and that genetic factors may play a role in explaining variance in HPA axis activity (Steptoe et al. 2009), one study tested the interaction between parent-infant attachment and the minor allele of the haplotype of FKBP5 (rs1360780) in predicting cortisol reactivity during the SS (mean age = 14 months; Luijk et al. 2010). A main contribution of FKBP5 rs1360780 on cortisol reactivity was found. Moreover, a significant interaction was found between insecure-resistant attachment and FKBP5 rs1360780 in predicting the same outcome. This indicates a double-risk for heightened cortisol reactivity levels during the SS in infants carrying the minor allele of the FKBP5 and an insecure-resistant attachment relationship with their mother (Luijk et al. 2010). Thus, the early development of physiological arousal in stressful social situations is mediated by the additive *and* joint contributions of genotype and early interactions with caregivers.

2.12.11 COMT

Carrying the minor allele of numerous dopaminergic system genes has been linked to infant difficult temperament (see Ebstein 2006) and ADHD (e.g., Faraone and Khan 2006). Although temperament was not found related to attachment security, it might be involved in infants' activity levels during interactions with caregivers. Moreover, a protective effect has been reported for COMT heterozygotes for Val/Met alleles (vs. homozygotes Val/Met) in the form of dopamine levels associated with optimal neonatal neurobehavioral features (Wahlstrom et al. 2010). Neonatal neurobehavioral organization, as assessed via examiner-rated scales, was linked to more secure attachment (Grossmann et al. 1985) and less attachment disorganization (Spangler et al. 1996). Therefore, distal associations between COMT genotype and infant attachment make this gene a potential candidate for the study of complex G-E interactions involving early parenting.

One study found a significant interaction between COMT and parental sensitivity in explaining variance on infant disorganization (mean age = 14 months; Luijk et al. 2011a, b). Yet, this finding could not be replicated across samples (Luijk et al. 2011a, b). Another study found that COMT significantly interacted with infant-parent attachment in the SS to predict alpha amylase basal levels (ages = 12–18 months; Frigerio et al. 2009). These findings suggest that COMT polymorphism may increase infant susceptibility to specific environments in predicting dysregulation of arousal during a stressful situation (i.e., SS) and basal stress level (i.e., alpha amylase).

Taken together, findings from linkage studies indicate that specific variations in infant dopaminergic and serotoninergic genotype, possibly through their contribution to infant temperament and self-regulation, moderate the association of early parenting with several behavioral, physiological and cognitive child outcomes; thereby creating a positive—or otherwise adverse—context for the child's later psychological adjustment. Several of these G-E interactions were found in the context of the differential susceptibility hypothesis.

On the other hand, variations in parental genotype, also involving genes from the dopaminergic and serotoninergic systems linked to self-regulatory capacities and emotional resilience in stressful situations, moderate the association of parental personal history, contextual stressors and infant characteristics with specific parental practices. Such G-E interactions may explain why some parents are more and others less impacted by child characteristics and stressful contexts in responding adaptively to their offspring's signals. Thus, robust evidence points to the importance of child *and* parental genetic risk (as well as their interaction) as moderators of putative risk factors in the child and in the parent.

While evidence of significant contributions of child genotype to features of the early caregiving environment suggests evocation processes in the form of child effects, evidence of unique parental genetic contribution to early parenting suggests partial independence of such child-driven evocations and point to the importance of parents' own characteristics in predicting their child-rearing practices. However, the reader should still bear in mind that infant and parent genetic risk are not necessarily independent, as shared genetic vulnerabilities with the child could account for parental involvement in specific practices. Thus, while evidence of child effects and independent contributions of maternal genotype represent distinct etiological processes, both may often coincide and jointly predict parental behavior, thereby shaping G-E interactions.

2.13 Conclusion

As this review shows, behavioral-genetics studies can provide comprehensive understanding of the nature of early parenting and its contribution to infant socio-emotional development. By disentangling genetic from environmental variance, important theoretical questions about the developmental role of early parenting and its interaction with child and parental genetic risk may be investigated. Also, genetically informed data helps specifying the direction and magnitude of developmental associations between early parenting and child development. In the context of the present chapter, we showed that *both* child and parental genotypes have a unique contribution *and* interact to predict multiple developmental problems in the child. However, a wide range of adaptive parenting practices are accounted for by parental characteristics and life experiences, whereas child heritable characteristics may account for specific, negative parenting practices in the normative range (but not in the more severe range). Such genetically informed studies may inform the prevention of child developmental problems, as well as early interventions promoting adaptive parent-infant interactions. The actual evidence suggests that interventions promoting sensitive and warm parenting should mainly focus on parental characteristics, perceptions and behaviors, while interventions promoting positive parent-child interactions may center both on child temperamental characteristics (and their purported effects on the caregiving environment) and parent training. Still, there are several methodological limitations inherent to the reviewed studies that should be considered for future research. We close this chapter by outlining these methodological caveats. Finally, we examine future directions in genetically sensitive studies of early parenting.

Many behavioral-genetics studies involving early parenting and child outcomes rely on parent or teacher reports to facilitate data collection and to reach the power necessary to detect complex genetic and environmental contributions. This approach, however, may produce biased results with respect not only to the nature of parent-child interactions (i.e., there is considerable inter-individual variability in perceptions of interpersonal relationships), but also in terms of their underlying etiology (e.g., parents may overestimate similarity between identical twins, thus artificially bolstering heritability estimates; Gervai 2009). The use of parent reports may also be problematic when the same person assesses both twins of the same family, as such assessments may be more informative of the parents' cognitive biases and beliefs toward their children—or else, of features of the family environment—than of children's actual characteristics. Thus, the fact that the characteristics of the twins and the features of the environment are not assessed independently for each twin may yield biased estimates. Despite the potentially high expenses and logistic challenges, twin studies might benefit from the use of multiple and independent assessments of child behavior (Gervai 2009). For that matter, observational data, which allows the investigation of micro-processes within parent-child interactions, may be used more systematically in future research. However, as stated earlier, investigators should take into account the fact that within experimental parent-infant interactions, parental behavior may be somewhat programmed by experimental context; thus falsely enhancing shared environmental variance (De Wolff and van IJzendoorn 1997). One adequate, although burdensome, way to estimate within-family differences in such contexts would be to conduct more than one assessment of observed parenting within a single design in order to control for contextual factors (e.g., unfamiliarity, experimental stress).

Another important problem with existing research, especially twin studies of infant attachment, concerns the limitations in statistical power (Roisman and Fraley 2006). Reliance on small samples is typical of the methodology based on detecting significant differences between twin correlations, although it considerably precludes generalization of results. Low statistical power is also an inherent methodological problem in molecular linkage studies, as the polygenic nature of inheritance of complex interpersonal phenotypes and small samples tend to limit replication of positive findings linking single gene variations with parenting practices or attachment behavior. As stated earlier, negative findings may reflect the low power of individual studies that, when combined, could yield a significant albeit small effect. Thus, although meta-analytic reviews of linkage studies of early parenting and infant attachment would be an appealing option for future research, researchers would also benefit from the long-term use of pooled genotyped samples, gathered from molecular genetics studies conducted across multiple countries. Combining samples from a variety of populations would partly control for population stratification (i.e., selection bias) and increase statistical power of future studies.

Additionally, although G-E correlations are hypothesized to cumulate and thereby become more important as the child grows up (Scarr and McCartney 1983), there is little evidence to support this assertion because of the lack of genetically informative longitudinal data on early parenting. Thus, future twin, adoption and linkage studies should use longitudinal designs with multiple measurements through infancy and early childhood in a more systematic fashion, in order to assess this purportedly increasing magnitude of G-E correlations. In addition, bivariate twin designs can show evidence of G-E correlations, but cannot discriminate between passive and evocative G-E correlations; which considerably restricts interpretation of results. Only adoption studies—through the assessment of birth parent(s)-adoptive parent(s)-adopted infant triads—can answer this question, but such designs cannot statistically disentangle shared and non-shared environmental sources of inter-individual variance, which also limits interpretation of results. To overcome this obstacle, future genetically sensitive studies of early parenting may, for instance, focus on validating G-E correlation evidence by assessing more systematically concordance of parent and infant genotype of interest or inferred genetic risk (e.g., indicators of emotional dysregulation).

2.13.1 *Future Directions*

Exciting avenues for future research on G-E processes in the context of early parenting may be offered by the better identification of the specific "causal factors" that lead low risk (i.e., sensitive, secure, positive) parents to have secure and well-adjusted children, given that observations of parenting quality are only moderately correlated with ratings of infant attachment security (e.g., Bokhorst et al. 2003; De Wolff and van IJzendoorn 1997; Fearon et al. 2006), as well as evidence that non-shared environmental processes account for a substantial proportion of the variation in infant attachment security (e.g., O'Connor and Croft 2000) and its covariation with parenting quality (Roisman and Fraley 2008). Certain authors (e.g., Fearon et al. 2006; Roisman and Fraley 2008) concur that these findings present a challenge to attachment and early parenting researchers who, until recently, have almost solely focused on identifying the antecedents of attachment security in variations in parenting quality assumed to be largely shared within families. These specific "causal factors" could be identified (and then incorporated in genetically sensitive designs), for a start, through the study other theoretical models of parent-infant interactions than attachment theory; for instance, parental reflective functioning (e.g., Grienenberger et al. 2005; Slade 2006), parenting styles (e.g., Cheah et al. 2009), etc.

On the other hand, as economic and psychosocial hardship and adversity may create considerable contextual stress and thus provide a negative background for early parent-infant interactions (Lee et al. 2011; McConnell et al. 2011; McLoyd 1998) which may be otherwise adaptive, "causal factors" may be investigated more systematically within behavior-genetics designs in the form of psychosocial risk indicators (but for a few exceptions, see Asbury et al. 2006; Ouellet-Morin et al. 2008) such as family income, social support, familial history of substance abuse disorders or parents' education. Identification of specific mechanisms through which psychosocial adversity may affect early interactions with the infant (e.g., creation of a sense of hopelessness in the parent, parental anger/frustration or depressive symptoms, lesser amount of time spent with the infant, fatigue) would also be necessary. That is to say, a more competent identification of such factors of environmental nature may considerably impact results from studies investigating G-E processes in the context of early parent-infant interactions, as the genetic-environmental etiology of the covariance between two modestly related phenotypes may differ from the genetic-environmental etiology of the covariance between two strongly related phenotypes. Precisely, there may be a positive association between the adequacy of a "causal factor" embedded in the infant's early environment and the odds of finding significant shared environmental mediation. For instance, the well-documented prominence of shared environmental variance on infant attachment behavior in univariate twin studies (e.g., Bakermans-Kranenburg and van IJzendoorn 2004; Roisman and Fraley 2006) may potentially be consolidated if more proficient environmental factors were identified and incorporated into bivariate twin designs.

As this review shows, both parental and infant genotype—interfaced with environmental putative risk factors—may play a role in the development of early parental practices and various child outcomes. Although such findings are stimulating, specific pathways through which genotype contributes to interpersonal behavior in parents and children during infancy often remain elusive. For that matter, the detection of different functional variants of specific genes is a first step to examine causal differences among alleles. However, these functional differences may be embedded in a complex, multifaceted system involving at least three levels of analysis: (1) the functional activity of the gene product itself; (2) the levels of its expression in different brain circuits or at different times during early child-hood, and (3) its differential expression to environmental risk factors (e.g., Chen et al. 2011). A deeper understanding of these neurochemical processes may help clarify the G-E processes involved in early parent-infant interactions that contrib-ute to individual differences in both adaptive and maladaptive socio-emotional development (Chen et al. 2011).

Moreover, recent human and animal research (though it cannot be translated directly to humans) suggests involvement of early parenting in epigenetic regulation in the human brain (Gervai 2009). For instance, the methylation pattern of the pro-moter of glucocorticoid receptor gene in the hippocampus of suicide victims with a history of childhood abuse differs from that of suicide victims with no childhood abuse (McGowan et al. 2009). Thus, there is a good reason for hypothesising that epigenetic modification of gene expression plays a role in the development of early parent-infant relationship, even if the study of such processes in human infants is currently not feasible. This avenue may offer an opportunity to gain deeper knowl-edge on intergenerational transmission of attachment and parenting (Gervai 2009).

In conclusion, behavior-genetics research on early parenting is still under development, as new compelling evidence surfaces monthly. The initial reports clearly show the potential of behavior genetics to deepen our understanding of the determinants and consequences of early parenting. Through the continuing study of G-E processes, this line of research will likely help to clarify the unique con-tribution of early parenting and child-parent attachment relationship to child func-tioning. The main challenge now lies in our capacity to use these powerful tools to understand the complex G-E interplay underlying the multifaceted aspects of early parenting and infant attachment in a fully developmental perspective.

References

Ainsworth, M. D. (1985). Attachments across the life span. *Bulletin of the New York Academy of Medicine, 61*, 792–812.

Ainsworth, M. D., Blehar, M. C., Waters, E., & Wall, S. (1978). *Patterns of attachment: A psy-chological study of the strange situation*. Hillsdale, IN: Erlbaum.

Alemany, S., Rijsdijk, F. V., Haworth, C. M., Fañanás, L., & Plomin, R. (2013). Genetic origin of the relationship between parental negativity and behavior problems from early childhood to adolescence: A longitudinal genetically sensitive study. *Development and Psychopathology, 25*, 487–500.

Asbury, K., Dunn, J. F., Pike, A., & Plomin, R. (2003). Nonshared environmental influences on individual differences in early behavioral development: A monozygotic twin differences study. *Child Development, 74*, 933–943.

Asbury, K., Wachs, F. P., & Plomin, R. (2005). Environmental moderators of genetic influence on verbal and nonverbal abilities in early childhood. *Intelligence, 33*(6), 643–661.

Asbury, K. J., Wachs, F. P., & Plomin, R. (2006). Birthweight-discordance and differences in early parenting relate to monozygotic twin differences in behavior problems and academic achievement at age 7. *Developmental Science, 9*, F22–F31.

Atkinson, L., Chisholm, V. C., Scott, B. G., Goldberg, S., Vaughn, B. E., Blackwell, J., et al. (1999). Maternal sensitivity, child functional level, and attachment in Down syndrome. *Monographs of the Society for Research in Child Development, 64*(3), 45–66.

Avinun, R., Ebstein, R. P., & Knafo, A. (2012). Human maternal behaviour is associated with arginine vasopressin receptor 1A gene. *Biology Letters, 8*, 894–896.

Bakermans-Kranenburg, M. J., & van IJzendoorn, M. H. (2004). No association of the dopamine D4 receptor (DRD4) and -521 C/T promoter polymorphisms with infant attachment disorganization. *Attachment & Human Development, 6*(3), 211–222.

Bakermans-Kranenburg, M. J., & van IJzendoorn, M. H. (2006). Gene-environment interaction of the dopamine D4 receptor (DRD4) and observed maternal insensitivity predicting externalizing behavior in preschoolers. *Developmental Psychobiology, 48*(5), 406–409.

Bakermans-Kranenburg, M. J., & van IJzendoorn, M. H. (2008). Oxytocin receptor (OXTR) and serotonin transporter (5-HTT) genes associated with observed parenting. *Social Cognitive and Affective Neuroscience, 3*(2), 128–134.

Bakermans-Kranenburg, M. J., van IJzendoorn, M. H., Bokhorst, C. L., & Schuengel, C. (2004). The importance of shared environment in infant–father attachment. A behavioral genetic study of the Attachment Q-Sort. *Journal of Family Psychology, 18*, 545–549.

Bakermans-Kranenburg, M. J., van IJzendoorn, M. H., Mesman, J., Alink, L. R., & Juffer, F. (2008a). Effects of an attachment-based intervention on daily cortisol moderated by dopamine receptor D4: A randomized control trial on 1- to 3-year-olds screened for externalizing behavior. *Development and Psychopathology, 20*(3), 805–820.

Bakermans-Kranenburg, M. J., van IJzendoorn, M. H., Pijlman, F. T. A., Mesman, J., & Juffer, F. (2008b). Experimental evidence for differential susceptibility: Dopamine D4 receptor polymorphism (DRD4 VNTR) moderates intervention effects on toddlers' externalizing behavior in a randomized controlled trial. *Developmental Psychology, 44*(1), 293–300.

Barry, R. A., Kochanska, G., & Philibert, R. A. (2008). G x E interaction in the organization of attachment: mothers' responsiveness as a moderator of children's genotypes. *Journal of Child Psychology and Psychiatry and Allied Disciplines, 49*(12), 1313–1320.

Becker, K. D., Ginsburg, G. S., Domingues, J., & Tein, J.-Y. (2010). Maternal control behavior and locus of control: Examining mechanisms in the relation between maternal anxiety disorders and anxiety symptomatology in children. *Journal of Abnormal Child Psychology, 38*, 533–543.

Belsky, J. (1984). The determinants of parenting: A process model. *Child Development, 55*, 83–96.

Belsky, J., & Jaffee, S. R. (2006). The multiple determinants of parenting. In D. Cicchetti & D. J. Donald (Eds.), *Developmental Psychopathology, Vol. 3: Risk, Disorder, and Adaptation* (pp. 38–85). Hoboken, NJ: Wiley.

Belsky, J., & Pluess, M. (2009). Beyond diathesis stress: Differential susceptibility to environmental influences. *Psychological Bulletin, 135*(6), 885–908.

Benjamin, J., Li, L., Patterson, C., Greenberg, B. D., Murphy, D. L., & Hamer, D. H. (1996). Population and familial association between the D4 dopamine receptor gene and measures of Novelty Seeking. *Nature Genetics, 2*(1), 81–84.

Boeldt, D. L., Rhee, S. H., DiLalla, L. F., Mullineaux, P. Y., Schulz, H. R., et al. (2012). The association between positive parenting and externalizing behavior. *Infant and Child Development, 21*(1), 85–106.

Boivin, M., Pérusse, D., Dionne, G., Saysset, V., Zoccolillo, M., Tarabulsy, G., et al. (2005). Parent's perceptions and self-assessed behaviors toward their 5-month-old infants in a large twin and singleton sample. *Journal of Child Psychology and Psychiatry and Allied Disciplines, 46*, 612–630.

Bokhorst, C. L., Bakermans-Kranenburg, M. J., Fearon, R. M., van IJzendoorn, M. H., Fonagy, P., & Schuengel, C. (2003). The importance of shared environment in mother–infant attachment security: A behavioral genetic study. *Child Development, 74*, 1769–1782.

Bornstein, M. C. (2002). *Handbook of parenting* (Vol. 1–4). Mahwah, NJ: Erlbaum.

Bowlby, J. (1982). *Attachment and loss (vol. 1): Attachment* (2nd ed.). London, UK: Hogarth Press.

Bradley, R. H., & Corwyn, R. F. (2007). Externalizing problems in fifth grade: Relations with productive activity, maternal sensitivity, and harsh parenting from infancy through middle childhood. *Developmental Psychology, 43*(6), 1390–1401.

Braungart-Rieker, J. M., Fulker, D. W., & Plomin, R. (1992). Genetic mediation of the home environment during infancy: A sibling adoption study of the HOME. *Developmental Psychology, 28*(6), 1048–1055.

Braungart-Rieker, J. M., Rende, R. D., Plomin, R., & DeFries, J. C. (1995). Genetic mediation of longitudinal associations between family environment and childhood behavior problems. *Development and Psychopathology, 7*, 233–245.

Bretherton, I., & Waters, E. (Eds.). (1985). *Growing points of attachment theory and research.* Monographs of the Society for Research in Child Development. Serial No. 209.

Brown, S. M., & Hariri, A. R. (2006). Neuroimaging studies of serotonin gene polymorphisms: Exploring the interplay of genes, brain, and behavior. *Cognitive, Affective & Behavioral Neuroscience, 6*(1), 44–52.

Brunner, H. G., & Nelen, M. (1993). Abnormal behavior associated with a point mutation in the structural gene for monoamine oxidase A. *Science, 262*(5133), 578–781.

Campbell, S. B., Brownell, C. A., Hugerford, A., Spieker, S. I., Mohan, R., & Blessing, J. S. (2004). The course of maternal depressive symptoms and maternal sensitivity as predictors of attachment security at 36 months. *Development and Psychopathology, 16*, 231–252.

Campbell, S. B., Matestic, P., von Stauffenberg, C., Mohan, R., & Kirchner, T. (2007). Trajectories of maternal depressive symptoms, maternal sensitivity, and children's functioning at school entry. *Developmental Psychology, 43*(5), 1202–1215.

Canli, T., & Lesch, K.-P. (2007). Long story short: The serotonin transporter in emotion regulation and social cognition. *Nature Neuroscience, 10*, 1103–1109.

Carpenter, L. L., Shattuck, T. T., Tyrka, A. R., Geracioti, T. D., & Price, L. H. (2011). Effect of childhood physical abuse on cortisol stress response. *Psychopharmacology (Berl), 214*, 367–375.

Caspi, A., McClay, J., Moffitt, T. E., Mill, J., Martin, J., Craig, I. W., et al. (2002). Role of genotype in the cycle of violence in maltreated children. *Science, 297*, 851–854.

Caspi, A., Moffit, T. E., Morgan, J., Rutter, M., Taylor, A., Arseneault, L., et al. (2004). Maternal expressed emotion predicts children's antisocial behavior problems: Using monozygotic-twin differences to identify environmental effects on behavioral development. *Developmental Psychology, 40*, 149–161.

Champagne, F. A., Chretien, P., Stevenson, C. W., Zhang, T. Y., Gratton, A., & Meaney, M. J. (2004). Variations in nucleus accumbens dopamine associated with individual differences maternal behavior in the rat. *Journal of Neuroscience, 24*, 4113–4123.

Cheah, C. S. L., Leung, C. Y. Y., Tahseen, M., & Schultz, D. (2009). Authoritative parenting among immigrant Chinese mothers of preschoolers. *Journal of Family Psychology, 23*(3), 311–320.

Chen, F. S., Barth, M. E., Johnson, S. L., Gotlib, I. H., & Johnson, S. C. (2011). Oxytocin receptor (OXTR) polymorphisms and attachment in human infants. *Frontiers in Psychology, 2*, 200–206.

Cicchetti, D., Rogosch, F. A., & Toth, S. L. (2011). The effects of child maltreatment and polymorphisms of the serotonin transporter and dopamine D4 receptor genes on infant attachment and intervention efficacy. *Development and Psychopathology, 23*(2), 357–372.

Collins, W. A., Maccoby, E. E., Steinberg, L., Hetherington, E. M., & Bornstein, M. H. (2000). Contemporary research on parenting: The case for nature and nurture. *American Psychologist, 55*, 218–232.

Comings, D. E. (1998). Polygenic inheritance and micro/minisatellites. *Molecular Psychiatry, 3*(1), 21–32.

Comings, D. E., Gonzalez, N., Wu, S., Gade, R., Muhleman, D., Saucier, G., et al. (1999). Studies of the 48 bp repeat polymorphism of the DRD4 gene in impulsive, compulsive, addictive behaviors: Tourette syndrome, ADHD, pathological gambling, and substance abuse. *American Journal of Medical Genetics, 88*(4), 358–368.

Cummings, E. M., & Davies, P. T. (2002). Effects of marital conflict on children: recent advances and emerging themes in process-oriented research. *Journal of Child Psychology and Psychiatry and Allied Disciplines, 43*(1), 31–64.

Deater-Deckard, K. (2000). Parenting and child behavioral adjustment in early childhood: A quantitative genetic approach to studying family processes. *Child Development, 71*, 468–484.

Deater-Deckard, K. (2001). Nonshared environmental processes in social-emotional development: an observational study of identical twin differences in the preschool period. *Developmental Science, 4*(2), F1–F6.

Deater-Deckard, K., Dunn, J., O'Connor, T. G., Davies, L., & Golding, J. (2001). Using the step-family genetic design to examine gene-environment processes in child and family functioning. *Marriage and Family Review, 33*, 131–156.

Deater-Deckard, K., & O'Connor, T. G. (2000). Parent–child mutuality in early childhood: Two behavioral genetic studies. *Developmental Psychology, 36*, 561–570.

Derijk, R. H., & de Kloet, E. R. (2008). Corticosteroid receptor polymorphims: determinants of vulnerability and resilience. *European Journal of Pharmacology, 583*(2–3), 303–311.

De Wolff, M. S., & van IJzendoorn, M. H. (1997). Sensitivity and attachment: A meta-analysis on parental antecedents of infant attachment. *Child Development, 68*, 571–591.

DiLalla, L. F., & Bishop, E. G. (1996). Differential maternal treatment of infant twins: Effects of infant behaviors. *Behavior Genetics, 26*(6), 535–542.

DiLalla, L. F., Elam, K. K., & Smolen, A. (2009). Genetic and gene-environment interaction effects on preschoolers' social behaviors. *Developmental Psychobiology, 51*(6), 451–464.

Dodge, K. A. (1990). Nature versus nurture in childhood conduct disorder: Is it time to ask a different question? *Developmental Psychology, 26*, 698–701.

Dozier, M., Stovall, K. C., Albus, K. E., & Bates, B. (2001). Attachment for infants in foster care: The role of caregiver state of mind. *Child Development, 72*, 1467–1477.

Drabant, E. M., Hariri, A. R., Meyer-Lindenberg, A., Munoz, K. E., Mattay, V. S., Kolachana, B. S., et al. (2006). Catechol O-methyltransferase val(158)met genotype and neural mechanisms related to affective arousal and regulation. *Archives of General Psychiatry, 63*, 1396–1406.

Easterbrooks, M. A., & Goldberg, W. A. (1984). Toddler development in the family: Impact of father involvement and parenting characteristics. *Child Development, 55*, 740–752.

Ebstein, R. P. (2006). The molecular genetic architecture of human personality: Beyond self-report questionnaires. *Molecular Psychiatry, 11*, 427–445.

Elam, K. K., Harold, G. T., Neiderhiser, J. M., Reiss, D., Shaw, D. S., Natsuaki, M. N., Gaysina, D., Barrett, D., & Leve, L. D. (2013, December 23). Adoptive parent hostility and children's peer behavior problems: Examining the role of genetically informed child attributes on adoptive parent behavior. *Developmental Psychology*. Advance online publication. doi: 10.1037/a0035470.

Enoch, M.-A., Steer, C. D., Newman, T. K., Gibson, N., & Goldman, N. (2012). Early life stress, MAOA, and gene-environment interactions predict behavioral disinhibition in children. *Genes, Brain, and Behavior, 9*(1), 65–74.

Faraone, S. V., & Khan, S. A. (2006). Candidate gene studies of attention-deficit/hyperactivity disorder. *The Journal of Clinical Psychiatry, 67*(Suppl 8), 13–20.

Faraone, S. V., Perlis, R. H., Doyle, A. E., Smoller, J. W., Goralnick, J. J., Holmgren, M. A., & Sklar, P. (2005). Molecular genetics of attention-deficit/hyperactivity disorder. *Biological Psychiatry, 57*(11), 1313–1323.

Fearon, R. M., van IJzendoorn, M. H., Fonagy, P., Bakermans-Kranenburg, M. J., Schuengel, C., & Bokhorst, C. L. (2006). In search of shared and nonshared environmental factors in security of attachment: A behavior-genetic study of the association between sensitivity and attachment security. *Developmental Psychology, 42*(6), 1026–1040.

Feldman, R., Zagoory-Sharon, O., Weisman, O., Schneiderman, I., Gordon, I., Maoz, R., et al. (2012). Sensitive parenting is associated with plasma oxytocin and polymorphisms in the OXTR and CD38 genes. *Biological Psychiatry, 72*(3), 175–181.

Finkel, D., & Matheny, A. P. (2000). Genetic and environmental influences on a measure of infant attachment security. *Twin Research, 3*, 242–250.

Finkel, D., Wille, D. E., & Matheny, A. P. (1998). Preliminary results from a twin study of infant-caregiver attachment. *Behavior Genetics, 28*(1), 1–8.

Forget-Dubois, N., Boivin, M., Dionne, G., Pierce, T., Tremblay, R. E., & Pérusse, D. (2007). A longitudinal twin study of the genetic and environmental etiology of maternal hostile-reactive behavior during infancy and toddlerhood. *Infant Behavior & Development, 30*(3), 453–465.

Frigerio, A., Ceppi, E., Rusconi, M., Giodra, R., Raggi, M. E., & Fearon, P. (2009). The role played by the interaction between genetic factors and attachment in the stress response in infancy. *Journal of Child Psychology and Psychiatry, 50*(12), 1513–1522.

Garai, E. M., Forehand, R. L., Colletti, C. J. M., Reeslund, K., Potts, J., & Compas, B. (2009). The relation of maternal sensitivity to children's internalizing and externalizing problems within the context of maternal depressive symptoms. *Behavior Modification, 33*(5), 559–582.

Gervai, J. (2009). Environmental and genetic influences on early attachment. *Child and Adolescent Psychiatry and Mental Health, 25*(3), 1–12.

Gervai, J., Nemoda, Z., Lakatos, K., Ronai, Z., Toth, I., Ney, K., & Sasvari-Szekely, M. (2005). Transmission disequilibrium tests confirm the link between DRD4 gene polymorphism and infant attachment. *American Journal of Medical Genetics. Part B, Neuropsychiatric Genetics: The Official Publication of the International Society of Psychiatric Genetics, 132B*(1), 126–130.

Gervai, J., Novak, A., Lakatos, K., Toth, I., Danis, I., et al. (2007). Infant genotype may moderate sensitivity to maternal affective communications: Attachment disorganization, quality of care, and the DRD4 polymorphism. *Social Neuroscience, 2*(3–4), 307–319.

Gilissen, R., Bakermans-Kranenburg, M. J., van IJzendoorn, M. H., & Linting, M. (2008). Electrodermal reactivity during the Trier Social Stress Test for children: interaction between the serotonin transporter polymorphism and children's attachment representation. *Developmental Psychobiology, 50*(6), 615–625.

Gray, C., Carter, R., & Silverman, W. (2011). Anxiety symptoms in African American Children: Relations with ethnic pride, anxiety sensitivity, and parenting. *Journal of Child and Family Studies, 20*(2), 205–213.

Grienenberger, J., Kelly, K., & Slade, A. (2005). Maternal reflective functioning, mother-infant affective communication and infant attachment: Exploring the link between mental states and observed caregiving behavior in the intergenerational transmission of attachment. *Attachment and Human Development, 7*(3), 299–311.

Grossmann, K., Grossmann, K. E., Spangler, G., Suess, G., & Unzner, L. (1985). Maternal sensitivity and newborns orientation responses as related to quality of attachment in northern Germany. *Monographs of the Society for Research in Child Development, 50*, 233–256.

Guimond, F.-A., Brendgen, M., Forget-Dubois, N., Dionne, G., Vitaro, F., Tremblay, R. E., & Boivin, M. (2012). Associations of mother's and father's parenting practices with children's observed social reticence in a competitive situation: A monozygotic twin difference study. *Journal of Abnormal Child Psychology, 40*(3), 391–402.

Jaffee, S. R., Caspi, A., Moffitt, T. E., Dodge, K. A., Rutter, M., Taylor, A., & Tully, L. (2005). Nature X nurture: Genetic vulnerabilities interact with child maltreatment to promote behavior problems. *Development and Psychopathology, 17*, 67–84.

Jaffee, S. R., Caspi, A., Moffit, T. E., Polo-Tomas, M., Price, T., & Taylor, A. (2004a). The limits of child effects: Evidence for genetically mediated child effects on corporal punishment but not on physical maltreatment. *Developmental Psychology, 40*, 1047–1058.

Jaffee, S. R., Caspi, A., Moffitt, T. E., & Taylor, A. (2004b). Physical maltreatment victim to antisocial child: Evidence of an environmentally mediated process. *Journal of Abnormal Psychology, 113*, 44–55.

Kaitz, M., Shalev, I., Sapir, N., Devor, N., Samet, Y., Mankuta, D., & Ebstein, R. P. (2010). Mothers' dopamine receptor polymorphism modulates the relation between infant fussiness and sensitive parenting. *Developmental Psychobiology, 52*(2), 149–157.

Kiff, C., Lengua, L., & Zalewski, M. (2011). Nature and nurturing: Parenting in the context of child temperament. *Clinical Child and Family Psychology Review, 14*(3), 251–301.

Kim, E. (2011). Korean American parental depressive symptoms and parental acceptance-rejection and control. *Issues in Mental Health Nursing, 32*(2), 114–120.

Klahr, A. M., & Burt, A. S. (2013). Elucidating the etiology of individual differences in parenting: A meta-analysis of behavior genetic research. *Psychological Bulletin, 140,* 544–586. doi:10.1037/a0034205.

Kochanska, G., Philibert, R. A., & Barry, R. A. (2009). Interplay of genes and early mother-child relationship in the development of self-regulation from toddler to preschool age. *Journal of Child Psychology and Psychiatry and Allied Disciplines, 50*(11), 1331–1338.

Knafo, A., & Plomin, R. (2006). Parental discipline and affection and children's prosocial behavior: Genetic and environmental links. *Journal of Personality and Social Psychology, 90*(1), 147–164.

Lakatos, K., Nemoda, Z., Birkas, E., Ronai, Z., Kovacs, E., Ney, K., et al. (2003). Association of D4 dopamine receptor gene and serotonin transporter promoter polymorphisms with infants' response to novelty. *Molecular Psychiatry, 8*(1), 90–98.

Lakatos, K., Nemoda, Z., Toth, I., Ronai, Z., Ney, K., Sasvari-Szekely, M., & Gervai, J. (2002). Further evidence for the role of dopamine D4 receptor (DRD4) gene in attachment disorganization: interaction of the exon III 48-bp repeat and the -521 C/T promoter polymorphisms. *Molecular Psychiatry, 7*(1), 27–31.

Lakatos, K., Toth, I., Nemoda, Z., Ney, K., Sasvari-Szekely, M., & Gervai, J. (2000). Dopamine D4 receptor (DRD4) gene polymorphism is associated with attachment disorganization in infants. *Molecular Psychiatry, 5*(6), 633–638.

Larsson, H., Viding, E., Risjdijk, F. V., & Plomin, R. (2008). Relationships between parental negativity and childhood antisocial behavior over time: a bidirectional effects model in a longitudinal genetically informative design. *Journal of Abnormal Child Psychology, 36*(5), 633–645.

Lee, S. S., Chronis-Tuscano, A., Keenan, K., Pelham, W. E., Loney, J., et al. (2010). Association of maternal dopamine transporter genotype with negative parenting: Evidence for gene x environment interaction with child disruptive behavior. *Molecular Psychiatry, 15,* 548–558.

Lee, C.-Y. S., Lee, J., & August, G. J. (2011). Financial stress, parental depressive symptoms, parenting practices, and children's externalizing problem behaviors: Underlying processes. *Family Relations: An Interdisciplinary Journal of Applied Family Studies, 60*(4), 476–490.

Lesch, K.-P., & Bengel, D. (1996). Association of anxiety-related traits with a polymorphism in the serotonin transporter gene regulator. *Science, 274*(5292), 1527–1532.

Leve, L. D., Harold, G. T., Ge, X., Neiderhiser, J. M., Shaw, J. M., et al. (2009). Structured parenting of toddlers at high versus low genetic risk: two pathways to child problems. *Journal of the American Academy of Child and Adolescent Psychiatry, 48*(11), 1102–1109.

Leve, L. D., Kerr, D. C. R., Shaw, D. S., Ge, X., Neiderhiser, J. M., Scaramella, L. V., et al. (2010). Infant pathways to externalizing behavior: Evidence of genotype x environment interactions. *Child Development, 31,* 340–356.

Lieb, R., Wittchen, H.-U., Höfer, M., Fuetsch, M., Stein, M. B., & Merikangas, K. R. (2000). Parental psychopathology, parenting styles, and the risk of social phobia in offspring: A prospective-longitudinal community study. *Archives of General Psychiatry, 57*(9), 859–866.

Lipscomb, S. T., Leve, L. D., Shaw, D. S., Neiderhiser, J. D., et al. (2012). Negative emotionality and externalizing problems in toddlerhood: Overreactive parenting as a moderator of genetic influences. *Development and Psychopathology, 24*(1), 167–179.

Losoya, S. H., Callor, S., Rowe, D. C., & Goldsmith, H. H. (1997). Origins of familial similarity in parenting: A study of twins and adoptive siblings. *Developmental Psychology, 33*(6), 1012–1023.

Lucki, I. (1998). The spectrum of behaviors influenced by serotonin. *Biological Psychiatry, 44*(3), 151–162.

Luijk, M. P. C. M., Roisman, G. I., Haltigan, J. D., et al. (2011a). Dopaminergic, serotonergic, and oxytonergic candidate genes associated with infant attachment security and disorganization? In search of main and interaction effects. *Journal of Child Psychology and Psychiatry, 52*(12), 1295–1307.

Luijk, M. P. C. M., Tharner, A., Bakermans-Kranenburg, M. J., van IJzendoorn, M. H., Jaddoe, V. W., Hofman, A., et al. (2011b). The association between parenting and attachment security is moderated by a polymorphism in the mineralocorticoid receptor gene: Evidence for differential susceptibility. *Biological Psychology, 88*(1), 37–40.

Luijk, M. P. C. M., Velders, F. P., Tharner, A., et al. (2010). FKBP5 and resistant attachment predict cortisol reactivity in infants: Gene–environment interaction. *Psychoneuroendocrinology, 35*(10), 1454–1461.

Lytton, H. (1977). Do parents create, or respond to, differences in twins? *Developmental Psychology, 13*(5), 456–459.

Lytton, H., Martin, N. G., & Eaves, L. (1977). Environmental and genetical causes of variation in ethological aspects of behavior in two-year-old boys. *Social Biology, 24*(3), 200–211.

Main, M., & Goldwyn, R. (1998). *Adult attachment classification system*. London, UK: University College.

Martini, T. S., Root, C. A., & Jenkins, J. M. (2004). Low and middle income mother's regulation of negative emotion: Effects of children's temperament and situational emotional responses. *Social Development, 13*(4), 515–530.

Martins, C., & Gaffan, E. A. (2000). Effects of early maternal depression on patterns of infant-mother attachment: A meta-analytic investigation. *Journal of Child Psychology and Psychiatry and Allied Disciplines, 41*(6), 737–747.

Matheny, A. P., Wilson, R. S., & Nuss, S. M. (1984). Toddler temperament: Stability across settings and over ages. *Child Development, 55*, 1200–1211.

McConnell, D., Breitkreuz, D., & Savage, A. (2011). From financial hardship to child difficulties: main and moderating effects of perceived social support. *Child: Care, Health and Development, 37*(5), 679–691.

McElwain, N. L., & Booth-LaForce, C. (2006). Maternal sensitivity to infant distress and nondistress as predictors of infant-mother attachment security. *Journal of Family Psychology, 20*, 247–255.

McGowan, P. O., Sasaki, A., D'Alessio, A. C., Dymov, S., Labonté, B., Szyf, M., et al. (2009). Epigenetic regulation of the glucocorticoid receptor in human brain associates with childhood abuse. *Nature Neuroscience, 12*, 342–348.

McLoyd, V. C. (1998). Children in poverty: Development, public policy, and practice. In W. Damon (Ed.), *Handbook of child psychology* (5th ed., Vol. 4, pp. 4–135). New York, NY: Wiley.

Mileva-Seitz, V., Kennedy, J., Atkinson, L., Steiner, M., Levitan, R., et al. (2011). Serotonin transporter allelic variation in mothers predicts maternal sensitivity, behavior and attitudes toward 6-month-old infants. *Genes, Brain and Behavior, 10*, 325–333.

Mills-Koonce, W. R., Propper, C. B., Gariepy, J.-L., Blair, C., Garrett-Peters, P., & Cox, M. J. (2007). Bidirectional genetic and environmental influences of mother and child behavior: The family system as the unit of analyses. *Development and Psychopathology, 19*(4), 1073–1087.

Morley, T., Bailey, H., Pederson, D., & Moran, G. (2011). Exploring the development of adolescent mother-infant attachment relationships: The contribution of ecological factors". *Psychology Presentations*. Paper 31. http://ir.lib.uwo.ca/psychologypres/31.

Murray, L. (1992). The impact of postnatal depression on infant development. *Journal of Child Psychology and Psychiatry, 33*, 543–561.

Nair, H., & Murray, A. D. (2005). Predictors of attachment security in preschool children from intact and divorced families. *Journal of Genetic Psychology, 166*(3), 245–263.

Natsuaki, M. N., Ge, X., Leve, L. D., Neiderhiser, J. D., et al. (2010). Genetic liability, environment, and the development of fussiness in toddlers: The roles of maternal depression and parental responsiveness. *Developmental Psychology, 46*(5), 1147–1158.

Nishikawa, S., Sundbom, E., & Hägglöf, B. (2010). Influence of perceived parental rearing on adolescent self-concept and internalizing and externalizing problems in Japan. *Journal of Child and Family Studies, 19*(1), 57–66.

Noble, E. P., Ozkaragoz, T. Z., Ritchie, T., Zhang, X., Bekin, T. R., & Sparkes, R. S. (1998). D2 and D4 dopamine receptor polymorphisms and personality. *American Journal of Medical Genetics, 81*, 257–267.

Numan, M. (2007). Motivational systems and the neural circuitry of maternal behavior in the rat. *Developmental Psychobiology, 49*, 12–21.

O'Connor, T. G., & Croft, C. M. (2000). A twin study of attachment in preschool children. *Child Development, 72*, 1501–1511.

Oosterman, M., & Schuengel, C. (2007). Physiological effects of separation and reunion in relation to attachment and temperament in young children. *Developmental Psychobiology, 49*, 119–128.

Oscar-Berman, M., & Bowirrat, A. (2005). Genetic influences in emotional dysfunction and alcoholism-related brain damage. *Neuropsychiatric Disease and Treatment, 1*(3), 211–229.

Ouellet-Morin, I., Boivin, M., Dionne, G., Lupien, S. J., Arsenault, L., Barr, R. G., et al. (2008). Variations in heritability of cortisol reactivity to stress as a function of early familial adversity among 19-month-old twins. *Archives of General Psychiatry, 65*(2), 211–218.

Ouellet-Morin, I., Dionne, G., Pérusse, D., Lupien, S. J., Arsenault, L., Barr, R. G., et al. (2009). Daytime cortisol secretion in 6-month-old twins: Genetic and environmental contributions as a function of early familial adversity. *Biological Psychiatry, 65*(5), 409–416.

Palusci, V. J., & Loeb, F. L. (2011). Risk factors and services for child maltreatment among infants and young children. *Children and Youth Services Review, 33*(8), 1374–1382.

Parrish, J. W., Young, M. B., Perham-Hester, K. A., & Gessner, B. D. (2011). Identifying risk factors for child maltreatment in Alaska: A population-based approach. *American Journal of Preventive Medicine, 40*(6), 666–673.

Pauli-Pott, U., Friedel, S., Hinney, A., & Hebebrand, J. (2009). Serotonin transporter gene polymorphism (5-HTTLPR), environmental conditions, and developing negative emotionality and fear in early childhood. *Journal of Neural Transmission, 116*(4), 503–512.

Petrill, S. A., & Deater-Deckard, K. (2004). Task orientation, parental warmth and SES account for a significant proportion of the shared environmental variance in general cognitive ability in early childhood: Evidence from a twin study. *Developmental Science, 7*(1), 25–32.

Plomin, R., DeFries, J. C., & Loehlin, J. C. (1977). Genotype-environment interaction and correlation in the analysis of human behavior. *Psychological Bulletin, 84*(2), 309–322.

Posner, M. I., Rothbart, M. K., & Sheese, B. E. (2007). Attention genes. *Developmental Science, 10*(1), 24–29.

Propper, C., Moore, G. A., Mills-Koonce, W. R., Halpern, C., Hill-Soderlund, A. L., Calkins, S., et al. (2008). Gene–environment contributions to the development of infant vagal reactivity: The interaction of dopamine and maternal sensitivity. *Child Development, 79*, 1377–1394.

Propper, C., Willoughby, M., Halpern, C. T., Carbone, M. A., et al. (2007). Parenting quality, DRD4, and the prediction of externalizing and internalizing behaviors in early childhood. *Developmental Psychobiology, 49*(6), 619–632.

Putnam, S. P., Sanson, A. V., & Rothbart, M. K. (2002). Child temperament and parenting. In M. H. Bornstein (Ed.), *Handbook of parenting: Vol. 1: Children and parenting* (2nd ed). Mahwah, NJ: Erlbaum.

Rhoades, K. A., Leve, L. D., Harold, G. T., Neiderhiser, J. D., et al. (2009). Longitudinal pathways from marital hostility to child anger during toddlerhood: genetic susceptibility and indirect effects via harsh parenting. *Journal of Family Psychology: JFP: Journal of the Division of Family Psychology of the American Psychological Association (Division 43), 25*(2), 282–291.

Ricciuti, A. E. (1993). Child–mother attachment: A twin study. *Dissertation Abstracts International, 54*, 3364.

Roisman, G. I., & Fraley, R. C. (2006). The limits of genetic influence: A behavior—genetic analysis of infant—caregiver relationship quality and temperament. *Child Development, 77*, 1656–1667.

Roisman, G. I., & Fraley, R. C. (2008). A behavior-genetic study of parenting quality, infant attachment security, and their covariation in a nationally representative sample. *Developmental Psychology, 44*(3), 831–839.

Rubin, K. H., Burgess, K. B., Dwyer, K. M., & Hastings, P. D. (2003). Predicting preschoolers' externalizing behaviors from toddler temperament, conflict, and maternal negativity. *Developmental Psychology, 39,* 164–176.

Rutter, M., & Silberg, J. (2002). Gene-environment interplay in relation to emotional and behavioral disturbance. *Annual Review of Psychology, 53,* 463–490.

Scarr, S., & McCartney, K. (1983). How people make their own environments: A theory of genotype-environment effects. *Child Development, 54,* 424–435.

Sheese, B. E., Voelker, P. M., Rothbart, M. K., & Posner, M. I. (2007). Parenting quality interacts with genetic variation in dopamine receptor D4 to influence temperament in early childhood. *Development and Psychopathology, 19,* 1039–1046.

Shih, J. C., Chen, K., & Ridd, M. J. (1999). Monoamine oxidase: From genes to behavior. *Annual Review of Neuroscience, 22*(1), 197–218.

Slade, A. (2006). Reflective parenting programs: Theory and development. *Psychoanalytic Inquiry, 26,* 640–657.

Smolka, M. N., Schumann, G., Wrase, J., Grüsser, S. M., et al. (2005). Catechol-*O*-methyltransferase *val*^{158}*met* genotype affects processing of emotional stimuli in the amygdala and prefrontal cortex. *The Journal of Neuroscience, 25*(4), 836–842.

Spangler, G., Fremmer-Bombik, E., & Grossmann, K. (1996). Social and individual determinants of infant attachment security and disorganization. *Infant Mental Health Journal, 17,* 127–139.

Spangler, G., Johann, M., Ronai, Z., & Zimmerman, P. (2009). Genetic and environmental influence on attachment disorganization. *Journal of Child Psychology and Psychiatry and Allied Disciplines, 50*(8), 952–961.

Spielewoy, C., Roubert, C., Hamon, M., Nosten-Bertrand, M., Betancur, C., & Giros, B. (2000). Behavioral disturbances associated with hyperdopaminergia in dopamine-transporter knockout mice. *Behavioral Pharmacology, 11,* 279–290.

Spinath, F. M., & O'Connor, T. G. (2003). A Behavioral Genetic Study of the Overlap Between Personality and Parenting. *Journal of Personality, 71*(5), 785–811.

Spokas, M., & Heimberg, R. G. (2009). Overprotective Parenting, Social Anxiety, and External Locus of Control: Cross-sectional and Longitudinal Relationships. *Cognitive Therapy & Research, 33*(6), 543–551.

Steele, M., Hodges, J., Kaniuk, J., Hillman, S., & Henderson, K. (2003). Attachment representations and adoption: associations between maternal states of mind and emotion narratives in previously maltreated children. *Journal of Child Psychotherapy, 29*(2), 187–205.

Steptoe, A., van Jaarzveld, C. H., Semmler, C., Plomin, R., & Wardle, J. (2009). Heritability of daytime cortisol levels and cortisol reactivity in children. *Psychoneuroendocrinology, 34,* 273–280.

Suhara, T., Yasuno, F., Sudo, Y., Yamamoto, M., Inoue, M., Okubo, Y., & Suzuki, K. (2001). Dopamine D2 receptors in the insular cortex and the personality trait of novelty seeking. *Neuroimage, 13,* 891–895.

Summer, G., & Spietz, A. L. (1995). *NCAST caregiver/parent-child interaction teaching manual* (2nd ed.). Seattle, WA: NCAST Publications.

Tarabulsy, G. M., Pascuzzo, K., Moss, E., St-Laurent, D., Bernier, A., Cyr, C., & Dubois-Comtois, K. (2008). Attachment-based intervention for maltreating families. *American Journal of Orthopsychiatry, 78*(3), 322–332.

Turkheimer, E., & Waldron, M. (2000). Statistical analysis, experimental method, and causal inference in developmental behavioral genetics. *Human Development, 43*(1), 51–52.

van IJzendoorn, M. H., & Bakermans-Kranenburg, M. J. (2006). DRD47-repeat polymorphism moderates the association between maternal unresolved loss or trauma and infant disorganization. *Attachment & Human Development, 8*(4), 291–307.

van IJzendoorn, M. H., Bakermans-Kranenburg, M. J., & Mesman, J. (2008). Dopamine system genes associated with parenting in the context of daily hassles. *Genes, Brain, and Behavior, 7*(4), 403–410.

Vaughn, B. E., & Waters, E. (1990). Attachment behavior at home and in the laboratory: Q-Sort observations and strange situation classifications of one-year-olds. *Child Development, 61,* 1965–1973.

Verissimo, M., & Salvaterra, F. (2006). Maternal secure-base scripts and children's attachment security in an adopted sample. *Attachment & Human Development, 8,* 261–273.

Vondra, J. S., Sysko, J., & Belsky, J. (2005). Developmental origins of parenting: Personality and relationship factors. In T. Luster & L. Okagaki (Eds.), *Parenting: An ecological perspective.* Mahwah, NJ: Erlbaum.

Wahlstrom, D., White, T., & Luciana, M. (2010). Neurobehavioral evidence for changes in dopamine system activity during adolescence. *Neuroscience and Biobehavioral Reviews, 34,* 631–648.

Watson, K. K., Ghodasra, J. H., & Platt, M. L. (2009). Serotonin transporter genotype modulates social reward and punishment in rhesus macaques. *PLoS ONE, 4,* e4156.

Zohsel, K., Buchmann, A. F., Blomeyer, D., Hohm, E., Schmidt, M. H., Esser, G., et al. (2014). Mothers' prenatal stress and their children's antisocial outcomes—a moderating role for the Dopamine D4 Receptor (DRD4) gene. *Journal of Child Psychology and Psychiatry, 55,* 69–76. doi: 10.1111/jcpp.12138.

Zuckerman, M. (1999). Diathesis-stress models. In M. Zuckerman (Ed.), *Vulnerability to psychopathology: A biosocial model.* Washington, DC: American Psychological Association.

Chapter 3
Parenting in Childhood

Alison Pike and Bonamy R. Oliver

Genetically sensitive studies of parenting now have a 30-year history, offering sufficient data to draw some conclusions about how genes and the environment influence parenting. The majority of this chapter will focus on summarizing this literature. The key aim of the current review is to look at potential differences in genetic and environmental influences as a function of informant, study design, age of children, and parenting dimension. Across studies, we demonstrate higher heritability for negative versus positive aspects of parenting, and contextualize these findings by outlining traditional socialization theories of parenting, enabling us to draw out the implications of the behavioral genetic findings for the field of parenting more generally. Future directions for research are also suggested.

3.1 Theories of Parenting

From early childhood onwards, the theoretical perspective that has gained the most research attention has been Baumrind's parenting styles. At its core, Baumrind's perspective has focused attention on two key aspects of parenting—responsiveness/warmth and demandingness/control (Baumrind 1973). The literature linking dimensions of warmth and control to children's outcomes indicates a robust, modest to moderate association (Parke and Buriel 2006). Certainly, the importance of parenting in children's lives is supported by thousands of research studies as well as common sense, moral, and legal expectations that parents are responsible for their children's behavior.

A. Pike (✉) · B.R. Oliver
School of Psychology, University of Sussex, Brighton BN1 9RH, UK
e-mail: alisonp@sussex.ac.uk

© Springer Science+Business Media New York 2015
B.N. Horwitz and J.M. Neiderhiser (eds.), *Gene-Environment Interplay in Interpersonal Relationships across the Lifespan*,
Advances in Behavior Genetics 3, DOI 10.1007/978-1-4939-2923-8_3

In 1984, Jay Belsky argued that the importance of parenting quality was well understood, and that it was time to focus attention on to the influences on parenting itself. In proposing his "Determinants of Parenting" framework, Belsky proposed that three factors influence normative (as well as abusive) variations in parenting quality—parents' personal resources, contextual sources of stress and support, and child characteristics. In addition, Belsky proposed that parenting is a buffered system, such that not every factor must be optimal for adequate functioning, and that there is a hierarchy to the order of importance of the three factors. Specifically, Belsky proposed that parental personal resources are the most important, and child characteristics the least important.

The past 30 years of research has largely supported Belsky's original propositions (see Belsky and Jaffee 2006 for a recent review). Briefly, parents' psychological resources have been shown to be of primary importance for parenting and for child outcomes, at least during infancy. Maternal mental health, and particularly maternal depression, has a marked effect on maternal sensitivity and subsequent child behavior (e.g., Campbell et al. 2007). Contextual sources of stress and support have also received ample empirical support. The marital relationship (or partnership) is usually the primary context (Erel and Burman 1995), although spillover between work and family life is also well documented (Crouter and McHale 2005). Finally, children's characteristics play a key role in parenting, especially at older ages (Karraker and Coleman 2005), with robust evidence linking child gender and parenting (Leaper 2002) and temperament and parenting (Putnam et al. 2002).

Genetically sensitive designs of parenting offer unique insights into the importance of child versus parent contributions to parenting (e.g., Kendler and Baker 2007; McGuire et al. 2012). In the studies reviewed here, the principal approach is to use measures of parenting as the phenotype, and traditional univariate behavioral genetic analysis is conducted thereon (see Chap. 10). In brief, parenting is assessed among relatives of varying genetic relatedness, and if parenting is found to be more similar among relatives that share more of their segregating genes, genetic influence is implied. The majority of such studies are child-based designs, in which the children are identical (monzygotic: MZ) or fraternal (dizygotic: DZ) twins. For example, a mother may be observed interacting with each of her twin children in turn, and these interactions coded for the degree of warmth shown by the mother. If mothers of identical twins treat them with more similar degrees of warmth than do mothers of fraternal twins, it is suggestive of genetically influenced traits of the children (e.g., temperament) being reflected in parental behavior. These studies also yield estimates of shared and nonshared environmental influence. Shared environmental influence refers to similarity in parenting across members of the twin pair, *once genetic similarity has been accounted for*. That is, finding shared environmental influence indicates consistency in parenting across children in a family. In contrast, nonshared environmental influence refers to differences in parenting the children, *once genetic differences have been accounted for*, that is, it is an indication of differential parental treatment.

A minority of the studies reviewed here used parent-based rather than child-based designs. In these cases, the aforementioned relatives of differing genetic relatedness used as units for analyses are the parents themselves. Specifically, the parenting of adult twin mothers is compared; if the parenting shown by identical twin mothers towards their children is more similar than the parenting shown by fraternal twin mothers, this indicates genetic influence. Finding significant genetic influence when using parent-based designs indicates that genetically influenced traits of the parent (e.g., maternal depression) are reflected in parent behavior towards the child. Again, shared and nonshared environmental estimates are also produced by studies of this kind. Shared environmental influence refers to similarity in parenting between twin mothers beyond that which would be expected by their genetic similarity, and would be interpreted as an effect of their own shared rearing environment. Nonshared environmental influence refers to disparity in parenting beyond that expected by genetic dissimilarity. This is most easily understood in the context of identical twin mothers, who would parent in an identical way to one another if their genetic propensities were entirely responsible for their parenting behaviors.

In a somewhat indirect fashion, then, behavioral genetic results have implications for our understanding of the determinants of parenting. We can gain insights into parent and child contributions to parenting by interpreting the meaning of influences found, in the context of the research design.

We now turn to the main business of this chapter, a review of genetically sensitive studies of parenting during childhood. A key aim of the current review was to look at potential differences in genetic and environmental parameter estimates as a function of informant, design, age, and parenting dimension.

3.2 Literature Review

It is exciting to be at a point where a substantial body of work can be summarized. Considering the research in an integrated fashion yields a comprehensive picture that is more than the sum of its parts. While many of the conclusions may come as no surprise (e.g., adoption studies have yielded smaller heritability estimates than have twin studies), a couple of the key messages have not previously been highlighted.

3.2.1 Study Selection

A first review of genetic influences on environmental measures, including parenting, was published some two decades ago (see Plomin and Bergeman 1991). For the purposes of the current review, we selected empirical studies that examined the genetic and environmental aetiology of aspects of parenting, excluding abstracts, reviews, dissertations, and designs based exclusively on differences between monozygotic (MZ) twins. We further excluded studies with sample sizes less than 100, and restricted our review to studies of childhood, that is under the age of 12 years, in order not to overlap with Chap. 10. We included all appropriate designs (i.e., twin children, twin parents, adoption, and retrospective designs). We acknowledge that the list of references included here (see "Appendix") may not be exhaustive, though we made our best attempt to capture all relevant empirical works.

3.3 Approach

Many of the references included in this review are studies that examine the aetiology of several parenting measures in one sample, several informants in one sample, or even more than one sample within the reference. Each line of the table represents a single measure, informant type, age, and sample (design). A key aim of the current review was to look at potential differences in genetic and environmental parameter estimates as a function of informant, design, age, and parenting dimension. We coded informant as 'mother or parent' (note that there were no studies that used fathers as sole informant), 'coded' which included observer coded parent-child interaction data from audio/video tapes or from live interaction, 'interviewer', 'retrospective twin-report', and 'retrospective parent-report'. Study design was coded as twin, adoption, twin retrospective reporting, and twin mothers (as opposed to twin children). Finally, based on Baumrind's parenting styles and subsequent conceptualizations of important parenting dimensions, we categorized each parenting measure as either possessing or not possessing any control of the child's behavior, as well as categorizing every measure's 'valence' as either positive or negative. Control and valence were blind coded from study measurement descriptions by both authors. Rater agreement for both valence and control was well over 80 %; in cases of disagreement we referred to the original article and discussed until an agreement was reached.

In order to glean a reasonable evaluation of genetic and environmental parameter estimates across all the empirical studies included here, we calculated A (additive genetic), C (shared environment), and E (non-shared environment) parameters from each study weighted by the size of the sample in each case. That is, for example, the first study in our reference list (Boivin et al. 2005; see "Appendix") is a twin study of 510 twin pairs, based on the Quebec Newborn Twin Study, and parameter estimates were thus weighted by the sample size (510) before inclusion in our summary. After weighting all studies on this basis we estimated the extent to which the mean proportion of parenting measure variance was accounted for by A, C, and E, first across all studies, and then as a function of various factors: informant, design, valence, control, and age, as well as two key interactions, namely valence and control, and design and informant.

3.4 Summary of Studies

3.4.1 Overall

As seen in Table 3.1, on average across the range of measures, ages, informants, designs, and other factors, parenting was accounted for by approximately one-third genetic, one-third shared environmental and one-third non-shared environmental influences. This indicates that parenting partly reflects genetically influenced characteristics of children, is partially a consistent behavior of parents towards all of their children, and partly reflects idiosyncratic interactions between parents and their individual children. However, these overall results belie differences in mean A, C, and E parameter estimates as a function of a variety of factors, and it is these differences that we argue are of potential importance.

Table 3.1 Average parameter estimates for each grouping

Factor	Level	A	C	E
All	All	32	36	32
Informant	Mother/parent	39	33	27
	Coded	17	37	45
	Interviewer	15	68	18
	Retrospective twin report	26	25	49
	Retrospective parent report	11	74	14
Design	Twin children	39	36	26
	Adoption	25	64	11
	Twin children (retrospective)	22	37	41
	Twin mothers	17	10	73
Age	<1 year	11	52	38
	Preschool (1–5 years)	42	36	22
	5–12 years	38	35	27
Control	No control element	32	31	37
	Control element	32	36	31
Valence	Positive valence	25	43	32
	Negative valence	38	29	33
Valence + control	Positive/no control	26	32	42
	Positive/control	25	48	27
	Negative/no control	44	29	27
	Negative/control	37	29	34
Design + informant	Twin + parent report	41	35	24
	Twin + coded interaction	16	34	50
	Twin + interview	15	68	18

3.4.2 Informant

We compared five different parenting information sources. The most "objective" measures were the videotaped, and subsequently coded, parent-child interactions. These yielded relatively low heritability estimates (averaging 17 %), moderate shared environment (37 %), and a relatively high estimate of nonshared environment (45 %). Coded interactions are considered by many to be the "gold standard" parenting measure because rater bias is removed and parenting behavior can be assessed during a standardized semi-structured task/interaction. We would argue, however, that there are key limitations to the methodology that are reflected in these genetic and environmental estimates. First, such interactions are typically only 5–10 min, on one particular day when the child (or parent) may be affected by illness, a poor night's sleep, good or bad news, and so on. In other words, the video may not capture a "typical" period of interaction. The specific task demands, although standardized, may also affect the children in individual ways. If present, such limitations increase measurement error, which is captured in the nonshared environment component. It seems likely that this is the case because of the relatively large nonshared environment estimate. The standardized nature of the interactions may also be responsible for the relatively low heritability. Such interactions may not be sensitive to the systematic adjustments that parents make in their parenting in response to genetically influenced traits of their children.

Next in line in terms of objectivity are the ratings made by interviewers. Unlike self-report questionnaires, this method combines (potentially biased) information by parents with standardized judgments made by interviewers. In this case, the heritability estimate was again relatively low (15 %), as was the nonshared environmental component (18 %), whereas the shared environmental component was very high (68 %). It is therefore possible that this method is particularly sensitive to between-family differences in parenting, at the expense of identifying parenting practices unique to each child in the family.

Parent self-report questionnaires are ostensibly the least objective of the three informants. The method is often criticized because of the risk of rater-bias, which is accentuated for designs involving one parent rating their own behavior towards two children. Parents are more likely to present themselves in a positive light (e.g., Gaylord et al. 2003), and also to report being more consistent in their behavior towards their children than they actually are (Pike et al. 1996). The summary estimates in "Appendix" are not congruent with these criticisms. The shared environment estimate is moderate in magnitude (33 %)—not inflated as would be expected if parents were over-estimating the consistency of their behavior. In addition, the moderate genetic and nonshared environmental estimates indicate that parent ratings do reflect both systematic, genetically-influenced traits of the children, as well as more idiosyncratic within-family differences in parenting. We would not argue that self-report parenting questionnaires should be adopted as the new gold standard. Rather, we make these points in reaction to the bad press that this methodology so often attracts. In fact, we argue that there is no single best measurement strategy—multiple methods and perspectives are necessary to capture the complexities of parenting.

Finally, we included studies of parenting that asked participants to answer questions about how they were parented as children (retrospective twin report), and studies that asked older adults to answer questions about how they had parented their now-adult children (retrospective parent report). As such, these methods include the standard limitations of rater bias, as well as being affected by memory failure and current parent-child relationship quality. It is therefore not surprising that these informant techniques seem to yield biased estimates. When asked retrospectively the parents report far more consistency than is the case when parents provide concurrent reports—as evidenced by a shared environmental estimate of 74 % versus 33 %. In contrast, the retrospective twin reports yielded a relatively large nonshared environment estimate (49 %), indicating individualized memories of childhood that may underestimate similarities in sibling experience.

3.4.3 Design

We found potentially important differences in parameter estimates as a function of study design. In terms of twin designs, the parameter estimates were, at first glance, remarkable in their similarity to the results across all studies. In terms of studies that employed the adoption design, while the average heritability was lower than for the twin design studies (25 %), the shared environmental component for this study design (0.64). Although it is tempting to be led to think of the difference between what we found from the twin and adoption designs as reflective of non-additive genetic variance, the twin studies formed the majority of all studies examined. In contrast, the adoption study findings are on the basis of just 8 % of the studies assessed, and are all from one sample, the Colorado Adoption Project (CAP). In terms of design, we also isolated twin children studies that relied on retrospective reports. These findings were described under Informant, above. Finally, for a few of the studies the twins were the mothers rather than the children. As shown in Table 3.1, genetic and shared environmental influences were modest (17 and 10 %, respectively) whereas nonshared environmental influences were substantial (73 %). The interpretation here is quite different than for child-based designs. That is, here, the modest genetic influence refers to genetically influenced characteristics of the *mothers* rather than the children. In this way, this thus-far limited research base indicates that children's genetic propensities are a more important determinant of parenting than are mothers' genetic propensities.

3.4.4 Age

The children in the studies reviewed ranged in age from 5 months to 12 years. We categorized the studies into those involving infants (children less than 12 months old), preschoolers (from one to five years old), and school-aged children (from five to 12 years-old) (see Table 3.1). A striking pattern emerged. During the first year of life, heritability estimates were modest (averaging 11 %). However, for the preschoolers and school-aged children, heritability was moderate (42 and 38 % respectively). This pattern suggests, perhaps unsurprisingly, that inherent characteristics of children seem not to influence variations in parenting infants; however, such child characteristics seem to exert moderate influence by preschool age. We explored this further by looking at genetic and environmental influences from 12 to 60 months, and found that the pattern was consistent throughout. In fact, it is striking that these child effects (as estimated by heritability of parenting) rapidly rise to moderate levels by preschool, and remain at this seeming "ceiling" level throughout childhood. We interpret this as meaning that parents are soon sensitized to the individual characteristics of their children, so that their parenting reflects—at least to a moderate degree—genetically influenced traits of the children.

Alongside this trend for heritability, estimates of shared environmental influence were substantial in infancy (52 %), dropping to more moderate levels by preschool. This is further indicative of parenting infants as quite different from parenting school-aged children, preschoolers, or even toddlers. Finding substantial shared environmental influence here suggests that the different ways that parents manage the early months of parenthood is relatively consistent among children in families, likely due to parental philosophies of child-rearing that alter little as a function of individual characteristics of the infant, and perhaps people's different proclivities for the newborn period.

3.4.5 Control

Reviews of behavioral genetic studies of parenting (e.g., Kendler and Baker 2007; Plomin 1994) have stated that control aspects of parenting yield low estimates of heritability, as opposed to affective aspects of parenting that yield moderate heritability estimates. It was thus remarkable to find identical, moderate summary estimates for measures with and without a control element (32 %). Previous reviews have focused on early studies of adolescents, with a preponderance of adolescent twin reports of parenting (e.g., Plomin et al. 1994; Rowe 1981). Such studies have overshadowed the fact that differential parental control during childhood *is* influenced by children's genetic propensities, as well as by shared and nonshared environmental factors. That is, children's genetically influenced traits are a significant factor that is reflected in the degree to which parents exhibit controlling versus non-controlling parenting.

3.4.6 Valence

Recent findings have alluded to different patterns of influence for positive versus negative aspects of parenting (Oliver et al. 2014; Rasbash et al. 2011). Here, in our summary of empirical findings we support this notion, finding heritability for positive valence (25 %) to be lower than for negative valence (38 %). On the flip-side, shared environment was greater for the positive versus negative valence measures (43 and 29 %, respectively). We interpret this as meaning that positive aspects of parenting such as warmth and affection vary largely on a family-by-family basis, perhaps stemming from shared environmental aspects such as the quality of the marital relationship, wellbeing of the parents, and attitudes towards parenting. Negative aspects of parenting such as harsh discipline and rejection may instead be more reactive to child characteristics. That is, a young child with a high activity level and prone to negative emotionality may elicit more frustration and harsh discipline than would a more easy-going child. While heritability and shared environment are important components to both positive and negative aspects of parenting, it is interesting to note this distinction—positive aspects may be more parent-led, and negative aspects more reflective of children's individual characteristics.

Important to note, in categorising measures of parenting into positive versus negative valence, we did not include maltreatment. That is, the pattern we report includes aspects of harsh discipline such as smacking, but it does not include abusive forms of parenting. One study that explicitly looked at this distinction found that harsh discipline was moderately genetically influenced (25 %), but physical maltreatment was not (7 %) (Jaffee et al. 2004). This provides concrete evidence of what we already know—abuse of children is not a reflection of the characteristics of the children, but of the parents.

3.4.7 Interactions

In order to check some of the key results, two interactions were examined to rule out confounds as an explanation for the pattern of results described above. First, the four possible combinations of control and valence were assessed. As can be seen in Table 3.1, this sub-division replicated the previous pattern—differences in heritability were due to negative versus positive valence, rather than control. Second, in order to check that the pattern of findings by informant was not due to confounds with design, we sub-divided by informant for the largest design type, twin children. Again, this confirmed the pattern described earlier—parent reports yielded much higher heritability estimates than did the coded interactions of interviews.

3.4.8 Bivariate

About half of the studies reviewed also contained a bivariate component. Basic genetic analyses are univariate; they decompose observed variance of a single measure into genetic and environmental components. Bivariate genetic analysis

Univariate Genetic Analysis   Bivariate Genetic Analysis

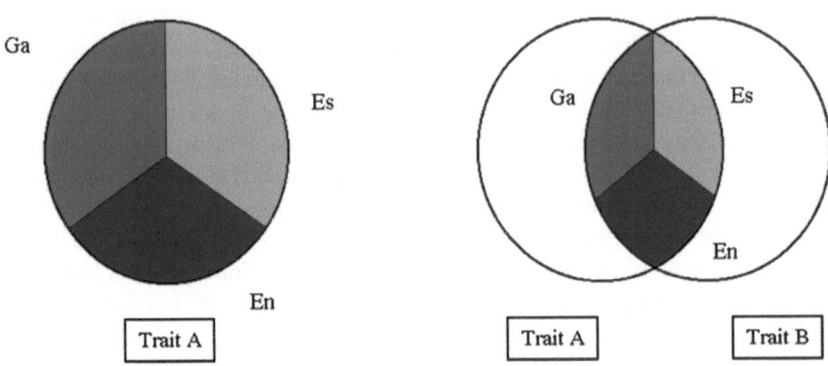

Fig. 3.1 Univariate genetic analysis decomposes the variance of one trait into its genetic (*Ga*), shared environmental (*Es*), and nonshared environmental (*En*) components. Bivariate genetic analysis decomposes the covariance between two traits into its genetic (*Ga*), shared environmental (*Es*), and nonshared environmental (*En*) components

focuses on the correlation between traits, decomposing this into its genetic and environmental components (see Fig. 3.1). Such bivariate analysis can tell us, for example, whether a link between parental treatment and children's behavior is due to the parenting environment (shared with siblings or nonshared), and/or whether common genetic influences linking parental treatment and child outcome play a part in the observed covariance.

The studies reviewed covered many known correlates of parenting such as child temperament (Boivin et al. 2005), child prosocial behavior (Knafo and Plomin 2006), and children's academic skills (Roisman and Fraley 2012). There is not yet a large enough body of research to summarise the findings from these studies, with one exception. Six studies report behavioral genetic findings for the association between parenting and children's externalizing problems. Table 3.2 contains a summary of the genetic and environmental contributions to these associations.

In the first study of this type, several moderate correlations emerged between parental behaviors and the children's behavior problems (Deater-Deckard 2000), as shown in Table 3.2. Bivariate genetic analyses were then conducted for these associations to determine the degree of genetic versus environmental mediation. The pattern of results was clear—the lion's share of associations was due to genetic mediation. This finding suggests that genetically influenced traits of these children were being reflected not only in their behavioral difficulties, but also in the treatment elicited from their parents.

The additional five studies have replicated this pattern; the link between parenting and children's externalizing problems is mainly due to genetic influence. In all cases, these studies utilized child-based designs, so the interpretation is that genetically influenced traits of the children are reflected in both the parenting

Table 3.2 Genetic and
environmental contributions
to correlations between
parenting and children's
externalizing problems

Reference	Covariance tested	Contributions to covariance (%)		
		A	C	E
Boeldt et al. (2012)	Maternal positivity and externalizing problems	59	32	9
Burt et al. (2003)	Parent-child conflict and externalizing problems	52	39	9
Deater-Deckard (2000)	Negative affect and problem behavior	76	0	24
	Positive affect and problem behavior	66	0	34
	Harsh discipline and problem behavior	75	0	25
Jaffee et al. (2004)	Corporal punishment and antisocial behavior	86	<1	13
Larsson et al. (2008)	Parental negativity and antisocial behavior (age 4)	94	–	–
	Parental negativity and antisocial behavior (age 7)	46	8	46
Roisman and Fraley (2012)	Supportive parenting and externalizing behavior	73	27	0

they receive and their levels of externalizing problems. Returning to Belsky's Determinants of Parenting framework (1984), the findings are strong evidence for the importance of child effects. As first described by Bell (1968), child effects highlight that parenting is an interactional process in which children's propensities influence parental behavior as well as the other way around (see also Scarr 1992).

3.5 Implications for Socialization Research

3.5.1 Univariate Findings

In order to contextualize the implications of the studies reviewed, we return to Belsky's Determinants of Parenting framework (Belsky 1984). The moderate genetic influence found for child-based designs is readily interpretable. This is clear evidence of the importance of child effects—at least after the youngest months of infancy. In concrete terms, these results suggest that mothers (and probably fathers) adjust their parenting in accordance with genetically influenced characteristics of the child, for example, temperament. This lends support to Bell's proposition that children influence parents as well as the other way around (Bell 1968). We note, however, that finding moderate genetic influence is also indicative that parents are being *sensitive* to the individual needs of their children.

Next, we consider the moderate estimate of shared environment from child-based designs. This is an indication that parents are moderately consistent in their behavior towards their children, once the genetic similarities of the children have been accounted for. This supports the importance of both parental personal resources and contextual support, as the expectation is that these factors will lead to parental consistency. For example, a depressed mother will show less responsive parenting to all of her children, and effects of a satisfying and supportive marriage will spillover to all parent-child relations.

The moderate influence of nonshared environmental factors is more difficult to interpret. This contribution cannot be readily attributable to any single determinant of parenting, and it is worth noting that the nonshared environmental contribution also contains measurement error. Philosophically, this chunk of idiosyncratic, relationship specific variance, serves as a reminder that human relationships may not be fully explained by psychological measurement.

Finally, the few twins-as-mothers designs are interpreted quite differently. As mentioned above, finding modest genetic influence indicates that genetically influenced traits of mothers are *not* substantial determinants of parenting. The scant shared environmental component means that co-twin mothers are not very consistent in their parenting practices. This is somewhat surprising, given the evidence for inter-generational transmission of parenting (Van Ijzendoorn 1992). The large nonshared contribution indicates that these twin mothers are behaving quite differently to one another. The reasons for this seem likely to include that the mothers are parenting different children, tying in with findings of the importance of child characteristics for parentings. In addition, the mothers may have different partners influencing their parenting, and have had idiosyncratic experiences that may impact their well-being and attitudes (parental personal resources).

3.5.2 Bivariate Findings

Finding evidence of genetic mediation between measures of parenting and children's outcomes suggests that non-genetic studies be interpreted with caution. Much of the developmental research relating parenting to children's adjustment is interpreted to mean that the parent's behavior is causing the child's behavior. In other words, parenting is hypothesized to be a functionally effective environmental factor. The bivariate findings reviewed here suggest that this is not always the case. Instead, it is the children's genes that are reflected in both the parent's behavior and in the child's adjustment. In terms of process, it is quite plausible that a child's genetic propensities that lead to adjustment difficulties would also lead to displays of negativity from parents.

This is certainly not the first critique of parental socialization put forward by behavioral geneticists. The most potent of these was put forward by Scarr in her presidential address to the Society for Research in Child Development in 1991 (Scarr 1992). Scarr utilized behavioral genetic evidence to make the case that children's experiences are driven by their genetic propensities. That is, children are seen as active agents in their own socialization, and this active selection and creation of environmental experiences is genetically determined. Perhaps most controversially, Scarr made the claim that "average" parents are "good enough." Scarr argued that within the species-normal range of environments, parents do not have a differential impact on their children's development. This idea runs counter to the traditional socialization theorists' claim that parents are the key socialization agents for young children, and Scarr's theory created lively debate and criticism (Baumrind 1993; Jackson 1993). We would argue that the findings concerning links between parenting and child adjustment serve to bolster Scarr's thesis.

3.6 Future Directions

Parents form the lynchpin for children's early environmental experience and socialization, and for every child the early years have critical implications for mental health and wellbeing. Indeed, parenting is one of the most commonly studied influences on child development, as well as the most common 'environmental' influence studied within a genetically sensitive context. As this review has shown, within the behavioral genetic context, parenting has been studied across a diverse age range, with numerous methods, measures and study designs. From these available studies there are conclusions to be drawn, and we have attempted to highlight some of those here. However, despite the plethora of research in the parenting field, there is by no means one well-worn path, rather, there is much to understand of existing lines of research, and many new directions to explore.

Perhaps most salient of the gaps in research of this kind is the examination of fathers' role in the parenting environment. Many studies use the term 'parents', yet mostly this is simply to cover reports collected from one or other carer, and is most usually the mother. The fundamental truth is that the mother-child relationship is at the heart of most genetically informed studies of parenting with maternal reports still the most universally used measures of parenting. Moreover, other methods such as parent-child coded interactions also have not had have fathers as their focus. Historically, this issue has been highlighted as important, but with the caveat that its significance was secondary to maternal parenting due to the then generic bias toward much more maternal than paternal child-care (Lamb 2004). While it is still true to say that on average mothers spend more time with their children than do fathers (Craig and Powell 2012), the role of fathers in families is one that has changed dramatically over the years (see Lamb and Tamis-LeMonda 2004), such that, while it remains maternally biased, father input to the parenting environment has more and more potential to influence children's development (Kochanska et al. 2008). Moreover, the increasing role of fathers implicates another critical, as yet small research field, that of coparenting (e.g., Feinberg 2003; Feinberg and Kan 2008). To our knowledge, this latter research area has been neglected by the behavioral genetic field. Genetically informed studies that aim to examine father-child as well as mother-child parenting will bolster knowledge of the family and its impact on children's wellbeing.

A similarly glaring void in the genetically informed parenting literature in childhood involves sex differences. Previous findings have shown that boys and girls may be exposed to rather different parenting styles (e.g., McKee et al. 2007), but very little is known within a behavioral genetic context. In part this may be due to a lack of statistical power to formally test such differences. In addition, all of the studies we review involve samples from North America or Western Europe, despite the fact that cultural differences are known to be important issues in the parenting field (Pachter and Dumont-Mathieu 2004). These areas are important new directions for genetically informed research in parenting.

The significance of our experiences lies, at least in part, in our perception of them. This has been shown to be true in many fields, including children's perceptions of their family environment (e.g., Atzaba-Poria and Pike 2008; Coldwell et al. 2006; Pike et al. 2005). Yet, in progressing our review we noticed that, although there is considerable focus on self-report in the adolescent literature (see Chap. 10), we found no behavioral genetic studies using child-reports of parenting earlier in development. Standardized methods designed to assess children's perception of the parenting they receive and the relationship they have with their parents (see Pike et al. 2006) show considerable validity, and have important potential in this field.

Behavioral genetic studies inform the extent to which individual differences in parenting are attributable to individual genetic and environmental differences in the population studied (i.e., the sample in question). In this way, such studies tell us about 'what is', but little about 'what could be', which lies in the hands of prevention and intervention scientists. However, although traditionally distanced, these two fields are far from at odds. Genetically informed research has the potential to broaden the horizons prevention and intervention science—and ultimately practice –in several ways. For example, intervention approaches that target parenting strategies have been shown to be useful for a considerable array of child outcomes (NICE 2006), yet transforming the emphasis of prevention and intervention from improving outcomes at a mean level, to examining individual differences in response would bridge the gap between idiographic clinical knowledge and randomized controlled designs. Doing so, has the potential to better identify those who benefit from the intervention—say, due to genetically influenced characteristics—and those who do not. Moreover, genetically informed research can help identify environmental vulnerabilities and protective factors important for response to parenting interventions, within the context of, or controlling for, genes. Finally, emerging findings that environmental influences have the potential to change the expression of genetic influence are very encouraging. Research that examines these possibilities within the parenting field has the potential to empower parents to change the behavior of their children while acknowledging that individual—genetically influenced—characteristics of children may be part of the initial catalyst for intervention need.

These new directions may be expensive, but therein may lie the key to truly understanding the heart of parent-child interaction and its importance for children's long term health or illness and how to reduce psychopathological difficulties and improve well-being.

Appendix

Reference	Children's age	N (pairs)	Design	Dataset	Informant	Construct	A	C	E
Boeldt et al. (2012)[b]	7–36 months (longitudinal)	374	Twin children	Twin infant project and longitudinal twin study	Coded m–c interaction, average over 5 time-points	Composite positivity (respect for child's autonomy, quality of instruction, sensitivity to child cues, warmth)	0.20	0.30	0.48
Boivin et al. (2005)[b]	5 months	510	Twin children	Quebec newborn twin study (QNTS)	Mother	Parental self-efficacy[a]	0.00	0.76	0.24
					Mother	Perceived parental impact[a]	0.00	0.69	0.31
					Mother	Parental hostile-reactive behaviour	0.31	0.53	0.16
					Mother	Parental overprotection	0.00	0.86	0.14
Burt et al. (2003)[b]	10–12 years	808	Twin children	Minnesota twin family study	Average of child and mother	Parent-child conflict	0.38	0.35	0.27
Deater-Deckard et al. (1999)	10–12 years	191	Adoption	Colorado adoption project (CAP)	Parents	Negativity	0.38	0.59	0.03
						Inconsistency	0.04	0.77	0.19
						Warmth	0.26	0.67	0.06
Deater-Deckard (2000)[b]	3 years	120	Twin children	TRACKS	Mother	Negative effect	0.55	0.23	0.22
						Positive affect	0.46	0.37	0.17
					Coded interaction	Negative affect	0.06	0.16	0.78
						Positive affect	0.00	0.51	0.49
						Negative control	0.00	0.47	0.53
						Positive control	0.00	0.49	0.51
						Responsiveness	0.49	0.02	0.51
					Interviewer	Harsh discipline	0.12	0.62	0.26

(Continued)

(Continued)

Reference	Children's age	N (pairs)	Design	Dataset	Informant	Construct	A	C	E
Deater-Deckard and O'Connor (2000)	3 years	120	Twin children	TRACKS	Coded interaction	Mother-child mutuality	0.59	0.00	0.41
	3 years	102	Adoption	CAP	Coded interaction	Mother-child mutuality	0.50	0.00	0.50
DiLalla and Bishop (1996)	7 months	186	Twin children	Twin infant project	Coded interaction	Respect for child autonomy[a]	0.00	0.22	0.78
						Quality of instruction[a]	0.42	0.11	0.47
						Sensitivity to cues from child[a]	0.56	0.01	0.44
						Warmth[a]	0.04	0.47	0.49
	9 months					Respect for child autonomy[a]	0.06	0.27	0.67
						Quality of instruction[a]	0.10	0.41	0.49
						Sensitivity to cues from child[a]	0.00	0.34	0.66
						Warmth[a]	0.00	0.50	0.50
Eley et al. (2010)[b]	8 years	265	Twin children	Emotions, cognitions, heredity and outcome (ECHO)	Coded interaction	Maternal control	0.63	0.00	0.37
Forget-Dubois et al. (2007)[b]	5 months	292	Twin children	QNTS	Parent	Parental hostile-reactive behaviours	0.31	0.50	0.19
	18 months						0.00	0.71	0.29
	30 months						0.26	0.56	0.18
Fujisawa et al. (2012)[b]	6.72 years (1.9 SD)	1677	Twin children	Tokyo twin cohort project and Tokyo twin cross sectional study	Parent	Negative parenting	0.39	0.46	0.15

(Continued)

(Continued)

Reference	Children's age	N (pairs)	Design	Dataset	Informant	Construct	A	C	E
Harlaar et al. (2008)	–	2334	Twin children (retrospective)	Finnish genetics of sex and aggression study	Twin retrospective reports	Maternal physical affection	0.30	0.24	0.46
						Maternal responsiveness	0.25	0.28	0.47
						Maternal control	0.25	0.22	0.53
						Maternal abuse	0.27	0.29	0.44
						Paternal physical affection	0.25	0.38	0.37
						Paternal responsiveness	0.20	0.35	0.45
						Paternal control	0.17	0.28	0.55
						Paternal abuse	0.17	0.44	0.39
Jaffee et al. (2004)[b]	5 years	1116	Twin children	The environmental risk (E-risk) longitudinal twin study	Interviewer	Corporal punishment	0.25	0.66	0.10
						Physical maltreatment	0.07	0.88	0.05
Kendler (1996)	–	1033	Twin children (retrospective)	Virginia twin registry	Twin retrospective reports	Paternal warmth	0.47	0.24	0.29
						Maternal warmth	0.63	0.00	0.37
						Paternal over-protectiveness	0.29	0.17	0.53
						Maternal over-protectiveness	0.29	0.23	0.47
						Paternal authoritarianism	0.24	0.19	0.57
						Maternal authoritarianism	0.18	0.29	0.53

(Continued)

(Continued)

Reference	Children's age	N (pairs)	Design	Dataset	Informant	Construct	A	C	E
	–	1033	Twin children (retrospective)	Virginia twin registry	Co-twin retrospective reports	Paternal warmth	0.20	0.45	0.35
						Maternal warmth	0.33	0.32	0.35
						Paternal over-protectiveness	0.17	0.23	0.61
						Maternal over-protectiveness	0.29	0.16	0.55
						Paternal authoritarianism	0.46	0.00	0.54
						Maternal authoritarianism	0.25	0.19	0.56
	–	1033	Twin children (retrospective)	Virginia twin registry	Parental retrospective reports	Paternal warmth	0.00	0.85	0.15
						Maternal warmth	0.12	0.71	0.17
						Paternal over-protectiveness	0.07	0.81	0.12
						Maternal over-protectiveness	0.05	0.83	0.12
						Paternal authoritarianism	0.06	0.79	0.15
						Maternal authoritarianism	0.00	0.85	0.15
	–	1033	Twin mothers	Virginia twin registry	Parent (twin) report (Twins as parents within same study as twins as children)	Maternal warmth	0.38	0.00	0.62
						Maternal over-protectiveness	0.00	0.29	0.71
						Maternal authoritarianism	0.00	0.20	0.80
Kim-Cohen et al. (2004)[b]	5 years	1116	Twin children	E-risk	Coded 5 min speech sample	Warmth	0.12	0.50	0.37

(Continued)

(Continued)

Reference	Children's age	N (pairs)	Design	Dataset	Informant	Construct	A	C	E
Knafo and Plomin (2006)[b]	3 years	9319	Twin children	TEDS	Parent	Positivity (discipline and feelings)	0.29	0.51	0.20
						Negativity (discipline and feelings)	0.58	0.20	0.22
	4 years					Positivity (discipline and feelings)	0.30	0.50	0.20
						Negativity (discipline and feelings)	0.53	0.24	0.23
	7 years—boys					Positivity (discipline and feelings)	0.30	0.40	0.30
						Negativity (discipline and feelings)	0.55	0.18	0.27
	7 years—girls					Positivity (discipline and feelings)	0.19	0.52	0.30
						Negativity (discipline and feelings)	0.54	0.19	0.27
Larsson et al. (2008)[b]	4 years	6230	Twin children	TEDS	Parent report	Negativity—parental feelings questionnaire	0.48	0.30	0.22
	7 years	6230	Twin children	TEDS	Parent report	Negativity—parental feelings questionnaire	0.42	0.27	0.32
Lichtenstein et al. (2003)[b]	45 years (±4.5)	326	Twin children (retrospective)	Twin moms study	Twin retrospective reports	Maternal warmth	0.33	0.35	0.45
						Maternal protectiveness	0.19	0.36	0.45
						Maternal authoritarianism	0.10	0.45	0.45
						Paternal warmth	0.27	0.41	0.32
						Paternal protectiveness	0.00	0.49	0.51
						Paternal authoritarianism	0.00	0.43	0.57

(Continued)

(Continued)

Reference	Children's age	N (pairs)	Design	Dataset	Informant	Construct	A	C	E
Losoya et al. (1997)	Under 8 years	186	Twin/sib parents		Parent report	Positive support	0.60	0.00	0.40
						Negative control	0.24	0.12	0.64
Perusse et al. (1994)	56.5 years (17.3 years)	1117	Twin children (retrospective reports)	US sample	Twin retrospective reports	PBI: maternal Care	0.38	0.00	0.62
						PBI: maternal overprotection	0.30	0.00	0.70
						PBI: paternal care	0.23	0.00	0.77
						PBI: paternal overprotection	0.18	0.00	0.82
Rende et al. (1992)	OS: 7 years YS: 4.5 years	124	Adoption	CAP	Coded interaction (maternal behavior)	Control	0.31	0.67	0.00
						Affection	0.00	0.88	0.12
						Attention	0.61	0.38	0.01
						Responsiveness	0.00	0.93	0.07
Roisman and Fraley (2006)	9 months	505	Twin children	Twin subsample from early childhood longitudinal study	Coded interaction	Overall observer rated infant-caregiver relationship quality	0.01	0.40	0.59
Roisman and Fraley (2012)	24 months and PreK	485	Twin children	Twin subsample from early childhood longitudinal study	Coded interaction	Aggregate supportive parenting across both time-points	0.00	0.72	0.28

(Continued)

(Continued)

Reference	Children's age	N (pairs)	Design	Dataset	Informant	Construct	A	C	E
Wade and Kendler (2000)	–	1033	Twin children (retrospective)	Virginia twin registry	Twin retrospective reports	Paternal physical discipline	0.33	0.28	0.39
						Maternal physical discipline	0.40	0.21	0.39
						Paternal limit-setting	0.28	0.23	0.49
						Maternal limit-setting	0.17	0.30	0.53
		1033	Twin children (retrospective)	Virginia twin registry	Parental retrospective reports	Paternal physical discipline	0.21	0.62	0.17
						Maternal physical discipline	0.19	0.66	0.15
						Paternal limit-setting	0.24	0.64	0.12
						Maternal limit-setting	0.19	0.67	0.14
		1033	Twin children (retrospective)	Virginia twin registry	Co-parent retrospective reports	Paternal physical discipline	0.09	0.72	0.19
						Maternal physical discipline	0.10	0.64	0.26
						Paternal limit-setting	0.19	0.69	0.12
						Maternal limit-setting	0.00	0.80	0.20
	–	1033	Twin mothers	Virginia twin registry	Parent (twin) report	Maternal physical discipline	0.21	0.00	0.79
						Maternal limit-setting	0.27	0.00	0.73

Note [a]Estimates calculated by authors from simple MZ DZ intraclass correlations in cases where model fitting results not provided. [b]Multivariate results available

References

Atzaba-Poria, N., & Pike, A. (2008). Determinants of parental differential treatment: Parental and contextual factors during middle childhood. *Child Development, 79*, 217–232. doi:10.1111/j.1467-8624.2007.01121.x.

Baumrind, D. (1973). The development of instrument competence through socialization. In A. D. Pick (Ed.), *Minnesota symposia on child psychology* (Vol. 7, pp. 3–46). Minneapolis: University of Minnesota Press.

Baumrind, D. (1993). The average expectable environment is not good enough: A response to Scarr. *Child Development, 64*, 1299–1317.

Bell, R. Q. (1968). A reinterpretation of the direction of effects in socialization. *Psychological Review, 75*, 81–95.

Belsky, J. (1984). The determinants of parenting: A process model. *Child Development, 55*, 83–96.

Belsky, J., & Jaffee, S. R. (2006). The multiple determinants of parenting. In D. Cicchetti & D. Cohen (Eds.), *Developmental psychopathology: Risk, disorder and adaptation* (2nd ed., pp. 38–85). Hoboken, NY: Wiley.

Boeldt, D. L., Rhee, S. H., DiLalla, L. F., Mullineaux, P. Y., Schulz-Heik, J., Corely, R. P., et al. (2012). The association between positive parenting and externalizing behaviour. *Infant and Child Development, 21*, 85–106. doi:10.1002/icd.764.

Boivin, M., Perusse, D., Dionne, G., Saysset, V., Zoccolillo, M., Tarabulsy, G. M., & Tremblay, R. E. (2005). The genetic-environmental etiology of parents' perceptions and self-assessed behaviours toward their 5-month-old infants in a large twin and singleton sample. *Journal of Child Psychology and Psychiatry, 46*, 612–630. doi:10.1111/j.1469-7610.2004.00375.x.

Burt, S. A., Krueger, R. F., McGue, M., & Iacono, W. (2003). Parent-child conflict and the comorbidity among childhood externalizing disorders. *Archives of General Psychiatry, 60*, 505–513. doi:10.1001/archpsyc.60.5.505.

Campbell, S. B., Matestic, P., von Stauffenberg, C., Mohan, R., & Kirchner, T. (2007). Trajectories of maternal depressive symptoms, maternal sensitivity, and children's functioning at school entry. *Developmental Psychology, 43*, 1202–1215. doi:10.1037/0012-1649.43.5.1202.

Coldwell, J., Pike, A., & Dunn, J. (2006). Household chaos—Links with parenting and child behaviour. *Journal of Child Psychology and Psychiatry, 47*, 1116–1122. doi:10.1111/j.1469-7610.2006.01655.x.

Craig, L., & Powell, A. (2012). Dual-earner parents' work-family time: The effects of atypical work patterns and non-parental childcare. *Journal of Population Research, 29*(3), 229–247. doi:10.1007/s12546-012-9086-5. [EPub ahead of print].

Crouter, A. C., & McHale, S. M. (2005). The long arm of the job revisited: Parenting in dual-earner families. In T. Luster & L. Okagaki (Eds.), *Parenting: An ecological perspective* (2nd ed.). Mahwah, NJ: Lawrence Erlbaum Associates.

Deater-Deckard, K. (2000). Parenting and child behavioral adjustment in early childhood: A quantitative genetic approach to studying family processes. *Child Development, 71*, 468–484. doi:10.1111/1467-8624.00158.

Deater-Deckard, K., Fulker, D. W., & Plomin, R. (1999). A genetic study of the family environment in the transition to early adolescence. *Journal of Child Psychology and Psychiatry, 40*, 769–775. doi:10.1111/1469-7610.00492.

Deater-Deckard, K., & O'Connor, T. G. (2000). Parent-child mutuality in early childhood: Two behavioral genetic studies. *Developmental Psychology, 36*, 561–570. doi:10.1037/0012-1649.36.5.561.

DiLalla, L. F., & Bishop, E. G. (1996). Differential maternal treatment of infant twins: Effects on infant behaviors. *Behavior Genetics, 26*, 535–542. doi:10.1007/BF02361226.

Eley, T. C., Napolitano, M., Lau, J. Y. F., & Gregory, A. M. (2010). Does childhood anxiety evoke maternal control? A genetically informed study. *Journal of Child Psychology and Psychiatry, 51*, 772–779. doi:10.1111/j.1469-7610.2010.02227.x.

Erel, O., & Burman, B. (1995). Interrelatedness of marital relations and parent-child relations: A meta-analytic review. *Psychological Bulletin, 118*, 108–132. doi:10.1037/0033-2909.118.1.108.

Feinberg, M. (2003). The internal structure and ecological context of coparenting: A framework for research and intervention. *Parenting Science and Practice, 3*, 95–131. doi:10.1207/S15327922PAR0302_01.

Feinberg, M., & Kan, M. L. (2008). Establishing family foundations: Intervention effects on coparenting, parent/infant well-being, and parent–child relations. *Journal of Family Psychology, 22*, 253–263. doi:10.1037/0893-3200.22.2.253.

Forget-Dubois, N., Boivin, M., Dionne, G., Pierce, T., Tremblay, R. E., & Prusse, D. (2007). A longitudinal twin study of the genetic etiology of maternal hostile-reactive and environmental behavior during infancy and toddlerhood. *Infant Behavior and Development, 30*, 453–465. doi:10.1016/j.infbeh.2006.12.005.

Fujisawa, K. K., Yamagata, S., Ozaki, K., & Ando, J. (2012). Hyperactivity/inattention problems moderate environmental but not genetic mediation between negative parenting and conduct problems. *Journal of Abnormal Child Psychology, 40*, 189–200. doi:10.1007/s10802-011-9559-6.

Gaylord, N. K., Kitzmann, K. M., & Coleman, J. K. (2003). Parents' and children's perceptions of parental behavior: Associations with children's psychosocial adjustment in the classroom. *Parenting Science and Practice, 3*, 23–47. doi: 10.1207/S15327922PAR0301_02.

Harlaar, N., Santtila, P., Bjorklund, J., Alanko, K., Jern, P., Varjonen, M., & Sandnabba, K. (2008). Retrospective reports of parental physical affection and parenting style: A study of finnish twins. *Journal of Family Psychology, 22*, 605–613. doi:10.1037/0893-3200.22.3.605.

Jackson, J. F. (1993). Human behavioral genetics, Scarr's theory, and her views on interventions: A critical review and commentary on their implications for African American children. *Child Development, 64*, 1318–1332.

Jaffee, S. R., Caspi, A., Moffitt, T. E., Polo-Tomas, M., Price, T. S., & Taylor, A. (2004). The limits of child effects: Evidence for genetically mediated child effects on corporal punishment but not on physical maltreatment. *Developmental Psychology, 40*, 1047–1058. doi:10.1037/0012-1649.40.6.1047.

Karraker, K. H., & Coleman, P. K. (2005). The effects of child characteristics on parenting. In T. Luster & L. Okagaki (Eds.), *Parenting: An ecological perspective* (2nd ed., pp. 147–176). Mahwah, NJ: Lawrence Erlbaum Associates.

Kendler, K. S. (1996). Parenting: A genetic-epidemiologic perspective. *American Journal of Psychiatry, 153*, 11–20.

Kendler, K. S., & Baker, J. H. (2007). Genetic influences on measures of the environment: A systematic review. *Psychological Medicine, 37*(5), 615–626. doi:10.1017/S0033291706009524.

Kim-Cohen, J., Moffitt, T. E., Caspi, A., & Taylor, A. (2004). Genetic and environmental processes in young children's resilience and vulnerability to socioeconomic deprivation. *Child Development, 75*, 651–668. doi:10.1111/j.1467-8624.2004.00699.x.

Knafo, A., & Plomin, R. (2006). Parental discipline and affection and children's prosocial behavior: Genetic and environmental links. *Journal of Personality and Social Psychology, 90*, 147–164. doi:10.1037/0022-3514.90.1.147.

Kochanska, G., Aksan, N., Prisco, T. R., & Adams, E. E. (2008). Mother-child and father-child mutually responsive orientation in the first 2 years and children's outcomes at preschool age: mechanism of influence. *Child Development, 79*, 30–44. doi:10.1111/j.1467-8624.2007.01109.x.

Lamb, M. E. (Ed.). (2004). *The role of the father in child development* (4th ed.). Hoboken, NJ: Wiley.

Lamb, M. E., & Tamis-LeMonda, C. S. (2004). The role of the father: An introduction. In M. E. Lamb (Ed.), *The role of the father in child development* (4th ed., pp. 1–31). Hoboken, NJ: Wiley.

Larsson, H., Viding, E., Rijsdijk, F. V., & Plomin, R. (2008). Relationships between parental negativity and childhood antisocial behavior over time: A bidirectional effects model in a longitudinal genetically informative design. *Journal of Abnormal Child Psychology, 36*, 633–645. doi:10.1007/s10802-007-9151-2.

Leaper, C. (2002). Parenting girls and boys. In M. H. Bornstein (Ed.), *Handbook of parenting: Children and parenting*. 2nd edn., Vol. 1, (pp. 127–152). Mahwah, NJ: Erlbaum.

Lichtenstein, P., Ganiban, J., Neiderhiser, J. M., Pedersen, N. L., Hansson, K., Cederblad, M., & Reiss, D. (2003). Remembered parental bonding in adult twins: Genetic and environmental influences. *Behavior genetics, 33*(4), 397–408.

Losoya, S. H., Callor, S., Rowe, D. C., & Goldsmith, H. H. (1997). Origins of familial similarity in parenting: A study of twins and adoptive siblings. *Developmental Psychology, 33*, 1012–1023. doi:10.1037/0012-1649.33.6.1012.

McGuire, S., Sagal, N. L., & Hershberger, S. (2012). Parenting as a phenotype: A behavioral genetic approach to understanding parenting. *Parenting Science and Practice, 12*(2–3), 192–201.

McKee, L., Roland, E., Coffelt, N., Olson, A. L., Forehand, R., et al. (2007). Harsh discipline and child problem behaviors: The roles of positive parenting and gender. *Journal of Family Violence, 22*, 187–196. doi:10.1007/s10896-007-9070-6.

National Institute for Health and Clinical Excellence (2006). *Parent training/education programmes in the management of children with conduct disorders.* www.nice.org.uk/page.aspx?o=TA102.

Oliver, B. R., Trzaskowski, M., & Plomin, R. (2014). Genetics of parenting: The power of the dark side. *Developmental Psychology, 50*(4), 1233–1240. doi:10.1037/a0035388.

Pachter, L. M., & Dumont-Mathieu, T. (2004). Parenting in culturally divergent setting. In M. S. Hoghughi & N. Long (Eds.), *Handbook of parenting: Theory, research, and practice.* Newbury Park, CA: Sage.

Parke, R., & Buriel, R. (2006). Socialization in the family: Ethnic and ecological perspectives. In N. Eisenberg (Ed.), *The handbook of child psychology: Social, emotional, and personality development* (6th ed., Vol. 3, pp. 429–504). New York: Wiley.

Perusse, D., Eaves, L. J., & Kendler, K. S. (1994). Depression and parental bonding—cause, consequence, or genetic covariance. *Genetic Epidemiology, 11*, 503–522. doi:10.1002/gepi.1370110607.

Pike, A., Coldwell, J., & Dunn, J. (2005). Sibling relationships in early/middle childhood: Children's perspectives and links with individual adjustment. *Journal of Family Psychology, 19*, 523–532. doi:10.1037/0893-3200.19.4.523.

Pike, A., Coldwell, J., & Dunn, J. (2006). *Family relationships in middle childhood.* York, UK: York Publishing Services/Joseph Rowntree Foundation.

Pike, A., Reiss, D., Hetherington, E. M., & Plomin, R. (1996). Using MZ differences in the search for nonshared environmental effects. *Journal of Child Psychology and Psychiatry, 37*, 695–704. doi:10.1111/j.1469-7610.1996.tb01461.x.

Plomin, R. (1994). *Genetics and experience: The interplay between nature and nurture* (6th ed.). Thousand Oaks, CA: Sage.

Plomin, R., & Bergeman, C. S. (1991). The nature of nurture: Genetic influence on environmental measures. *Behavior and Brain Sciences, 14*, 373–427. doi:10.1017/S0140525X00070278.

Plomin, R., Reiss, D., Hetherington, E. M., & Howe, G. (1994). Nature and nurture: Contributions to measures of the family environment. *Developmental Psychology, 30*, 32–43. doi:10.1037/0012-1649.30.1.32.

Putnam, S. P., Sanson, A., & Rothbart, M. K. (2002). Child temperament and parenting. In M. Bornstein (Ed.), *Handbook of parenting* (2nd ed., Vol. 1, pp. 255–278). Mahwah, NJ: Erlbaum.

Rasbash, J., Jenkins, J., O'Connor, T. G., Tackett, J., & Reiss, D. (2011). A social relations model of observed family negativity and positivity using a genetically informative sample. *Journal of Personality and Social Psychology, 100*, 474–491. doi:10.1037/a0020931.

Rende, R. D., Slomkowski, C. L., Stocker, C., Fulker, D. W., & Plomin, R. (1992). Genetic and environmental-influences on maternal and sibling interaction in middle childhood—a sibling adoption study. *Developmental Psychology, 28*, 484–490. doi:10.1037/0012-1649.28.3.484.

Roisman, G. I., & Fraley, R. C. (2006). A behavior-genetic study of the legacy of early caregiving experiences: Academic skills, social competence, and externalizing behavior in kindergarten. *Child Development, 83*(2), 728–742. doi:10.1111/j.1467-8624.2011.01709.x.

Roisman, G. I., & Fraley, R. C. (2012). The limits of genetic influence: A behavior-genetic analysis of infant-caregiver relationship quality and temperament. *Child Development, 77*, 1656–1667. doi:10.1111/j.1467-8624.2006.00965.x.

Rowe, D. C. (1981). Environmental and genetic influences on dimensions of perceived parenting: A twin study. *Developmental Psychology, 17*, 203–208. doi:10.1037/0012-1649.17.2.203.

Scarr, S. (1992). Developmental theories for the 1990s: Development and individual differences. *Child Development, 63*, 1–19. doi:10.1111/j.1467-8624.1992.tb03591.x.

Van Ijzendoorn, M. H. (1992). Intergenerational transmission of parenting: A review of studies in nonclinical populations. *Developmental Review, 12*, 76–99. doi:10.1016/0273-2297(92)90004-L.

Wade, T. D., & Kendler, K. S. (2000). The genetic epidemiology of parental discipline. *Psychological Medicine, 30*, 1303–1313. doi:10.1017/S0033291799003013.

Chapter 4
The Sibling Relationship as a Source of Shared Environment

Shirley McGuire, Meenakshi Palaniappan and Taryn Larribas

Siblings are children's most common social companions during childhood and, for many, their relationship with their siblings is their longest lasting across development (Conger and Kramer 2010). Sibling research suggests that siblings do impact each other's social, emotional, and cognitive development (Brody 2004; Dunn 2007; McGuire and Shanahan 2010). Sibling rivalry has been associated with greater loneliness, depression, and feelings of low self-worth in children (e.g., Stocker 1994). While some aggression and teasing is common when siblings are young (e.g., Dunn and Munn 1986), high sibling conflict is linked to children's behavior problems and distressed family circumstances (e.g., Perlman et al. 2007). Positive sibling relationships, on the other hand, may foster well being and serve as a protective factor against developmental risks, such as stressful life events and peer problems (e.g., East and Rook 1992; Gass et al. 2007).

Work on the role of sibling relationships in children's development is increasing (McGuire and Shanahan 2010). Consequently, it is surprising that very little research has been conducted on the nature of sibling dynamics from a BG perspective. While the heritability of parenting literature shows significant heritability (McGuire, Segal, Hershberger 2012), sibling research shows significant share environment influences, indicating that it may be a key source of family environment influences. The goal of the current chapter is to: (1) review BG work studies

S. McGuire (✉)
College of Arts and Sciences, University of San Francisco,
San Francisco, CA 94117, USA
e-mail: mcguire@usfca.edu

M. Palaniappan
University of Florida, Gainesville, USA

T. Larribas
Los Angeles Unified School District, Los Angeles, USA

© Springer Science+Business Media New York 2015
B.N. Horwitz and J.M. Neiderhiser (eds.), *Gene-Environment Interplay in Interpersonal Relationships across the Lifespan*,
Advances in Behavior Genetics 3, DOI 10.1007/978-1-4939-2923-8_4

conducted on sibling relationship quality; (2) discuss studies on growing literature on sibling socialization effects, including BG studies; (3) outline ways that BG approaches can further the literature on sibling contributions to development and mental health.

4.1 Sibling Relationships as a Phenotype

Behavioral geneticists have not ignored siblings. On the contrary, siblings have played a prominent role in BG research. Many kinship studies include sibling pairs that vary in genetic relatedness, the most common being monozygotic (MZ) and dizygotic (DZ) twins (Plomin et al. 2008). Sibling differences are also the main focus of nonshared environment research (McGuire 2001). However, the sibling relationship itself has been the focus of only a handful of studies. Instead, the goal of kinship studies is to understand sibling similarity, not sibling relationship quality. The sibling dyads are a tool to understand the etiology of a behavior or trait and very few of these studies include any measure of sibling interaction or relationship quality. Sibling relationship quality dimensions, however, can also be treated as a measured behavior—or phenotype—in a kinship study. It is important to understand genetic and environmental contributions to this relationship that plays a prominent role in many people's lives.

The BG approach conceptualizes social relationships not as pure environmental phenomena, but as "phenotypes" that can be influenced by nature and nurture (McGuire et al. 2012). Phenotypes are measured behaviors or traits. BG studies on the "nature of nurture" show that many measures of the child's "environment" are heritable (Plomin 1994). Significant genetic contributions to "environmental" measures are conceptualized as genotype-environment correlation, which is a form of genotype-environment interplay (Plomin 1994). There are three types of rGE processes: passive, evocative, and active (McGuire et al. 2012). Passive rGE occurs when family members share both environments and genes. Evocative rGE processes refer to the extent to which genetically linked traits in individuals elicit responses from others in the environment. Active rGE processes refer to the extent to which people seek out environments that are correlated with their genetically linked traits. The BG literature devoted to understanding the etiology of parenting focuses on unraveling the nature of genetic contributions to theses social experiences (McGuire et al. 2012).

BG studies are designed to disentangle the environmental as well as genetic contributions to individual differences in phenotypes. Kinship designs allow researchers to separate phenotypic variability into three components of variance: h^2, c^2, and e^2 (see Plomin et al. 2008 for details about BG methods). The h^2 statistic represents heritability and estimates the degree to which individual differences in a phenotype are associated with individual differences in genotypes within a population. The c^2 statistic estimates the degree to which individual differences in a phenotype are associated with family similarity after controlling for genetic similarity. Researchers often refer to this statistic as "common environment" or "shared environment" because it was assumed to be due to shared family experiences. The e^2 statistic estimates the degree to which family members are not similar in a phenotype, and includes measurement error. Researchers often refer to this statistic as "nonshared environment" because it was assumed to be due to dissimilar family experiences. BG analyses of behaviors/traits, however, do not directly capture environmental processes. If family members respond differently to a "common" environment (e.g., parental divorce), the impact of the experience would be found in e^2, not c^2. If different circumstances lead family members to the same outcome (e.g., behavior problems), c^2 would be significant for the outcome, not e^2.

The kinship studies reviewed in this chapter involve twins and siblings of different degrees of genetic relatedness and the designs used have different strengths and weaknesses (see McGuire et al. 2012; Plomin et al. 2008). The twin design includes MZ twins and DZ twins. MZ twins result from the division of a single fertilized egg and share all genes. DZ twins result from the separate fertilization of two eggs by separate sperm and share (on average in the population) 50 % of their segregating genes. Family designs include full siblings (FS) who, like DZ twins, result from the separate fertilization of two eggs, but at different points in time, and share 50 % of their segregating genes (on average in the population). Stepfamily designs include full siblings who share both biological parents, half siblings (HS) who share only one biological parent, and are on average 25 % genetically related, and genetically-unrelated (UR) siblings. Adoption designs include either full siblings or adoptive siblings who are genetically unrelated. Kinship designs may compare similarity in different types of family relationships, such as parents and their children, cousins, and aunts/uncles and nieces/nephews. The studies reviewed in this chapter all used sibling dyads as the basis of the design.

Heritability is significant when kinship similarity in the measured phenotype is associated with genetic similarity. The basic unit of comparison is the sibling intraclass correlation, which is the correlation between one siblings' score with the other. Significant heritability (h^2) would be found if the sibling intraclass correlations followed the pattern MZ twins > DZ twin and adoption study would follow the pattern FS > UR siblings. If all of the correlations are high and do not differ significantly across the pairs, then shared environment (c^2) will be significant. The adoptive pair correlation, in particular, should be significant because adoptive sibling correlations are considered direct estimates of common environment. E^2 includes error, but will also be significant when intraclass correlations are not high across kinship pairs.

4.2 BG Studies of Siblings Relationships: Some h^2, but Mostly c^2

To our knowledge, only five studies have examined genetic and environmental contributions to sibling relationship quality using a BG design. These studies vary in several characteristics, such as the children's ages, the BG design employed, and the methods used to assess the sibling relationship (see Table 4.1). There were

Table 4.1 Sibling intraclass correlations for sibling negativity measures by sibling type

Variable	Reporter/method	Study	Sibling type by degree of genetic relatedness				
			MZ	DZ	FS	HS	UR
Instigating conflict	Parent: questionnaire	Lemery and Goldsmith (2001)	0.68 (n = 200)	SS = 0.55 (n = 159) OS = 0.42 (n = 156)			
Physical aggression	Child: interview	McGuire et al. (2000)[a]			0.42 (n = 59)		0.13 (n = 49)
Verbal aggression	Child: interview	McGuire et al. (2000)			0.27 (n = 59)		0.10 (n = 49)
Rivalry	Child: interview	McGuire et al. (2000)			0.35 (n = 59)		0.00 (n = 49)
Hostility	Child: questionnaire	Pike and Atzaba-Poria (2003)[b]	0.68 (n = 102)	0.50 (n = 110)			
Rivalry	Child: questionnaire	Pike and Atzaba-Poria (2003)	0.55 (n = 102)	0.24 (n = 110)			
Negativity	Child: questionnaire	Plomin (1994)[b]	0.64 (n = 93)	0.57 (n = 97)	ND = 0.37 (n = 94) Step = 0.45 (n = 179)	0.45 (n = 109)	0.47 (n = 129)
Negativity	Mother: questionnaire	Plomin (1994)	0.90 (n = 93)	0.74 (n = 97)	ND = 0.74 (n = 94) Step = 0.67 (n = 179)	.65 (n = 109)	0.74 (n = 129)
Negativity	Father: questionnaire	Plomin (1994)	0.87 (n = 93)	0.83 (n = 97)	ND = 0.81 (n = 94) Step = 0.86 (n = 179)	.77 (n = 109)	0.80 (n = 129)

(continued)

Table 4.1 (continued)

Variable	Reporter/method	Study	Sibling type by degree of genetic relatedness				
			MZ	DZ	FS	HS	UR
Conflict	Observations: structured	Rende et al. (1992)[a]			0.91 (n = 67)		0.85 (n = 57)
Competition	Observations: structured	Rende et al. (1992)			0.66 (n = 67)		0.46 (n = 57)
Negativity	Observations: unstructured	Rende et al. (1992)			0.79 (n = 67)		0.62 (n = 57)
Negativity	Mother: interview	Rende et al. (1992)			0.85 (n = 67)		0.65 (n = 57)

Note MZ monozygotic twins; *DZ* dizygotic twins; *FS* full siblings; *HS* half siblings; *UR* genetically unrelated siblings, in either adoptive families or in step-families; *SS* same-sex pairs; *OS* opposite-sex pairs; *ND* from never divorced and Step = from stepfamilies
[a]Data for sibling pairs in this study includes same-sex and opposite-sex pairs
[b]Data for sibling pairs in this study includes only same-sex pairs

two twin studies: Lemery and Goldsmith's (2001), which study included 534 pairs of 3- to 8-old twins, and Pike and Atzaba-Poria's (2003) study, which included 212 12- to 15-year old twins. Rende et al. (1992) report focused on the 57 adoptive and 67 non-adoptive pairs from the Colorado Adoption Project when the siblings were 4- and 7-years old. The McGuire et al. (2000) study used data from the longitudinal follow-up of the same study three years later when the siblings were 7- and 10-years old. Plomin (1994) paper uses data from the Nonshared Environment in Adolescence Development project, which included over 700 pairs of twins and siblings and in never-divorced and step families (Reiss et al. also discussed the results for this study in a book about the project in 2000). Given that there are very few studies, and that the studies are different from each other in important ways, a meta-analysis would not be appropriate. The results of these studies, however, are remarkably consistent and worthy of discussion.

Most research on sibling relationship quality focuses on three main dimensions: conflict, warmth, and relative power (Furman and Buhrmester 1985). Sibling conflict, which is often measured as the broader construct "sibling negativity", includes physical aggression, verbal aggression, competition, and rivalry. Sibling intraclass correlations for sibling negativity for the five studies are in Table 4.1. Sibling warmth, which is often measured as the broader construct "sibling positivity", includes closeness, intimacy, companionship, admiration, nurturance, and affection. The sibling intraclass correlations for sibling positivity for the fives studies are in Table 4.2. Relative power refers to the equity and asymmetry of the relationship; it may be accompanied by positive emotion, in the case of sibling caregiving, or by negative emotion, in the case of sibling control. The data for this dimension is not included in this chapter because it was only measured in the Colorardo Adoption Project (i.e., McGuire et al. 2000).

As Tables 4.1 and 4.2 show, there is evidence of modest genetic contributions to sibling negativity and sibling positivity. Similar to the heritability of parenting literature (McGuire et al. 2012), heritability appears to be lower for structured observations and higher for self-report data. More data are needed, of course, before conclusion can be drawn about differences in the heritability of sibling relationship quality across dimensions, context, and ages. It is interesting to note that both Lemery and Goldsmith (2001) and Pike and Atzaba-Poria (2003) found that temperament explained much of the genetic contributions to sibling negativity. Lemery and Goldsmith found that temperament explained some of the heritability of parents' reports of sibling conflict in young children. Pike and Atzaba-Poria found that temperament explained some of the heritability of teenagers' reports of their hostility and rivalry toward each other, respectively. As Lemery and Goldsmith discuss, the link between children's temperament characteristics and sibling conflict may be either the result of a "lack of fit" between siblings' temperaments, with two dissimilar temperaments leading to greater conflict, or the result of a "buffering effect", with two adaptive temperament styles leading to less conflict. They found evidence to support both hypotheses. Sibling pairs where both children had difficult temperaments had the highest levels of conflict, followed by pairs with one child with an easy temperament. Not surprisingly, pairs with two

Table 4.2 Sibling intraclass correlations for sibling positivity measures by sibling type

Variable	Reporter/method	Study	Sibling type by degree of genetic relatedness				
			MZ	DZ	FS	HS	UR
Instigating cooperation	Parent: questionnaire	Lemery and Goldsmith (2001)	0.64 (n = 200)	SS = 0.57 (n = 159) OS = 0.60 (n = 156)			
Sharing personal possessions	Child: interview	McGuire et al. (2000)[a]			0.54 (n = 59)		0.00 (n = 49)
Sharing common items	Child: interview	McGuire et al. (2000)			0.21 (n = 59)		0.12 (n = 49)
Affection	Child: questionnaire	Pike and Atzaba-Poria (2003)[b]	0.70 (n = 102)	0.55 (n = 110)			
Positivity	Child: questionnaire	Plomin (1994)[b]	0.74 (n = 93)	0.54 (n = 97)	ND = 0.38 (n = 94) Step = 0.39 (n = 179)	0.50 (n = 109)	0.40 (n = 129)
Positivity	Mother: questionnaire	Plomin (1994)	0.90 (n = 93)	0.70 (n = 97)	ND = 0.84 (n = 94) Step = 0.73 (n = 179)	0.75 (n = 109)	0.73 (n = 129)
Positivity	Father: questionnaire	Plomin (1994)	0.94 (n = 93)	0.81 (n = 97)	ND = 0.80 (n = 94) Step = 0.83 (n = 179)	0.82 (n = 109)	0.86 (n = 129)

(continued)

Table 4.2 (continued)

Variable	Reporter/method	Study	Sibling type by degree of genetic relatedness				
			MZ	DZ	FS	HS	UR
Cooperation	Observations: structured	Rende et al. (1992)[a]			0.76 (n = 67)		0.81 (n = 57)
Positivity	Observations: unstructured	Rende et al. (1992)			0.85 (n = 67)		0.63 (n = 57)
Positivity	Mother: interview	Rende et al. (1992)			0.81 (n = 67)		0.62 (n = 57)

Note MZ monozygotic twins; *DZ* dizygotic twins; *FS* full siblings; *HS* half siblings; *UR* genetically unrelated siblings, in either adoptive families or in stepfamilies; *SS* same-sex pairs; *OS* opposite-sex pairs; *ND* from never divorced and Step = from stepfamilies

[a]Data for sibling pairs in this study includes same-sex and opposite-sex pairs

[b]Data for sibling pairs in this study includes only same-sex pairs

children with the easy temperaments showed lower conflict and more prosocial behavior. These results tentatively suggest that there may be passive and evocative genotype-environment processes at work in sibling relationship dynamics, particularly for sibling negativity.

The most striking finding in Tables 4.1 and 4.2 is that the majority of the variance in sibling relationship dimensions is shared environment. McGuire et al. (2000) seems to be the only exception, with lower adoptive sibling intraclass correlations for physical aggression, verbal aggression and rivalry than would be expected if shared environment is significant. The findings of this study may differ from the others because of the measure used to collect the data. In McGuire et al., the siblings were asked to discuss the last fight they had with their sibling and to explain why they fight with their siblings during a semi-structured interview. The method was designed to capture the child's perspective and, consequently, it is not surprising that the sibling intraclass correlations suggest genetic and non-shared environment contributions. The other studies used methods that may have tapped into the quality of the dyadic relationship. Rende et al. (1992) discuss the role of methodology in the high levels of shared environment they found for the structured observations in their study. They suggested that the constrained context might have increased or encouraged synchronicity in social interaction. Given that all of the studies used sibling dyads as the basis of the kinship design, significant sibling intraclass correlations are expected and it could be that the high percentage of shared environment found is just an artifact of the design method. Parents and observers may see the dyad as "a unit" when asked about their relationship with each other, even if the siblings treat each other differently. New research in the sibling socialization area, however, suggests that shared sibling experiences exist and contribute to children's development. Behavioral geneticists are beginning to test if the sibling relationship acts as a moderator of sibling similarity in important developmental outcomes, particularly in the delinquency area.

4.3 Sibling Socialization Effects in Development

Studies of siblings do suggest that some sibling pairs are socializing each other in antisocial behaviors, such as substance abuse and smoking (e.g., Kramer and Conger 2009; Slomkowski et al. 2001). For instance, Slomkowski et al. (2005) reviewed an emerging literature providing strong support for older siblings' influences over younger siblings' smoking behavior. Work with families of antisocial children show that siblings sometimes colluding together and reinforcing antisocial behavior in each other. The influence of older siblings' behavior on the younger sibling, both prosocial and antisocial, has been documented in single-parent families and teenage child-bearing families. While some researchers argue that sibling socialization is due to social learning (Whiteman et al. 2007), behavior genetic studies suggest that sibling relationships processes may be interacting with genetic similarity to create similarity in risky or delinquent behavior.

Rowe and Gulley (1992) argued two decades ago that sibling pairs who had a warm relationship might become "partners in crime" by teaching each other risky and deviant behaviors. That is, the sibling relationship dimension warmth would moderate sibling similarity in delinquent behavior. They suggested using BG designs to tap into these processes while controlling for genetic similarity. A few researchers have used this strategy with some success. Using twin and sibling data from the US National Longitudinal Study of Adolescent Health (Add Health), Slomkowski et al. (2005) and Rende et al. (2005) studied the role of "sibling connectedness" (i.e., amount of sibling contact and number of mutual friendships) on sibling similarity in smoking and drinking behavior. They found evidence to suggest that social connectedness moderated sibling similarity in smoking and drinking behavior after controlling for the genetic relatedness of the pairs. That is, sharing friends and high social contact between the siblings helped explain shared environmental, but not genetic, contributions to smoking behavior in teens. Also using the Add Health data, McHale, Bissell and Kim (2009) studied the effect of family-level warmth and sibling closeness on sibling similarity in sexual risk. They found that sibling similarity in risky sexual attitudes was higher in dyads high in sibling closeness. They also found that family warmth and sibling closeness mediated genetic contributions to sexual risk. These reports for the Add Health study provide further support of genotype-environmental associations between sibling relationship processes and genetic factors. BG researchers should build on this research to investigate the role of sibling relationship as a moderator of sibling similarity in many important developmental outcomes that show significant shared environment, including development psychopathology (Burt 2009).

4.4 Future Directions

Behavioral geneticists and family researchers should consider building on previous research on genetic and environmental contributions to sibling relationships. There is evidence of modest heritability, particularly for sibling negativity. Heritabilities of sibling relationship dimensions may show the same pattern as heritabilities of parenting dimensions, with h^2 changing across dimension, development, and context (McGuire et al. 2012). Adding sibling relationship measures to BG studies of family dynamics would help further this literature.

The significant shared environment found for sibling relationship measures suggest that psychologists and sociologists in the clinical, family, and developmental areas should consider investigating sibling socialization effects. Rowe and Gulley (1992) may be correct that siblings that share a close relationship may be influencing each other in antisocial ways. Sibling researchers, however, are moving beyond the basic three dimensions—negativity, positivity, and relative power—to investigate sibling trust (McGuire et al. 2010), sibling attachment

(Tancredy and Fraley 2006), and sibling self-disclosure (Howe et al. 2001), cooperation, and coordination. More research needs to be conducted exploring process such as observational learning (e.g., Brody et al. 2003; Whiteman et al. 2007), coercion, collusion, and reinforcement (Slomkowski et al. 2001). This work may be useful in the detection of sibling processes that lead to greater sibling similarity in developmental and mental health outcomes.

Scientists should also examine the joint contributions of sibling and peer influences. Behavioral geneticists have explored genetic and environmental contributions to friendship, peer group characteristics, and peer experiences; they found significant heritability and nonshared environment (e.g., Manke et al. 1995; Pike and Atzaba-Poria 2003). These studies, however, focused on children's perceptions of their peer experiences in adolescence. More relevant to the research reviewed in this chapter, Burt and Klump (2013) found that delinquent peer affiliation moderated shared environmental effects on childhood delinquency. Their study differed from others conducted during adolescence that found that exposure to peer deviance interacted with genetic contributions to antisocial behavior. As Burt and Klump note, childhood may be a prime time to examine peer influences on development. It may be, however, the best time to examine the joint effects of siblings and peers on children's development. In fact, McGuire and Segal (2013) found that school-aged siblings overlapped by 53 % in their peer networks, on average, in a study of over 300 pairs of school-aged twins and siblings. Interestingly, sibling similarity in number of friends was high and consistent across different pairs, suggesting shared environmental influence. The percentage of peer network overlap for the dyad, on the other hand, differed by genetic relatedness, sex composition, age difference, degree of relationship intimacy, and time spent together at school. MZ twins showed the highest overlap (82 %) and opposite-sex full sibling pairs showed the lowest overlap (27 %). The more personal characteristics and social experiences siblings had in common, the more friends they had in common. These findings suggest that peer and sibling influences on risky and delinquent behavior may not be happening independent of each other.

While only a handful of studies have been conducted on genetic and environmental contributions to sibling relationships, the majority of the data show significant shared environmental contributions to sibling negativity and positivity. The sibling relationship may be a prime candidate for the discovery of shared environmental influences in the family. A few researchers have, in fact, found that sibling connectedness and closeness moderate adolescent risky behavior. Expanding the sibling relationship dimensions beyond "positivity" and "negativity" to include more specific processes such as trust, self-disclosure, collusion, and coercion, will further this literature. Investigating the overlap between sibling and peer influences, particularly in childhood, and how they interact with genetic contributions to behavior will help us understanding siblings' role in important development.

References

Burt, S. A. (2009). Rethinking environmental contributions to child and adolescent psychopathology: A meta-analysis of shared environmental influences. *Psychological Bulletin, 135*, 608–637.

Burt, S. A., & Klump, K. L. (2013). Delinquent peer affiliation as an etiological moderator of childhood delinquency. *Psychological Medicine, 43*, 1269–1278

Brody, G. H. (2004). Siblings'direct and indirect contributions to child development. *Current Directions in Psychological Science, 13*, 124–212.

Brody, G. H., Kim, S., Murry, V. B., & Brown, A. C. (2003). Longitudinal direct and indirect pathways linking older sibling competence to the development of younger sibling competence. *Developmental Psychology, 39*, 618–628.

Conger, K. J., & Kramer, L. (2010). Introduction to the special section: Perspectives on sibling relationships: Advancing child development research. *Child Development Perspectives, 4*(2), 69–71.

Dunn, J. (2007). Siblings and socialization. In J. Grusec & P. D. Hastings (Eds.), *Handbook of socialization: Theory and research* (pp. 309–327). New York: Guilford.

Dunn, J., & Munn, P. (1986). Sibling quarrels and maternal intervention: Individual differences in understanding and aggression. *Journal of Child Psychology and Psychiatry, 27*, 583–595.

East, P. L., & Rook, K. S. (1992). Compensatory support among children's peer relationships: a test using friends, nonschool friends and siblings. *Developmental Psychology, 28*, 163–172.

Furman, W., & Buhrmester, D. (1985). Children's perceptions of the qualities of sibling relationships. *Child Development, 56*, 448–461.

Gass, K., Jenkins, J., & Dunn, D. (2007). Are sibling relationships protective? A longitudinal study. *Journal of Child Psychology and Psychiatry, 48*, 167–175.

Horwitz, B., Marceau, K., & Neiderhiser, J. M. (2011). Family relationship influences on development: What can we learn from genetic research? In K. Kendler, S. Jaffee, & D. Romer (Eds.), *The dynamic genome and mental health: The role of genes and environments in development* (pp. 128–144). NY: Oxford University Press.

Howe, N., Aquan-Assee, J., Bukowski, W., Lehoux, P., & Rinaldi, C. (2001). Siblings as confidants: Emotional understanding, relationship warmth, and sibling self-disclosure. *Social Development, 10*(4), 439–454. doi:10.1111/1467-9507.00174.

Kramer, L., & Conger, K. J. (Eds.). (2009). *Siblings as agents of socialization: New directions in child and adolescent development* (Vol. 126). San Francisco: Jossey-Bass.

Lemery, K. S., & Goldsmith, H. H. (2001). Genetic and environmental influences on preschool sibling cooperation and conflict: Associations with difficult temperament and parenting style. In K. Deater-Deckard & S. Petrill (Eds.), *Gene-environment process in social behaviors and relationships*. Philadelphia: Haworth Press.

Manke, B., McGuire, S., Reiss, D., Hetherington, E. M., & Plomin, R. (1995). Genetic contributions to adolescents' extrafamilial interactions: Teachers, best friends, and peers. *Social Development, 4*, 238–256.

McGuire, S. (2001). Nonshared environment research: What is it and where is it going? *Marriage and Family Review, 33*, 31–57.

McGuire, S. & Segal, N. L. (2013). Peer network overlap in twin, sibling, and friend dyads. *Child Development, 84*, 500–511. doi:10.1111/j.1467-8624.2012.01855.x

McGuire, S., Segal, N. L., & Hershberger, S. (2012). Parenting as phenotype: a behavioral genetic approach to understanding parenting. *Parenting, Science and Practice, 12*(2–3), 192–201. doi:10.1080/15295192.2012.683357

McGuire, S., Segal, N. L., Gill, P., Whitlow, B., & Clausen, J. M. (2010). Siblings and trust. In K. Rotenberg (Ed.), *Interpersonal trust during childhood and adolescence* (pp. 133–154). Cambridge, UK: Cambridge University Press.

McGuire, S., & Shanahan, L. (2010). Sibling relationships in evolving family contexts. *Child Development Perspectives, 4*, 72–79.

Plomin, R. (1994). *Genetics and experience: The interplay between nature and nurture.* Thousand Oaks, CA: Sage.

Plomin, R., DeFries, J. C., McClearn, G. E., & McGruffin, P. (2008). *Behavioral genetics* (5th ed.). New York: Worth.

Perlman, M., Garfinkel, D., & Turrell, S. L. (2007). Sibling and parent influences on the quality of children's conflict behaviours across the preschool period. *Social Development, 16,* 619–641.

Pike, A., & Atzaba-Poria, N. (2003). Do sibling and friend relationships share the same temperamental origins? A twin study. *Journal of Child Psychology and Psychiatry, 44,* 598–611.

Reiss, D., Neiderhiser, J. M., Hetherington, E. M., & Plomin, R. (2000). *The relationship code.* Cambridge, MA: Harvard University Press.

Rende, R.D., Slomkowski, C.L., Stocker, C., Fulker, D., & Plomin, R. (1992). Genetic and environmental influences on maternal and sibling behavior in middle childhood: A sibling adoption study. *Developmental Psychology, 28,* 484–490.

Rowe, D. C., & Gulley, B. (1992). Sibling effects on substance use and delinquency. *Criminology, 30,* 217–233.

Slomkowski, C., Rende, R., Conger, K., Simons, R., & Conger, R. (2001). Sisters, brothers, and delinquency: Evaluating social influence during early and middle adolescence. *Child Development, 72*(1), 271–283. doi:10.1111/1467-8624.00278.

Slomkowski, C., Rende, R., Novak, S., Richardson, E., & Niaura, R. (2005). Sibling effects on smoking in adolescence: Evidence for social influence from a genetically-informative design. *Addiction, 100,* 430–448.

Snyder, J., Bank, L., & Burraston, B. (2005). The consequences of antisocial behavior in older male siblings for younger brothers and sisters. *Journal of Family Psychology, 19,* 1–11.

Stocker, C. (1994). Children's perceptions of relationships with siblings, friends, and mothers: compensatory processes and links with adjustment. *Journal of Child Psychology and Psychiatry, 35,* 1447–1459.

Tancredy, C., & Fraley, R. (2006). The nature of adult twin relationships: An attachment-theoretical perspective. *Journal of Personality and Social Psychology, 90*(1), 78–93. doi:10.1037/0022-3514.90.1.78.

Whiteman, S., McHale, S., & Crouter, A. (2007). Competing processes of sibling influence: Observational learning and sibling deidentification. *Social Development, 16*(4), 642–661. doi:10.1111/j.1467-9507.2007.00409.x.

Chapter 5
Gene-Environment Transactions in Childhood and Adolescence: Problematic Peer Relationships

Mara Brendgen and Michel Boivin

For a large part of the twentieth century, theoretical perspectives on human devel-opment have mostly focused on the role of environmental experiences in explain-ing interindividual differences in social adjustment. Accumulating evidence suggests, however, that genetic as well as environmental factors contribute to shape human development (Turkheimer 2000). Although genetic and environmen-tal influences were initially posited to influence development in an independent fashion, it is now clear that these forces are likely to interact and to reciprocally influence each other through various mechanisms of gene-environment interplay (Allen et al. 1961; Plomin et al. 1977; Scarr and McCartney 1983). In this con-text, the vast majority of studies have examined how genetic factors work together with family-related factors (e.g., parental behavior, socioeconomic status, physi-cal or sexual maltreatment) to shape developmental outcomes (e.g., Bakermans-Kranenburg and van IJzendoorn 2006; Kim-Cohen et al. 2006; Thapar et al. 2007). Already very early in life, however, children spend many hours in daycare settings or schools, away from their families and in the company of peers. Prior to school entry, experiences with peers are thus likely to play a fundamental role in chil-dren's development, and they assume an even greater importance as youngsters make the transition from childhood to adolescence (Bukowski et al. 2006). Indeed, given that even identical twins often do no affiliate with the same friends (Thorpe and Gardner 2006), experiences with peers may be a particularly important source of nonshared environmental influence during childhood and adolescence.

M. Brendgen (✉)
Department de Psychologie, Université du Québec à Montréal, Montreal, Canada
e-mail: brendgen.mara@uqam.ca

M. Brendgen · M. Boivin
Université Laval, Québec City, Canada

© Springer Science+Business Media New York 2015
B.N. Horwitz and J.M. Neiderhiser (eds.), *Gene-Environment Interplay in Interpersonal Relationships across the Lifespan*,
Advances in Behavior Genetics 3, DOI 10.1007/978-1-4939-2923-8_5

Based on the works of Piaget (1932) and Sullivan (1953), Youniss (1980) argued that the unique characteristic of peer experiences is grounded in the symmetrical reciprocity of the relation between children. Contrary to relations with adults, which can be described as asymmetrical and unilateral in power, children's peer relations are considered as being balanced and egalitarian. Through interaction with peers as co-equals, children have the opportunity to discuss and negotiate conflicting ideas and multiple perspectives, and to decide whether to accept or reject the notions held by others. In so doing, the children presumably acquire fundamental socio-cognitive skills like social perspective-taking, cooperation, and methods of conflict resolution, which are the basis for successful social relations later in life. Once established, these peer relations offer both social support and security, which, in turn, enhance the development of a healthy self-concept (Boivin et al. 2005). However, despite their many positive features, experiences with peers are not always based on egalitarian and supportive relationships, and can in some instances also seriously impede children's healthy development. This could occur, for example, when youths are exposed to chronic peer victimization or when they affiliate with antisocial friends who may reinforce inappropriate behaviors such as aggression. Youngsters who experience such problematic peer relations are not only deprived of the opportunity to learn basic skills required to establish and maintain future social relations, but they are also at substantial risk of suffering undesirable developmental outcomes such as depression, academic failure, delinquency, or substance abuse (Boivin et al. 2005; Bukowski et al. 2006).

Compared with studies examining gene-environment interplay with respect to family-related risk factors or stressful life events, behavioral genetic studies that investigate problematic peer relations are still few in number. Nevertheless, in the past few years several studies have been published that provide important insights into how genetic factors work together with perhaps the most worrisome peer-related experiences—i.e., peer victimization and affiliation with deviant peers—in influencing child development. In this chapter, we offer a synopsis of the current state of knowledge in this context, including research on both children and adolescents and covering diverse adjustment outcomes. We will first illustrate the different types of gene-environment interplay relevant to our understanding of how deviant peer affiliation and peer victimization relate to behavioral and emotional maladjustment. We then provide a brief description of prominent theoretical perspectives of deviant peer affiliation and peer victimization, respectively. This will be followed by a review of the extant empirical evidence from behavioral genetic studies on deviant peer affiliation and peer victimization in childhood and adolescence. The chapter concludes with a discussion of future directions in genetically informed research on problematic peer relations.

5.1 Problematic Peer Relations: Deviant Peer Affiliation

Notwithstanding the importance of friendship relations for youngsters' healthy development, youths who affiliate with deviant peers are a major concern for parents, teachers, and clinical practitioners alike. While mostly considered a problem during adolescence, there is evidence that some children associate with deviant peers already during the preschool period (Estell et al. 2002; Farver 1996; Snyder et al. 1997, 2005). Although external circumstances such as ineffective parenting and neighborhood disadvantage have been shown to predict deviant peer affiliation (Ary et al. 1999; Dodge et al. 2006), most scholars concur that affiliation with aggressive and antisocial peers is also due to an active selection process based on the similarity of attitudes and behaviors (Cairns et al. 1998; Hektner et al. 2000; Lahey et al. 1999; Poulin and Boivin 2000; Urberg et al. 1998). This active selection of deviant friends begins by mid-childhood (or possibly earlier), accelerates thereafter, and peaks by early adolescence.

Numerous studies have documented that deviant peer affiliation, either within dyadic friendships or within larger cliques of deviant peers, is related to a subsequent increase in externalizing problems such as violent and nonviolent delinquency or substance use (Elliott et al. 1985; Kim et al. 1999; Patterson et al. 2000; Scaramella et al. 2002; Simons et al. 2001). These findings are in line with the *peer influence model,* which posits that social learning via modeling and reinforcement of antisocial attitudes and behaviors by deviant peers is the main cause of delinquent behavior in youth (Akers et al. 1979; Cohen 1977; Elliott et al. 1985; Johnson et al. 1987; Sutherland 1947). Not all studies provide unequivocal support for the peer influence hypothesis, however (Coie et al. 1995; Tremblay et al. 1995). For instance, Tremblay et al. (1995) found that affiliation with deviant peers and delinquency were only related because of their underlying common association with early antisocial behavior and that relation vanished once early antisocial behavior was controlled. Such findings concord with the *individual characteristics model,* which posits that deviant peer affiliation plays no causal role in the development of antisocial behavior, but is rather a mere epiphenomenon of youngsters' preexisting deviant tendencies (e.g., Cairns et al. 1988; Gottfredson and Hirschi 1990). The *social interactional or enhancement model* merges aspects of both the individual characteristics model and the peer influence model (Dishion 1990a, b; Patterson et al. 1989). According to this integrated model, a child's preexisting deviant tendencies work together with his or her peers' deviancy in an interactive and mutually enhancing fashion to promote later antisocial or delinquent behaviors. Support for this viewpoint comes from findings that affiliation with deviant peers has been linked to later delinquency more strongly in children with preexisting externalizing behavior than in children without such problematic behavioral tendencies (Vitaro et al. 1997).

5.2 Problematic Peer Relations: Peer Victimization

Peer victimization among children is a severe problem in many countries around the world (Smith et al. 1999). Although physical peer victimization tends to decline over the course of primary school (Brame et al. 2001), nonphysical forms of victimization such as rumor spreading or social exclusion actually increase with age, with a sizable peak between ages 11 and 15 when children experience puberty and enter high school (Archer and Côté 2005; Pellegrini and Long 2002). In addition, Internet or cell-phone-based peer victimization, called cyber-bullying, is becoming increasingly common among early adolescents (Li 2006). Whether peer victimization is physical or psychological, all forms likely carry severe risks for the victims. Indeed, there is considerable evidence based on singleton samples that many victims of peer victimization develop internalizing mental health problems, notably anxiety and depression symptoms and suicide ideation (Carney 2000; Fekkes et al. 2006; Hodges et al. 1999; Paul and Cillessen 2003; Vernberg et al. 1992; Vuijk et al. 2007). Peer victimization has also been linked to concurrent and later externalizing mental health problems, specifically angry-oppositional behavior and antisocial conduct (Arseneault et al. 2006; Hanish and Guerra 2002; Lamarche et al. 2007; Schwartz et al. 1998a, b). In addition, there is an association between peer victimization and later physical health problems, including respiratory infections, psycho-physiological disorders, skin disorders, and excessive weight gain (Brendgen and Vitaro 2008; Fekkes et al. 2006; Nishina et al. 2005; Rigby 1999; Slee 1995; Strauss et al. 1985; Sweeting et al. 2005; Williams et al. 1996).

Although peer victimization is a social stressor that may lead to negative health outcomes, preexisting externalizing or internalizing mental health problems can also put children at risk of peer victimization (Boivin et al. 2001; Hanish and Guerra 2000; Hodges et al. 1999; Hodges and Perry 1999; Lamarche et al. 2007; Paul and Cillessen 2003; Schwartz et al. 1999). Children who exhibit internalizing symptoms such as behavioral inhibition and withdrawal may be seen as "easy targets" that are unlikely to retaliate (Boivin and Hymel 1997; Hodges et al. 1999). Externalizing behavior symptoms such as screaming and angry outbursts may be considered annoying, thus fostering victimization from peers (Pope and Bierman 1999). Physical symptoms, such as obesity, may also increase children's risk of becoming a target of peer victimization (Cramer and Steinwert 1998).

5.3 Understanding the Developmental Impact of Problematic Peer Relations: The Usefulness of Behavioral Genetic Designs

As outlined above, numerous studies have documented significant associations between problematic peer relations, notably peer victimization and deviant peer affiliation, during childhood and adolescence and a variety of adjustment problems later in life (Bukowski et al. 2006). Moreover, the existing literature points to a bidirectional link between these peer relation difficulties and youngsters' behavioral and emotional problems. However, the vast majority of peer relation studies are based on a correlational design, which cannot provide a completely valid test of whether, for example, adolescents' aggressive behavior is the consequence or the cause of affiliation with deviant peers. This limitation is true even for transactional longitudinal studies, which are typically considered the most stringent type of longitudinal correlational design for testing the directionality of association between two variables (Boivin et al. 2010). To test causal hypotheses one would ideally use experimental designs. However, experimental studies are often prohibited due to ethical concerns, specifically when studying potentially harmful environmental influences, such as deviant peer affiliation. In some instances, researchers can make use of randomized intervention studies to test causal links between problematic peer relations and future maladjustment by trying to reduce the purported antecedents of these problems (Vitaro et al. 2001).

Another alternative is the use of genetically informed designs, notably behavioral genetic studies. Behavioral genetic studies are considered a type of quasi-experimental design, because the comparison of family members with differing degrees of genetic relatedness, such as MZ and DZ twins, constitutes a natural experiment that can pull apart processes that are typically confounded—in this case co-occurring genetic and environmental processes (Rutter et al. 2001). Thus, although behavioral genetic designs cannot offer conclusive proof of causation, they are useful for testing environmental causation beyond the mere identification of risk factors that is provided by traditional correlational studies based on singletons (Moffitt 2005). For instance, the *social homogamy* model would suggest that genetic factors do not play any role in the link between individuals' and their friends' deviant characteristics, and that any behavioral resemblance is explained by environmental conditions that foster deviancy in all individuals who are exposed to these conditions (Gilson et al. 2003; Neale and Cardon 1992). In contrast, the *child effects* model proposes that youngsters' involvement with deviant peers is explained by personal and potentially heritable characteristics inherent to the youngsters themselves. Evidence of genetic effects on deviant peer affiliation would support the child effects model and could suggest that an active selection process based on heritable personal characteristics is involved in adolescents' deviant peer affiliation. Moreover, an extension of such an analysis that involves the inclusion of a measure of child aggression would allow assessing to what extent heritable factors associated with aggressive behavior contribute to adolescents' affiliation with deviant peers. By disentangling genetic from nongenetic sources of interindividual variance, behavioral genetics can thus help test competing theoretical models of peer relation difficulties. Behavioral genetics designs also permit the test of other hypotheses about the association between peer experiences and child behavior. For instance, one might test whether a gene-environment interaction concordant with a diathesis-stress process can be observed, whereby peer victimization increases the likelihood that a genetic vulnerability for depression is expressed. The various mechanisms of gene-environment interplay that may help explain how problematic peer experiences relate to behavioral and emotional adjustment problems are described next.

5.4 Mechanisms of Gene-Environment Interplay

Although the term "gene-environment interplay" can refer to a variety of concepts (Rutter et al. 2006), the two genetic-environmental mechanisms relevant for our understanding of the developmental role of problematic peer experiences are *gene-environment correlations* and *gene-environment interactions*. **Gene-environment correlation (*rGE*)** refers to a situation where heritable factors are associated with specific environmental experiences. Of specific interest with respect to problematic peer experiences is the type of rGE that arises when a child's heritable traits (e.g., behaviors, cognitive abilities, physical characteristics) affect the kind of peer environment he or she will experience. Genetic factors can influence individuals' environmental experiences through either *passive, active,* or *evocative* rGE processes (Plomin et al. 1977; Scarr and McCartney 1983). A *passive rGE* would occur when parents' personal, heritable characteristics naturally influence the quality of the peer environment their children are exposed to. For example, parents with a history of delinquent behavior (a significantly heritable trait) may be more likely than others to settle in disadvantaged neighborhoods with a high crime rate, which increases their children's risk of being recruited by criminal gangs. As a consequence, the child's genotype, which is inherited from the parents, becomes correlated with the peer environment he or she is exposed to. This type of rGE is called passive because the child's environmental experience, in this case deviant peer affiliation, is not brought about by the child's own characteristics, but instead by parents' characteristics that are themselves genetically influenced and heritable.

Contrary to passive rGE, active rGE and evocative rGE involve a direct influence of a child's heritable characteristics on his or her environmental experiences. *Active* (sometimes also called *Selective) rGE* arises where an individual actively selects or shapes his or her own environment, based on genetically influenced personal characteristics. For example, active rGE would refer to a situation where delinquent adolescents (whose behavior may in part be genetically influenced) deliberately choose to affiliate with peers who show similar behavioral characteristics. Active rGE thus corresponds to a special case of the popular saying of "birds of a feather flock together," which would be based on partly heritable characteristics and could partly account for the often observed behavioral similarity between friends (Boivin et al. 2005). However, active rGE is unlikely to be implicated in explaining peer experiences that are not driven by active choice, most notably negative experiences such as being the target of peer victimization. Here, a more likely mechanism is evocative rGE. *Evocative (sometimes also called reactive) rGE* occurs when the child's genetically influenced characteristics elicit specific reactions from the peer environment. For example, a child who displays shy and anxious behavior (a moderately heritable phenotype) may be perceived as an easy target and, as a result, become the victim of frequent peer harassment. The environmental variable (in this case peer victimization) thus becomes correlated with the anxiogenic genotype.

The relative importance of the three types of rGE may change over the course of development (Scarr and McCartney 1983). Such a developmental change is especially likely in regard to genetic effects on children's peer-related environment. Passive rGE is particularly likely to occur in early childhood or during the early elementary school years. Indeed, passive rGE may be the primary mechanism underlying parents' influence on children's peer experiences through selecting, structuring, or modifying their child's physical or social environment (Ladd et al. 1992). As children grow older, however, passive rGE should decline and active rGE should increase as youth become more autonomous in shaping their social environment and selecting their peer affiliation. In contrast, evocative rGE probably plays a relatively consistent role in influencing youngsters' peer experiences, including negative ones such as peer victimization. It should also be mentioned that the presence of rGE due to an influence of genetic factors on problematic peer experiences does not mean that these problematic peer experiences do not, in turn, affect youngsters' behaviors and emotions. Instead, the presence of rGE means that at least part of the association between a problematic peer environment and a problematic developmental outcome is explained by genetic factors that underlie both.

In addition to a possible correlation, genetic factors may also interact with problematic peer experiences in shaping children's and adolescents' developmental outcomes. Such a *gene-environment interaction (GxE)* refers to any situation in which (a) the effect of a genetic disposition on a developmental outcome varies depending on environmental conditions, or (b) the effect of the environment on a developmental outcome varies according to an individual's genetic disposition (Shanahan and Hofer 2005). From a statistical point of view, it is irrelevant whether the environment moderates genetic effects or whether genetic effects moderate the effect of the environment. However, because genetic effects in quantitative studies are usually conceptualized as latent (i.e., non-measured) variables, researchers have mostly interpreted GxE in terms of environmental moderation of genetic effects (for an alternative approach, see Brendgen et al. 2008a; Jaffee et al. 2005).

Gene-environment interactions involving problematic peer experiences may take different forms or processes. A *contextual trigger* process of GxE occurs when the environment triggers or exacerbates a genetic predisposition for a specific behavior or other outcome, or when an environmental condition only leads to a specific developmental outcome in individuals with a genetic predisposition. When it involves negative environmental conditions, such as deviant peer affiliation or peer victimization, and negative adjustment outcomes, such as aggression or depression, a contextual trigger process is analog to a *diathesis–stress* model of psychopathology. This diathesis–stress process may be apparent, for example, when a genetic disposition for aggressive behavior is more likely to be manifested in youth who are frequently victimized by their peers or who affiliate with aggressive friends.

Contrary to a contextual trigger process, a *contextual suppression* process arises when the environment suppresses or at least reduces the influence of genetic factors on a developmental outcome, such that many individuals exhibit the same problem behavior or symptoms irrespective of their genetic disposition. Especially when the environment involves stressful conditions, a contextual suppression process thus often implies that interindividual variability with regard to the problem outcome is lower, whereas the overall mean level is higher in individuals exposed to the environmental stressor than in unexposed individuals. Such a suppression process of GxE would be suggested, for example, if genetic factors were less important and environmental factors more important for explaining depressed affect among victimized children than among non-victimized children.

Importantly, GxE and rGE processes may be related in the sense that the same environmental factors may simultaneously be involved in both types of processes with respect to a given developmental outcome (Eaves et al. 2003; Purcell 2002). Environmental features that are proximal to the child are the most likely to be simultaneously implicated in GxE and rGE (Shanahan and Hofer 2005). For example, adolescents with a genetic disposition for aggression have been found to experience more problematic relationships with their parents, thus suggesting rGE (Narusyte et al. 2007). At the same time, a problematic relationship with parents also fosters aggressive behavior in youths who have a genetic disposition for aggression, thus showing GxE (Feinberg et al. 2007). Purcell (2002) has shown that failure to account for such potential gene-environment correlations when investigating gene-environment interactions can lead to biased estimates and, hence, false conclusions. Indeed, as we will show, a similar feedback loop of rGE and GxE is likely with respect to youngsters' problematic experiences with peers, notably peer victimization and deviant peer affiliation.

5.5 Gene-Environment Interplay Linking Deviant Peer Affiliation and Development: Findings from Behavioral Genetic Studies

5.5.1 Gene-Environment Correlation

The vast majority of studies that examined gene-environment interplay with respect to deviant peer affiliation have focused on the adolescent period. In most of these studies, measures of peer deviancy were based on parents', teachers' or youth's reports about the peers' deviant behavior. In these cases, findings (almost) consistently reveal significant genetic effects on deviant peer affiliation. For example, using a population-based sample of 153 MZ and 558 DZ twins aged 11–18 years (Mean age $= 15.24$, $SD = 2.21$), Button and colleagues (2007) asked participants to indicate how many of their friends had participated in 13 delinquent behaviors, including alcohol and drug use, stealing, and violence, in the previous 6 months. Adolescents' own conduct problems were assessed with the self-report Diagnostic Interview Schedule for Children_IV (Shaffer et al. 1997), a structured interview that assesses *DSM_IV lifetime* symptoms for conduct disorder. Univariate ACE modeling revealed a significant genetic association with deviant peer affiliation, accounting for 21 % of the variance, leaving 40 and 30 % to shared environment and nonshared environment, respectively. Furthermore, bivariate modeling showed that common genetic influences explained most (i.e., 86 %) of the correlation between adolescents' own and their peers' deviancy, with nonshared environmental effects accounting for the remaining 14 %. Similar findings were reported in other studies, with estimates of genetic contributions varying between 20 and 42 % (Button et al. 2009; Dick et al. 2007; Fowler et al. 2007; Hicks et al. 2009; Rose 2002). Exceptions to this pattern of results come from two studies, which found no genetic effects on perceived peers' deviancy (Iervolino et al. 2002; Walden et al. 2004).

Adolescents are likely well informed about their peers' deviant behavior. Indeed, antisocial and delinquent acts are more often committed in the company of peers rather than in isolation (Dishion et al. 2000; Elliott et al. 1985). Nevertheless, halo effects may lead to a confound between individuals' own deviant behavior and their perception of their peers' deviant behaviors (Berndt and Keefe 1995). Such potential halo effects might be further accentuated if individuals are asked to rate the deviant behavior of the peer group as a whole instead of evaluating the behavior of a specific peer. Because deviant behaviors such as aggression, delinquency, and substance use are all to a considerable extent under genetic influence, this halo effect might thus suggest that affiliation with deviant peers is also partly genetically determined when in fact this may not be the case at all. However, findings from recent behavioral genetic studies that include the peers' own reports of their deviant behavior suggest that affiliation with deviant peers may indeed be partly genetically influenced—at least in adolescence.

Many of these studies used data from the National Longitudinal Study of Adolescent Health (ADDHealth), a nationally representative study of adolescent health and risk behaviors collected in the U.S. during the 1994–1995 academic year that includes a large subsample of twin and non-twin siblings in grades 7 through nine (mean age = 16 years). Participants of the ADDHealth study were asked to nominate up to five male and five female school friends, ranked by friendship closeness. Because the ADDHealth study collected data from whole schools, it was not only possible to have access to the nominated friends' self-reported deviant behavior but also to verify whether the friendship nomination was reciprocal. Based on these data, a significant genetic contribution was found to nominated friends' smoking and drinking, accounting for 64 % of interindividual variability in affiliation with substance using friends (Cleveland et al. 2005). A similar pattern of rGE was observed in another study with this sample, where only the closest same sex friend's substance use was considered (Harden et al. 2008). Not surprisingly, a considerable portion (i.e., around 40 %) of genetic influences on affiliation with substance using best friends was explained by genetic factors associated with adolescents' own substance (Harden et al. 2008). Similar evidence of genetic influences on deviant peer affiliation has been found with respect to friends' aggressive behavior in the ADDHealth sample (Guo 2006).

The considerable genetic effects on affiliation with deviant peers observed in mid- and late adolescence may not necessarily occur at younger ages, however. Some evidence to that effect comes from a study of 12-year-old identical and fraternal Finnish twins (Rose 2002). The results showed that boys' genetic make-up was unrelated to their nominated friends' self-reported alcohol and substance use, and only weak and inconsistent genetic effects were found for girls. Similarly, two other studies of still younger identical and fraternal twins (mean age = 10 years) found no rGE when using teacher-ratings and direct observations to assess children's and their close friends' antisocial behaviors (Bullock et al. 2006; Leve 2001). Finally, a similar conclusion can be drawn from findings from the Quebec Newborn Twin Study (QNTS). The QNTS involves a population-based longitudinal sample of identical and fraternal twins who have been followed since birth and for whom in-classroom peer nominations (including friendship nominations and behavior ratings of classmates) were obtained over multiple time points. Examining whether genetic factors contribute to deviant peer affiliation already in kindergarten, Van Lier and colleagues focused specifically on the twins' and their best friends' physically aggressive behavior as rated by teachers and classmates (Van Lier et al. 2007). As in other studies (DiLalla 2002), ACE modeling showed that the twins' own physical aggression was highly heritable, explaining 66 % of the variance. However, twins' genetic make-up explained only a small (i.e., 15 %) portion of their friends' physical aggression, with the remaining variance entirely explained by nonshared environmental effects. Furthermore, no association was found between the twins' genetic disposition for physical aggression and their friends' physically aggressive behavior, suggesting an absence of rGE. These results were the same for girls and boys.

A similar lack of genetic effect on children's affiliation with physically aggressive peers was also found in a follow-up study with the QNTS sample in Grade 1 (Brendgen et al. 2008a). Using again teachers' and classmates' evaluations of the twins' and their friends' aggressive behavior, ACE modeling revealed that genetic factors explained 58 % of the variance of the twins' own physical aggression, but were not associated with their friends' physically aggressive behavior. Instead, affiliation with physically aggressive friends was completely explained by nonshared environmental influences. Importantly, the study revealed that this lack of rGE did not only apply to children's affiliation with physically aggressive friends, but also to their affiliation with friends who used more sophisticated aggressive strategies such as rumor spreading, public humiliation, or social exclusion. Specifically, genetic factors accounted for a substantial portion (i.e., 43 %) of the variance of the twins' own social aggression, leaving the remaining variance to nonshared environmental factors. In contrast, affiliation with socially aggressive friends was explained mostly (87 %) by nonshared environmental factors, with the remaining variance explained by shared environmental factors.

Recent findings using grade four data from the QNTS indicate that this lack of rGE not only seems to concern the type of friends children affiliate with, but also whether they have friends at all, and how many (Brendgen et al. 2013a). Thus, despite their greater genetic similarity, MZ twins were not found to be more similar than DZ twins with respect to friendship participation (i.e., whether they had any reciprocal friends in the classroom or not). Moreover, among those with at least one reciprocal friend, MZ twins were no more similar to each other than were DZ twins with respect to the number of reciprocal friends.

Together, findings from behavioral genetic studies thus suggest that affiliation with antisocial friends is largely unrelated to individuals' genetic disposition prior to adolescence. However, the consistent pattern of genetic contribution to deviant peer affiliation found in older samples also suggests that rGE gradually appears in early-to mid-adolescence. Due to the largely cross-sectional design of the studies and the lack of measures of parents' deviant characteristics, it is not possible to draw definite conclusions about the specific type of gene-environment correlation at play. Nevertheless, if passive rGE reflecting some form of parental transmission affecting adolescents' peer affiliations were at play, rGE should also be observed in regard to young children's peer affiliations. The absence of rGE underlying the peer affiliations of young children thus suggests that the rGE found in adolescence likely reflects active rather than passive processes. This emerging rGE in adolescence is in line with the idea that youth show an increased propensity toward an active selection of their peer group—including deviant peers—based on heritable characteristics over the course of development (Scarr and McCartney 1983). Of course, the lack of rGE in regard in early- and mid-childhood does not mean that parents have no influence on their young children's peer affiliations. Rather, parents may represent a source of nonshared, child-specific environmental influence on their children's peer relations, for example, through parenting practices geared toward each child's specific needs. Clearly, more research is needed to obtain a more precise picture of the potential role of genetic factors in children's and adolescents' affiliations with deviant and nondeviant peers.

5.5.2 Gene-Environment Interaction

As mentioned previously, rGe and GxE can, and often do co-occur. In other words, genetic factors may not only influence whether youth befriend deviant peers, but these affiliations may in turn influence youngsters' own behavior. Moreover, the nature and extent of the influence of deviant peers may depend on, and interact with, youngsters' genetic disposition for deviant behavior. Again, many of the available behavioral genetic studies that have examined GxE in regard to deviant peer affiliation relied on youths' reports of their peers' deviant behavior. These studies all found evidence for an enhancement process of GxE, suggesting that genetic risk for deviant behavior is more likely to be expressed when adolescents are exposed to deviant peers (Button et al. 2007, 2009; Dick et al. 2007; Fowler et al. 2007; Hicks et al. 2009). Importantly, this pattern was also confirmed in studies that employed more stringent measures of peer deviancy. One example is the aforementioned study by Harden and colleagues (Harden et al. 2008). Indicative of a possibly active rGE, adolescents with a stronger genetic propensity for substance use (i.e., drinking and smoking) were more likely than others to affiliate with substance-using friends. However, even after controlling for rGE, affiliation with a drinking and smoking best friend was related to an increased rate of substance use. Dividing the sample into quartiles of genetic risk, the relation between friends' and adolescents' own smoking and drinking was then examined as a function of adolescents' genetic risk. Adolescents with the highest genetic liability showed the highest levels of substance use, even when their best friends reported minimal substance use. Moreover, indicative of an enhancement process of GxE, adolescents with a stronger genetic disposition for substance use (i.e., upper two quartiles of genetic risk) drank and smoked even more if their best friends did as well. In contrast, adolescents with little or no genetic risk for substance use (i.e., lower two quartiles of genetic risk) showed minimal substance use, regardless of their best friends' behavior. Using data from the same sample, a similar enhancement process of GxE was also found when examining specifically adolescents' and their friends' alcohol use (Guo et al. 2009).

Findings by Boardman and colleagues (2008) with the ADDHealth study data reveal that genetically vulnerable youth are not only influenced by their close friends' deviant behavior, but also by popular students in their group. Specifically, genetically liable adolescents were more likely to smoke when they attended schools where the most popular students also smoked. As previously mentioned, peers' tobacco and alcohol use was assessed through the peers' self-reports in the ADDHealth study and data were collected from all students in the participating schools. These findings lend further support to a possible enhancement process of GxE, whereby a deviant peer environment facilitates the expression of a genetic predisposition for such behavior.

Recent findings from the QNTS suggest that this GxE enhancement process is not necessarily limited to the adolescent period. Specifically, a first study used teacher- and peer ratings of the twins and of their three reciprocal classroom friends in kindergarten (Van Lier et al. 2007). Each twin's genetic risk for aggression was calculated as a function of his or her co-twin's level of aggression and the pair's zygosity (Andrieu and Goldstein 1998). A twin's genetic risk was considered highest in MZ twins (who are assumed to share 100 % of their genetic material) whose co-twin's aggression level was at or above the 80th sex-specific percentile. Genetic risk was considered moderately high in DZ twins (who are assumed to share on average 50 % of their genetic material) whose co-twin's aggression was at or above the 80th sex-specific percentile. Genetic risk was considered moderately low if the DZ co-twin was below the 80th percentile on aggression. Finally, genetic risk was considered lowest in MZ twins whose co-twin's level of aggression was below the 80th percentile of the distribution. In line with a contextual trigger process of GxE, aggression was most frequently observed in children who were at high genetic risk for such behavior **and** who also affiliated with highly aggressive friends. In contrast, children with very low genetic risk for aggressive behavior showed very low levels of aggression regardless of their friends' behavior.

A follow-up study conducted with data collected in grade 1 (Brendgen et al. 2008a) revealed that this trigger process of GxE held for the link between friends' and children's physical aggression (i.e., hitting, biting, kicking), but not for relational aggression. As mentioned, relational aggression is a more indirect type of aggression that includes social exclusion or malicious gossiping in order to harm the social status or the self-esteem of the victim. Similar to the previously reported results in kindergarten, a GxE reflective of a contextual trigger process was again observed for affiliation with physically aggressive friends in grade 1. In contrast, no evidence of GxE was found in regard to affiliation with socially aggressive friends. Instead, affiliation with socially aggressive friends seemed to foster relational aggression independently of, and in addition to the effect of children's genetic disposition for such behavior.

One possible explanation for these distinct genetic-environmental processes may be that relational aggression involves rather subtle behaviors that may carry a much lower risk of retribution than physical aggression. Because social aggression is often diffuse and ubiquitous, it is difficult to identify the initiator and punish him or her. Adults also feel less negative toward, and are less likely to intervene against children's use of social aggression compared to physical aggression (Colwell et al. 2002). Social aggression may also offer potential rewards for the perpetrators, as it has been shown to promote cohesiveness and closeness among friends who perpetrate these acts against a common 'enemy' (Werner and Crick 2004). Social aggression has even been linked to perceived popularity in the peer group (Cillessen and Mayeux 2004). Thus, children with a genetic disposition for relational aggression may use this behavior toward social gains irrespective of their friends' behavior. Furthermore, the lack of punishment and the potential benefits associated with relational aggression may render children without a genetic disposition susceptible to imitating their friends' behavior. In contrast, physical aggression usually entails negative sanctions from the social environment rather than positive consequences (Brendgen et al. 2006). Children who are genetically predisposed to physical aggression may be more impervious to such negative sanctions and thus more vulnerable to the influence of physically aggressive friends than children who lack a genetic predisposition to physical aggression.

Interestingly, the contribution of friends' social and physical aggression to child aggression was found only for the same type of aggressive behavior in the Brendgen et al. study (2008a). No crossover contributions from friends' physical aggression to children's social aggression, or vice versa, were found. These context-specific contributions of friends to social and physical aggressive behaviors are consistent with findings that social and physical aggression are driven by largely the same underlying genetic dispositions, but by mostly different environmental influences (Brendgen et al. 2005). Although replication studies are needed, these results emphasize the potentially diverse transactions at play between genetic factors and the peer environment at play in the development of different types of antisocial behaviors.

5.6 Gene-Environment Interplay in the Link Between Peer Victimization and Development: Findings from Behavioral Genetic Studies

5.6.1 Gene-Environment Correlation

In contrast to the (potentially active) rGE involved in deviant peer affiliation, which seems to emerge only during the transition to adolescence, evocative rGE with respect to problematic peer relations appears at a much younger age. Indeed, although relatively few genetically informed studies have been published in this context, the findings almost systematically suggest that heritable characteristics play a significant role in explaining why some children find themselves the target of rejection and negative treatment by the peer group.

Some evidence in this regard comes from a study with the QNTS sample conducted in kindergarten (Brendgen et al. 2009). Peer nominations by the twins' classmates were used to assess rejection in the peer group. Specifically, all children in the class were asked to nominate three classmates they most liked to play with (positive nominations) and three classmates they least liked to play with (negative nominations). Following widely used criteria for assessing peer rejection (Coie et al. 1982), the total number of received positive nominations was calculated for each participant and z-standardized within classroom to create a total Liked-Most-score. Similarly, the total number of received negative nominations was calculated for each participant and z-standardized within classroom to create a total Liked-Least-score. The Liked-Least-score was then subtracted from the Liked-Most-score to create a continuous scale, where low levels indicate rejection by peers. Univariate ACE modeling revealed that around 30 % of the interindividual differences in peer rejection was explained by heritable factors, another 15 % was accounted for shared environmental sources, and the remaining 55 % of the variance was due to nonshared environmental factors.

These initial findings suggest that genetic factors are associated with peer difficulties. However, they bear limitations. First, as stated earlier, peer rejection provides only an indirect assessment of peer difficulties, and actual peer victimization plays a central part in children's social adjustment problems (Boivin et al. 1995a, b; Dodge et al. 2003; McDougall et al. 2001; Rubin et al. 1998). Whereas peer rejection reflects negative feelings of the peer group toward a child, peer victimization refers to actual negative behavior manifested repeatedly by one or more peers toward specific children (Boivin et al. 2001). Thus, assessing these two aspects together would provide a more comprehensive coverage of various forms of peer difficulties. Second, this initial study relied on only one source of information, i.e., peers, to assess peer difficulties. Because peers participate in and witness the social arena, peer assessments can be seen as the gold standard when it comes to assessing peer relationship difficulties (Boivin et al. 2013a; Rubin et al. 2006). However, peer assessments may also be biased by social reputations and group dynamics. Other potentially useful sources of information include the child himself or herself, or a member of the social entourage such as the teacher or the mother. A potential shortcoming of self-reports is that they partly reflect the self-system (Boivin and Hymel 1997; Boivin et al. 1992), and they may therefore yield biased estimates of genetic and environmental contributions to actual peer difficulties. Mothers and teachers may only have limited information regarding the child's peer relations (Fekkes et al. 2005; Houndoumadi and Pateraki 2001). Thus, each measurement approach has its biases and limitations when used exclusively. One way to overcome these limitations would be to use multiple informants to establish more a reliable 'latent' measure of the phenomenon. Third, this first study and many others provide a limited view of developmental processes, as peer relationships were assessed only at a single point in time. There could be developmental changes in the relative strength of genetic contributions to peer difficulties (Boivin et al. 2013a). For instance, Ball et al. (2008) found a strong association between peer victimization and genetic factors in middle childhood, but this finding was not replicated among kindergarten children (Brendgen et al. 2008b) This differing pattern could be due to the use of different methods for assessing peer victimization (i.e., mother versus peer evaluations). It could also reflect a growing association with age between genetic factors and peer difficulties driven by the progressive establishment of an evocative rGE as described above.

Accordingly, a recent study with the QNTS sample adopted a longitudinal multiple-assessment approach to examine the genetic and environmental contributions to a more global construct of peer relationship difficulties during the early years of school (Boivin et al. 2013a). To this end, the study utilized a combination of peer-assessed peer rejection and peer, teacher and self-assessed peer victimization in kindergarten, grade 1, and grade 4. The multivariate results revealed that, starting in kindergarten and then in later grades, genetic factors accounted for most of the variance in peer difficulties, as indexed by a latent factor combining peer, teacher, and self-ratings. Specifically, genetic factors accounted for 73, 73, and 94 % of individual differences in the peer difficulties construct in kindergarten, grade 1, and grade 4, respectively, leaving 27, 27, and 6 % to nonshared environmental factors. In plain words, twins of the same family were highly similar in terms of their peer difficulties, as defined by combining all four assessments, and this similarity was mainly explained by their genetic relatedness. Genetic factors also accounted for the stability in peer difficulties in the early school years. Specifically, stability of the peer difficulty latent factor was high ($r = 0.73$ and 0.69, respectively), and these stability coefficients were mainly accounted for by shared genetic factors between the different time points.

The usefulness of the multiple-informant approach was clearly illustrated when the multivariate and univariate results were contrasted. Whereas the multivariate results indicated strong genetic contributions to initial and stable peer difficulties, those stemming from the univariate approach were qualified by age. Specifically for most measures, the pattern of familial aggregation in kindergarten was generally low and unstable. The results for the later grades were clearer, and the univariate findings pointed toward the progressive establishment of a G-E correlation regarding peer difficulties in the later grade school years. At the univariate level, the genetic contributions emerged progressively, as did a growing consensus among informants with respect to those who experienced peer difficulties.

Thus, through a multi-informant longitudinal approach, the study showed that the genetic factors underlying peer difficulties appeared to be enduring: as the negative experiences crystallized, the same children with the same genetic vulnerabilities tended to become chronically embroiled in a cycle of negative peer experiences (Boivin et al. 2013a). These results underline the need to intervene early and persistently, and to target the child and the peer context to prevent peer difficulties and their consequences. Although replication studies are needed to support the findings from these studies, together they suggest that—in line with an evocative rGE—children's heritable characteristics may indeed elicit dislike and negative treatment by the peer group, and quite substantially so when the sources of information are combined and children are assessed accordingly.

The significant genetic effects on peer rejection and victimization indicate the presence of (presumably evocative) rGE, suggesting that personal and heritable characteristics may at least in part explain why some children are subject to covert or even overt disdain from peers. The question thus arises what these heritable characteristics might be. A significant body of research supports the view that social behavior is one of the main sources of these difficulties, over and above atypical physical attributes such as speech problems and physical clumsiness (Boivin et al. 2005; Rubin et al. 2006). It is well documented that children who display disruptive behaviors, especially aggressive behaviors, are more likely to experience peer relation difficulties in school (Rubin et al. 2006) as well as in preschool (Boivin et al. 2005). The fact that early disruptive behaviors (such as aggression)—which are known to be influenced by genetic factors (Rhee and Waldman 2002)—predict peer relation difficulties strongly suggests that a significant part of the risk of peer difficulties could be traced back to genetic vulnerability for these behaviors in the child. Until recently, these alternative hypotheses had not been tested empirically.

The first published study to examine this issue (Ball et al. 2008) was based on data from the Environmental Risk (E-Risk) Longitudinal Twin Study, which tracks the development of a birth cohort of over 1000 identical and fraternal twin pairs in England and Wales. Participants' level of aggressive behavior (i.e., bullying behavior toward other pupils) and of peer victimization (i.e., being the target of verbal, relational, or physical bullying) was evaluated by mothers and teachers at age 10. Univariate ACE modeling revealed that almost three-quarters of the variance of peer victimization (i.e., 73 %) were explained by genetic factors, with the remaining variance explained by nonshared environmental influences. What is more, indicative of rGE, results from a subsequent bivariate Cholesky model showed that a significant if modest part (i.e., 14 %) of the genetic influences on peer victimization were those that also underlie aggressive behavior.

Similar results were obtained with grade four data from the QNTS sample, but this time based on children's self-reported peer victimization and peer nominated verbal and physical aggression, and using an ordinal index of genetic risk for aggression. Specifically, an elevated genetic risk for aggressive behavior was associated with a higher level of victimization by the peer group ($r = 0.32, p < 0.001$) (Brendgen et al. 2013a). Findings based on grade 1 data from the QNTS also suggest that a genetic propensity for aggression may be an important risk factor for peer victimization in younger children (Brendgen et al. 2011). The twins' levels of verbal and physical peer victimization, as well as their levels of verbally and physically aggressive behavior toward others were assessed through peer nominations from the twins' classmates. Univariate ACE modeling revealed that, together, genetic factors accounted for more than half (i.e., 53 %) of the variance of aggression, with the remaining variance explained by nonshared environmental factors. Genetic factors also explained a significant portion (25 %) of the variance of peer victimization, with the remaining variance (75 %) again explained by nonshared environmental factors. The finding that not only children's aggressive behavior but also their level of victimization by the peer group has some genetic basis suggested a possible gene-environment correlation (rGE) in the link between these two variables. In line with this notion, results from a Cholesky modeling showed that 100 % of the genetic influence on peer victimization was due to genetic factors that also influenced aggression.

These initial findings suggest that genetic factors are associated with peer difficulties, and that part of this association is accounted for by child characteristics, including aggressive behaviors, which suggests rGE. However, they were limited by the cross-sectional nature of their design, i.e., they assessed aggressive behaviors and peer difficulties at a single point in time. Indeed, implicit to the idea of rGE is the view that there is a "flow of influence" from a child characteristic, i.e., aggressive behavior in this case, to a social experience, i.e., peer difficulties (Boivin et al. 2013b). In such a context, the combined use of a twin design and a longitudinal approach provides a fine-grained description of the G-E dynamic linking disruptive behaviors and peer difficulties.

Accordingly, we recently tested the presence of rGE linking disruptive behaviors and peer relation difficulties from kindergarten to grade 1, again using QNTS data (Boivin et al. 2013b). To this end, we used a biometric cross-lagged design initially proposed by Burt et al. (2005). This design takes advantage of the developmental time lag and heuristics of the twin design to more precisely assess directional associations and their underlying G-E architecture linking disruptive behaviors and peer difficulties. In this study, the construct of disruptive behaviors was expanded to include impulsive/hyperactive behaviors, in addition to aggressive behaviors. Impulsive/hyperactive behaviors often imply rude and unpredictable responses aversive to peers, and have indeed been found to forecast negative peer perceptions and status in newly formed groups (Dodge 1983; Erhardt and Hinshaw 1994).

As predicted, disruptive behaviors were concurrently and predictively associated with peer relation difficulties, and genetic factors clearly accounted for a substantial part of these associations, thus confirming rGE. Interestingly, disruptive behaviors in kindergarten were found to independently contribute to peer difficulty in grade 1 (with proper constraints for preexisting associations), and this unique contribution was largely a function of genetic factors. Given that peer difficulties are an experience (i.e., not a child phenotype), these findings provide strong confirmation that an evocative rGE links disruptive behaviors and peer difficulties in the early years of school, perhaps through self-regulation processes. Most importantly, this rGE process is established in the first year of school, if not earlier for some children (Barker et al. 2008), and tends to persist over time. Hence, children displaying aggressive and impulsive/hyperactive tendencies, perhaps due to self-regulation problems, may not only provoke potential bullies to pick on them, but they may also elicit little sympathy and support from classmates when they are rejected and victimized. Indeed, already preschool children have been found to endorse the use of aggression against others they consider 'mean' (Giles and Heyman 2005). It is thus not surprising that many peers perceive aggressive-disruptive children who are victimized as largely responsible for their own fate and show little inclination to come to their defense, thus solidifying their reputation as socially rejected and "deserving victims" and contributing to a continued cycle of violence (Boivin et al. 2001; Schuster 2001; Teräsahjo and Salmivalli 2003).

However, there could be some contextual moderation of these general tendencies. Indeed, social behavior, and aggressive-disruptive behaviors in particular, were found to be variably associated with peer relation difficulties, depending on group norms (Boivin et al. 1995a, b; Stormshak et al. 1999). Therefore, one may question whether this moderation effect extends to the rGE underlying the association between disruptive behavior and peer difficulties. Recent findings from the QNTS suggest it does, as rGE linking aggressive behavior with peer victimization was found to vary according to the behavioral norms that prevail in the peer group (Brendgen et al. 2014). In the QNTS sample, these classroom norms were assessed in grade 4, based on the correlation between aggressive behavior and the level of popularity or rejection among classmates: A highly positive correlation indicated high acceptance of aggressive behavior (i.e., highly favorable aggression norms), whereas a highly negative correlation indicated high rejection of aggressive behavior (i.e., highly unfavorable aggression norms). In line with previous research (Henry et al. 2000), we found considerable variation between classrooms with respect to these 'injunctive' aggression norms (i.e., the level of acceptability of aggression), with some classrooms adopting a very favorable view and others clearly condemning such behavior. Moreover, using an ordinal index of genetic risk (Andrieu and Goldstein 1998), the findings showed that children with a genetic disposition for aggression were at increased risk of being victimized only when such behavior was disapproved by the peer group. In contrast, children with a genetic disposition for aggression were *less* likely to be victimized than others in classrooms where such behavior was highly accepted. These results, which applied to both physical and relational aggression, emphasize that the link between specific genetic susceptibilities and specific environmental experiences emerges in the context of other environmental conditions that may significantly influence rGE.

Early rGE may also involve other form of personal relationships. For instance, previous research with adolescents has shown that individuals with a genetic propensity for depression are at increased risk to experience a problematic parent–child relationship as well as negative social life events that are at least somewhat controllable, such as a rupture of a friendship or romantic relationship, academic failure, or job loss (Lau et al. 2008; Pike et al. 1996; Rice et al. 2003; Silberg et al. 1999). Findings from three studies, again from the QNTS sample, show that this environmental risk may already occur at earlier ages and extend to relationship difficulties with the peer group. The first evidence in this regard comes from the previously mentioned study by Brendgen and colleagues (2009), which utilized nominations by the twins' classmates to assess peer rejection in the peer group and teacher reports to evaluate the twins' depression symptoms in kindergarten. The results of a bivariate ACE Cholesky model revealed that, in line with rGE, a genetic propensity for depressive behavior as rated by teachers also increased the risk of rejection by peers. Specifically, about 54 % of the genetic variance of children's depressed behavior was explained by genetic factors that were also associated with peer acceptance/rejection. The second, more recent study conducted with grade 4 data from the QNTS shows that this pattern also applies to older children and extends to active expressions of peer dislike (Brendgen et al. 2013a). In that study, peer victimization and depression symptoms were measured using each child's self-reports. Using an ordinal index of genetic risk for depression (Andrieu and Goldstein 1998), the analyses revealed that a high genetic risk for depression symptoms was significantly associated with an elevated level of victimization by the peer group ($r = 0.16$, $p < 0.01$). Based on a bivariate ACE Cholesky model, similar results were obtained in another recent study with regard to the association between self-reported peer victimization and anxiety in grade 8 (Brendgen et al. 2015).

Together, these results thus corroborate previous findings that individuals with a genetic risk for depressive behavior are exposed to more negative social environmental experiences and suggest that, as early as kindergarten, these negative experiences also include peer rejection and victimization. Because peer rejection and victimization reflect the peer group's negative perception of the target child and is clearly not the result of a voluntary or active choice of the individual, these results most likely indicate an evocative, rather than active or passive rGE. Indeed, depressed children's peer interactions are characterized by more conflict and less collaboration than interactions of nondepressed children (Altmann and Gotlib 1988; Rudolph et al. 1994), which may at least in part explain the peers' negative perceptions and hostile actions.

5.6.2 Gene-Environment Interaction

A genetic disposition for internalizing or externalizing behaviors may not only put children at risk of being rejected or even victimized by their peers, but such traumatic experiences may also contribute to the maintenance and further increase of such problems. The effects of peer victimization on youngsters' emotional and behavioral adjustment may not necessarily be equivalent for all youth, however, but rather interact with their genetic vulnerabilities for internalizing and externalizing problems. For example, using an ordinal index of genetic risk (Andrieu and Goldstein 1998), findings based on kindergarten data from the QNTS study revealed that victimized girls showed high levels of generalized aggression mainly if they had a high genetic risk of being aggressive ($\beta = 1.00$, $p < 0.001$) (Brendgen et al. 2008b). The predictive effect of peer victimization on aggressiveness was much weaker, however, in girls with a very low genetic risk of aggressive behavior ($\beta = 0.17$, $p < 0.05$). Interestingly, a similar GxE in the link between peer victimization and aggression was not found in boys. Instead, a high level of peer victimization predicted a high level of aggression for all boys, independently of their genetic disposition to such behavior. These sex-specific patterns of GxE may at least in part be due to different socialization experiences in girls' compared to boys' peer groups. Girls rate aggressive behavior as more unacceptable than boys even when displayed in response to an aggressive provocation from peers (Goldstein et al. 2002). When faced with provocation and victimization from others, girls may therefore be most likely to behave aggressively if they are genetically disposed to do so. In contrast, because aggressive responses to hostile peer provocations may be more acceptable for boys, many victimized boys may retort with aggression regardless of whether they have a genetic disposition for such behavior or not.

At least for girls, these findings are in line with a Diathesis–Stress Hypothesis of psychopathology, according to which an environmental stressor such as peer victimization should lead to a greater risk for maladjustment in individuals with preexisting genetic vulnerabilities (Monroe and Simons 1991; Zuckerman 1999). Interestingly, physical abuse by parents has also been related to a very high probability of conduct disorder only in children with a very high genetic risk for such disorder, whereas children with a very low genetic risk show little to no conduct disorder, regardless of whether they were abused by their parents or not (Jaffee et al. 2005). Together, these findings suggest that experiences of abuse, whether from parents or peers, may constitute a risk factor for the development or maintenance of aggressive behavior, especially in children with a genetic disposition for such behavior. By the same token, the presence of positive environmental influences from other socializing agents, such as teachers, may help alleviate at least to some extent the negative consequences of peer victimization for genetically vulnerable children.

This notion is supported by findings from grade 1 data in the QNTS sample, utilizing peer nominations to assess both peer victimization and aggressive behavior and estimates based on latent genetic factors using ACE modeling (Brendgen et al. 2011). Statistical power constraints made it impossible to perform sex-limited analyses, which may explain why only additive but no interactive contributions of peer victimization and genetic risk to aggressive behavior were found. However, the results also showed that genetic contributions to aggression were significantly reduced in children who had a warm and conflict-free relationship with their teacher. This pattern of GxE is indicative of a compensation process, whereby the presence of a positive environment prevents or reduces the expression of a genetic disposition for problem behavior. The beneficial effect of a good relationship with the teacher may eventually also extend to genetically vulnerable children's risk of being harassed by their peers. Specifically, because genetically vulnerable children are less likely to be aggressive if they have a positive relationship with the teacher, these children may as a consequence more easily elicit sympathy from classmates who may then be more willing to defend them against peer victimization. Although such cascading processes remain speculative, the overall variance of peer victimization was indeed greatly reduced—and nonshared environmental influence on peer victimization was also less varied and more consistent between individuals—when children enjoyed a positive relationship with the teacher.

Stressful experiences with peers may also interact with children's genetic propensity to predict internalizing problems. The little available research in this regard suggests, however, that at least for young children the process underlying GxE for internalizing behaviors may differ from those observed for externalizing problems. Evidence to that effect comes from a study with the QNTS sample using classmates' nominations of liking and disliking and teacher ratings of twins' depression symptoms during kindergarten (Brendgen et al. 2009). Results from a multivariate Cholesky model (Purcell 2002) showed that genetic effects on depression symptoms in kindergarten varied significantly depending on the twin's level of rejection by peers. Inspection of the interaction revealed that interindividual differences in depression symptoms were *less* associated with genetic factors, and *more* related to environmental factors when children were highly rejected by their peers. In contrast, a preexisting genetic disposition appeared to play a stronger role in accounting for interindividual differences in depression symptoms in accepted children, who presumably face little peer-related stress. Interindividual variability of depressive behavior was also lower—whereas the mean level of depressive behavior was higher—when children were rejected by their peers than when they were accepted.

This finding is inconsistent with a diathesis–stress or a compensation process, and rather points to a suppression mechanism underlying GxE between peer rejection and depressive behavior. Similar suppression-based GxE processes have been reported for the link between family adversity and stress-reactivity in infancy (Ouellet-Morin et al. 2008) and for the link between a problematic parent–child relationship and negative emotionality during childhood (Burt 2008). They seem to signal an overriding programming effect of the environment over gene expression, perhaps associated with some form of social entrapment within an adverse environment. Peer rejection is indeed stable and it hurts; it has been shown to activate the same areas of the brain that register physical pain (Eisenberger et al. 2003; McDougall et al. 2001). Young children entering school may have relatively few opportunities to escape this hurtful social context. The more vulnerable of them may not have acquired the social interactional skills and coping strategies necessary to deal effectively with peer rejection; some tend to cry or run away when being harassed by peers, which is unlikely to remedy the situation and may accentuate their negative peer status (Smith et al. 2001).

These results suggest that, at least at younger ages, stressful experiences—including those with peers—may foster internalizing symptoms even in children without a genetic disposition for such problems. Further support for this notion comes from a recent follow-up study with the QNTS sample in grade 4 (Brendgen et al. 2013a). Here, both peer victimization and depression symptoms were assessed using the twins' self-reports and an ordinal index based on the pair's zygosity and the co-twin's depression was used to approximate genetic risk for depressive symptoms (Andrieu and Goldstein 1998). Regression analyses showed only main effects ($\beta = 0.42$, $p < 0.001$ for genetic risk and $\beta = 0.25$, $p < 0.001$ for peer victimization) but no interaction, indicating that peer victimization was related to increased levels of depressive feelings irrespective of youngsters' genetic risk for depression. It is still possible, however, that a GxE pattern signaling a contextual trigger process could operate in adolescence. Some adolescents may possess a greater arsenal of coping strategies and hence show greater variability in depression symptoms when faced with peer difficulties. Some evidence indeed suggests that the relative influence of genetic factors on depression increases from childhood to adolescence (Rice 2009). Moreover, another recent study with the QNTS sample suggests that the specific pattern of GxE (i.e., contextual trigger versus contextual suppression process) observed in the link between peer victimization and internalizing problems may depend on the degree of closeness with the tormentor (Brendgen et al. 2015). Thus, although most research has focused on victimization perpetrated by classmates or relatively unfamiliar peers, there is evidence that between 5 and 30 % of youth experience psychological or even physical abuse from their close friends (e.g., Crick and Nelson 2002; Waasdorp et al. 2009). Using participants' self-reports about their victimization by a close friend and by other peers and their anxiety level in grade 8, bivariate Cholesky modeling showed important differences in the specific mechanisms of gene-environment interplay that link the two victimization experiences with anxiety. Specifically, in line with the contextual suppression process of GxE observed at earlier ages with the QNTS, interindividual differences in anxiety were less explained by genetic factors in adolescents who were frequently victimized by peers they were not (or never were) friends with. In contrast, victimization by a close friend exacerbated the effect of genetic influences on anxiety, in line with a diathesis–stress process of GxE. Experimental research has revealed that anxious children show greater physiological reactivity and helplessness than their non-anxious counterparts when faced with social rejection by a friend (Gazelle and Druhen 2009). Compared to individuals without a genetic propensity for anxiety, highly anxious youth may thus be less likely to assert themselves against attacks from their friend. Anxious youth have also been found to possess fewer reciprocal friends than others (La Greca and Lopez 1998) and this lack of alternatives may make it especially difficult for victims who are prone to anxiety to end the relationship. Thus, as suggested by Crick and Nelson (2002), abusive friendships may in many ways resemble abusive romantic relationships. More research is necessary to clarify potential developmental changes and the influence of other contextual factors with respect to the nature of gene-environment interactions linking peer victimization experiences with mental health outcomes, most notably internalizing problems.

5.7 Conclusion

Genetically informed research on the role of problematic peer experiences in youngsters' development, including affiliation with deviant peers and peer victimization, is a rather recent phenomenon. The relative dearth of studies, the diversity of methods and conceptual content, and the resulting equivocal findings make it challenging to trace a global portrait. Nevertheless, the results from quantitative genetic studies so far suggest that genetic factors account for the degree of peer relationship problems children experience, and modulate how these problematic peer experiences may affect their psychosocial development. Quantitative genetic studies that statistically infer genetic risk have been, and remain instrumental in informing about the ways genetic factors as a whole work together with problematic peer experiences to shape youngsters' development.

Such genetically informed studies rely on a number of assumptions that may not always be tenable, however (Keller and Coventry 2005). For example, two of the potentially most problematic assumptions of the classical twin design are that MZ and DZ twin pairs are influenced to the same extent by their shared environment, and that there is no assortative mating by the parents on the trait under study. Violation of one or more of these assumptions will result in biased estimates and either over-or underestimations of the genetic and shared environmental variance components. A related issue refers to the fact that finding evidence of GxE in the presence of rGE typically requires very large sample sizes (Purcell 2002; Van Der Sluis et al. 2012), and many behavioral genetic studies, particularly those relating to the role of peer relations, have been relatively underpowered. Failure to uncover gene-environment interaction involving peers in some studies may thus at least in part be due to a lack of power rather than true absence of GxE. Another important limitation of quantitative genetic studies is related to the interpretation of the estimated latent genetic effects. These estimates reflect overall genetic influences that also include any existing non-measured gene-environment correlations and gene-environment interactions, as well as heritable epigenetic processes. Thus, findings from quantitative genetic studies should be followed by studies of genetic markers in order to specify the specific nature of the genes and biological processes involved in the various gene-environment interplays.

To date, less than a dozen molecular genetic studies have been published that examined potential rGe and GxE with respect to deviant peer affiliation and peer victimization. Yet, these studies offer a first glimpse of how interindividual differences at the genomic level might work together with peer relation difficulties in shaping development (for a review, see Brendgen 2012). One interesting contribution of molecular genetic studies rests in the ability to investigate specific patterns of gene-environment interactions that may be difficult to test in the context of quantitative genetic designs, most notably the differential susceptibility hypothesis of GxE. As previously mentioned, the diathesis–stress process of GxE involves genetic vulnerability factors that are particularly likely to lead to developmental maladjustment when individuals are exposed to a negative environment. However, several scholars have argued that the same genetic disposition may render an individual particularly sensitive to very negative and to very positive environments (Belsky and Pluess 2009; Ellis et al. 2011). In other words, the same individual that is genetically vulnerable to adverse conditions may also benefit most from exceptionally positive environments. This type of U-shaped contextual trigger GxE effect, which is also referred to as the Differential Susceptibility Hypothesis, may occur in individuals who are genetically predisposed to show greater physiological reactivity to experiences of both pleasure and displeasure and to respond more intensively to both reward and punishment. Because genetic factors are inferred rather than directly measured and involve overall heritable influences (including non-measured rGE and GxE as well as epigenetic processes) on a given phenotype, assessment of the Differential Susceptibility Hypothesis in the context of quantitative genetic studies poses a challenge. The few molecular genetic studies providing evidence for differential susceptibility have so far focused on experiences within the family context (e.g., Bakermans-Kranenburg and van IJzendoorn 2006; Bakermans-Kranenburg et al. 2008). Nevertheless, a similar process is conceivable with respect to problematic peer experiences. For example, genetic factors that make youth especially susceptible to the influence of deviant peers may also render these youth more likely than others to follow positive peer role models.

Despite such promising avenues in molecular genetic research, the past decade of research has shown how difficult it is to find genes that contribute to phenotypic variance, even in traits that are known to be highly heritable (Goldstein 2009). Few genetic markers that have been associated with social behaviors or traits so far have withstood replication. This situation is complicated further by the fact that genes are likely to interact with each other in addition to interacting with environmental influences, with each effect explaining only a very small portion of the overall variance. Perhaps even more important are recent findings that genome function—and ultimately the phenotype—is not so much defined by the individual genotype but rather the epigenotype (i.e., the individual pattern of gene activation) (Meaney 2010). Recent research suggest that the epigenotype can be modified by physical as well as social environmental influences, but most of the evidence to date comes from animal studies (for a review, see Champagne 2012). As such, quantitative genetic studies will likely still figure prominently in ongoing efforts to understand how genetic factors work together with problematic peer experiences in influencing children's and adolescents' development. In this context, the findings from quantitative genetic studies can help inform the selection of environmental variables for testing rGE and GxE in molecular genetic research (Moffitt et al. 2005). In turn, molecular genetic studies can help validate findings of gene-environment interplay from quantitative genetic studies. Much more research is needed, however, if we are to obtain a comprehensive picture of gene-by-peer-environment interplay in the link between problematic peer experiences and youngsters' development. For instance, quantitative genetic studies investigating rGE and GxE in regard to deviant peer affiliation have so far mainly focused on externalizing behaviors such as aggression and delinquency, despite findings from nongenetically informed research that internalizing problems such as low self-esteem and depressive feelings may both precede and result from deviant peer affiliation (e.g., Brendgen et al. 1998, 2000). Similarly, while findings from nongenetically informed studies that a peer victimization is associated with a variety of behavioral, emotional, somatic, and academic problems, quantitative genetic research has only just begun to examine the potential gene-environment processes underlying these links. In this regard, longitudinal research will be needed to help understand to what extent different types of rGE and GxE occur at different developmental periods. Such studies will be able to assess more clearly, for example, a possible emerging rGE with respect to deviant peer affiliation. Moreover, sufficiently powered studies are required to examine potential sex differences in rGE and GxE linking problematic peer experiences with psychosocial adjustment.

References

Akers, R. K., Krohn, M. D., Lanza-Kaduce, L., & Radosevich, M. (1979). Social learning and deviant behavior. A specific test of a general theory. *American Sociological Review, 44*, 636–655.

Allen, G., Knobloch, H., & Pasamanick, B. (1961). Intellectual potential and heredity. *Science, 133*(3450), 378–380.

Altmann, E. O., & Gotlib, I. H. (1988). The social behavior of depressed children: An observational study. *Journal of Abnormal Child Psychology, 16*(1), 29–44.

Andrieu, N., & Goldstein, A. M. (1998). Epidemiologic and genetic approaches in the study of gene-environment interaction: An overview of available methods. *Epidemiologic Reviews, 20*, 137–147.

Archer, J., & Côté, S. (2005). Sex differences in aggressive behavior: A developmental and evolutionary perspective. In R. E. Tremblay, W. W. Hartup, & J. Archer (Eds.), *Developmental origins of aggression* (pp. 425–443). New York: Guilford.

Arseneault, L., Walsh, E., Trzesniewski, K., Newcombe, R., Caspi, A., & Moffitt, T. E. (2006). Bullying victimization uniquely contributes to adjustment problems in young children: A nationally representative cohort study. *Pediatrics, 118*(1), 130–138.

Ary, D. V., Duncan, T. E., Duncan, S. C., & Hops, H. (1999). Adolescent problem behavior: The influence of parents and peers. *Behaviour Research and Therapy, 37*, 217–230.

Bakermans-Kranenburg, M. J., & van IJzendoorn, M. H. (2006). Gene-environment interaction of the dopamine D4 receptor (DRD4) and observed maternal insensitivity predicting externalizing behavior in preschoolers. *Developmental Psychobiology, 48*(5), 406–409.

Bakermans-Kranenburg, M. J., Van Ijzendoorn, M. H., Pijlman, F. T. A., Mesman, J., & Juffer, F. (2008). Experimental evidence for differential susceptibility: Dopamine D4 receptor polymorphism (DRD4 VNTR) moderates intervention effects on toddlers' externalizing behavior in a randomized controlled trial. *Developmental Psychology, 44*(1), 293–300.

Ball, H. A., Arseneault, L., Taylor, A., Maughan, B., Caspi, A., & Moffitt, T. E. (2008). Genetic and environmental influences on victims, bullies and bully-victims in childhood. *Journal of Child Psychology and Psychiatry, 49*(1), 104–112.

Barker, E. D., Boivin, M., Brendgen, M., Fontaine, N., Arseneault, L., Vitaro, F., et al. (2008). Predictive validity and early predictors of peer-victimization trajectories in preschool. *Archives of General Psychiatry, 65*(10), 1185–1192.

Belsky, J., & Pluess, M. (2009). Beyond diathesis stress: Differential susceptibility to environmental influences. *Psychological Bulletin, 135*(6), 885–908.

Berndt, T. J., & Keefe, K. (1995). Friends' influence on adolescents' adjustment to school. *Child Development, 66*, 1312–1329.

Boardman, J. D., Saint Onge, J. M., Haberstick, B. C., Timberlake, D. S., & Hewitt, J. K. (2008). Do schools moderate the genetic determinants of smoking? *Behavior Genetics, 38*(3), 234–246.

Boivin, M., Brendgen, M., Vitaro, F., Dionne, G., Girard, A., Pérusse, D., et al. (2013a). Strong gene-environment correlation in peer relation difficulties at school entry: Findings from a longitudinal twin study. *Child Development, 84*(3), 1098–1114.

Boivin, M., Brendgen, M., Vitaro, F., Forget-Dubois, N., Feng, B., Tremblay, R. E., et al. (2013b). Evidence of gene-environment correlation for Peer difficulties: Disruptive behaviors predict early school peer relation difficulties through genetic effects. *Development and Psychopathology, 25*, 79–92.

Boivin, M., Dodge, K. A., & Coie, J. D. (1995a). Individual-group behavioral similarity and peer status in experimental play groups: the social misfit revisited. *Journal of Personality and Social Psychology, 69*(2), 269–279.

Boivin, M., & Hymel, S. (1997). Peer experiences and social self-perceptions: A sequential model. *Developmental Psychology, 33*(1), 135–145.

Boivin, M., Hymel, S., & Burkowski, W. M. (1995b). The roles of social withdrawal, peer rejection, and victimization by peers in predicting loneliness and depressed mood in childhood. *Development and Psychopathology, 7*(4), 765–785.

Boivin, M., Hymel, S., & Hodges, E. V. E. (2001). Toward a process view of peer rejection and harassment. In *Peer harassment in school: The plight of the vulnerable and victimized.* (pp. 265–289): New York, NY, US: Guilford Press.

Boivin, M., Petitclerc, A., Feng, B., & Barker, E. D. (2010). The developmental trajectories of peer victimization in middle to late childhood and the changing nature of their behavioral correlates. *Merrill-Palmer Quarterly, 56*(3), 231–260.

Boivin, M., Vitaro, F., & Gagnon, C. (1992). A reassessment of the self-perception profile for children: Factor structure, reliability and convergent validity of a French version among second through sixth grade children. *International Journal of Behavioral Development, 15*(2), 275–290.

Boivin, M., Vitaro, F., & Poulin, F. (2005). Peer relationships and the development of aggressive behavior in early childhood. In R. E. Tremblay, W. W. Hartup, & J. Archer (Eds.), *Developmental origins of aggression* (pp. 376–397). New York: Guilford Press.

Brame, B., Nagin, D. S., & Tremblay, R. E. (2001). Developmental trajectories of physical aggression from school entry to late adolescence. *Journal of Child Psychology and Psychiatry and Allied Disciplines, 42*, 503–512.

Brendgen, M. (2012). Genetics and peer relations: A review. *Journal of research on adolescence, 22*, 419–437.

Brendgen, M., Boivin, M., Dionne, G., Barker, E. D., Vitaro, F., Girard, A., et al. (2011). Gene-environment processes linking aggression, peer victimization, and the teacher-child relationship. *Child Development, 82*(6), 2021–2036.

Brendgen, M., Boivin, M., Vitaro, F., Bukowski, W. M., Dionne, G., Tremblay, R. E., et al. (2008a). Linkages between children's and their friends' social and physical aggression: Evidence for a gene-environment interaction? *Child Development, 79*, 13–29.

Brendgen, M., Boivin, M., Vitaro, F., Dionne, G., Girard, A., & Pérusse, D. (2008b). Gene-environment interactions between peer victimization and child aggression. *Development and Psychopathology, 20*, 455–471.

Brendgen, M., Dionne, G., Girard, A., Boivin, M., Vitaro, F., & Pérusse, D. (2005). Examining genetic and environmental effects on social aggression: A study of 6-year-old twins. *Child Development, 76*(4), 930–946.

Brendgen, M., Girard, A., Vitaro, F., Dionne, G., & Boivin, M. (2014, Early View). Gene-environment correlation linking aggression and peer victimization: Do classroom behavioral norms matter? *Journal of Abnormal Child Psychology.* doi 10.1007/s10802-013-9807-z

Brendgen, M., Girard, A., Vitaro, F., Dionne, G., & Boivin, M. (2015). The Dark Side of Friends: A Genetically Informed Study of Victimization Within Early Adolescents' Friendships. Journal of Clinical Child and Adolescent Psychology. *Special Issue on Peer Victimization, 44*(3), 417-431.

Brendgen, M., & Vitaro, F. (2008). Predictive links between peer rejection and physical health problems in early adolescence. *Journal of Developmental and Behavioral Pediatrics, 29*(3), 183–190.

Brendgen, M., Vitaro, F., Barker, E. D., Girard, A., Dionne, G., Tremblay, R. E., et al. (2013a). Does other people's plight matter? A genetically informed twin study of the role of social context in the link between peer victimization and children's aggression and depression symptoms. *Developmental Psychology, 49*(2), 327–340.

Brendgen, M., Vitaro, F., Boivin, M., Girard, A., Bukowski, W. M., Dionne, G., et al. (2009). Gene-environment linkages between peer rejection and depressive symptoms in children. *Journal of Child Psychology and Psychiatry, 50*(8), 1009–1017.

Brendgen, M., Vitaro, F., & Bukowski, W. M. (1998). Affiliation with delinquent friends: Contributions of parents, self-esteem, delinquent behavior, and peer rejection. *Journal of Early Adolescence, 18*, 244–265.

Brendgen, M., Vitaro, F., & Bukowski, W. M. (2000). Deviant friends and early adolescents' emotional and behavioral adjustment. *Journal of Research on Adolescence, 10*(2), 173–189.

Brendgen, M., Vitaro, F., Bukowski, M., Dionne, G., Tremblay, R. E., & Boivin, M. (2013b). Can friends protect genetically vulnerable children from depression? *Development and Psychopathology, 25*, 277–289.

Brendgen, M., Wanner, B., & Vitaro, F. (2006). Verbal abuse by the teacher and child adjustment from kindergarten through grade 6. *Pediatrics, 117*(5), 1585–1598.

Bukowski, W. M., Brendgen, M., & Vitaro, F, (2006). Peer relations. In J. E. Grusec & P. D. Hastings (Eds.), *Handbook of socialization* (pp. 355–381). New York, NY: Guilford Press.

Bullock, B. M., Deater-Deckard, K., & Leve, L. D. (2006). Deviant peer affiliation and problem behavior: A test of genetic and environmental influences. *Journal of Abnormal Child Psychology, 34*(1), 29–41.

Burt, S. A. (2008). Gene-environment interactions and their impact on the development of personality traits. *Psychiatry, 7*(12), 507–510.

Burt, S. A., McGue, M., Krueger, R. F., & Iacono, W. G. (2005). How are parent-child conflict and childhood externalizing symptoms related over time? Results from a genetically informative cross-lagged study. *Development and Psychopathology, 17*(1), 145–165.

Button, T. M. M., Corley, R. P., Rhee, S. H., Hewitt, J. K., Young, S. E., & Stallings, M. C. (2007). Delinquent peer affiliation and conduct problems: A twin study. *Journal of Abnormal Psychology, 116*(3), 554–564.

Button, T. M. M., Stallings, M. C., Rhee, S. H., Corley, R. P., Boardman, J. D., & Hewitt, J. K. (2009). Perceived peer delinquency and the genetic predisposition for substance dependence vulnerability. *Drug and Alcohol Dependence, 100*(1–2), 1–8.

Cairns, R. B., Cairns, B. D., Neckerman, H. J., Gest, S. D., & Gariépy, J.-L. (1988). Peer networks and aggressive behavior: Social support or social rejection? *Developmental Psychology, 24*, 815–823.

Cairns R., Xie, H., & Leung, M. (1998). The popularity of friendship and the neglect of social networks: Toward a new balance. In W. M. Bukowski & A. H. Cillessen (Eds.), *Sociometry then and now: Building on six decades of measuring children's experiences with the paper group: No. 80. New directions for child development* (pp. 5–24). San Francisco, CA: Jossey-Bass.

Carney, J. V. (2000). Bullied to death. *School Psychology International, 21*(2), 213–223.

Champagne, F. A. (2012). Interplay between social experiences and the genome. Epigenetic consequences for behavior. *Advances in Genetics, 77*, 33–57.

Cillessen, A. H. N., & Mayeux, L. (2004). From censure to reinforcement: Developmental changes in the association between aggression and social status. *Child Development, 75*(1), 147–163.

Cleveland, H. H., Wiebe, R. P., & Rowe, D. C. (2005). Sources of exposure to smoking and drinking friends among adolescents: A behavioral-genetic evaluation. *Journal of Genetic Psychology, 166*(2), 153–169.

Cohen, J. (1977). *Statistical power analysis for the behavioral sciences.* New York: Academic Press.

Coie, J. D., Dodge, K. A., & Coppotelli, H. (1982). Dimensions and types of social status: A cross-age perspective. *Developmental Psychology, 18*, 557–570.

Coie, J., Terry, R., Akriski, A., & Lochman, J. (1995). Early adolescent social influences on delinquent behavior. In J. McCord (Ed.), *Coercion and punishment in long-term perspectives* (pp. 229–244). New York, NY: Cambridge University Press.

Colwell, M. J., Mize, J., Pettit, G. S., & Laird, R. D. (2002). Contextual determinants of mothers' interventions in young children's peer interactions. *Developmental Psychology, 38,* 492–502.

Cramer, P., & Steinwert, T. (1998). Thin is good, fat is bad: How early does it begin? *Journal of Applied Developmental Psychology, 19*(3), 429–451.

Crick, N. R., & Nelson, D. A. (2002). Relational and physical victimization within friendships: Nobody told me there'd be friends like these. *Journal of Abnormal Child Psychology, 30,* 599–607.

Dick, D. M., Pagan, J. L., Holliday, C., Viken, R., Pulkkinen, L., Kaprio, J., et al. (2007). Gender differences in friends' influences on adolescent drinking: A genetic epidemiological study. *Alcoholism, Clinical and Experimental Research, 31*(12), 2012–2019.

DiLalla, L. F. (2002). Behavior genetics of aggression in children: Review and future directions. *Developmental Review, 22,* 593–622.

Dishion, T. J. (1990a). The family ecology for boys' peer relations in middle childhood. *Child Development, 61,* 874–892.

Dishion, T. J. (1990b). Peer context of troublesome behavior in children and adolescents. In P. Leone (Ed.), *Understanding troubled and troublesome youth* (pp. 128–153). Beverly Hills, CA: Sage.

Dodge, K. A. (1983). Behavioral antecedents of peer social status. *Child Development, 54*(6), 1386–1399.

Dodge, K. A., Dishion, T. J., & Lansford, J. E. (2006). *Interventions and policies that aggregate deviant youth, and strategies to optimize outcomes.* New York, NY: Guilford Press.

Dodge, K. A., Lansford, J. E., Burks, V. S., Bates, J. E., Pettit, G. S., Fontaine, R., & et al. (2003). Peer rejection and social information-processing factors in the development of aggressive behavior problems in children. doi: 10.1111/1467-8624.7402004. *Child Development, 74*(2), 374–393.

Eaves, L., Silberg, J., & Erkanli, A. (2003). Resolving mutliple epigenetic pathways to adolescent depression. *Journal of Child Psychology and Psychiatry, 44*(7), 1006–1014.

Eisenberger, N. I., Lieberman, M. D., & Williams, K. D. (2003). Does rejection hurt? An fMRI study of social exclusion. *Science, 302*(5643), 290–292.

Elliott, D. S., Huizinga, D., & Ageton, S. S. (1985). *Explaining delinquency and drug use.* Beverly Hills, CA: Sage.

Ellis, B. J., Boyce, W. T., Belsky, J., Bakermans-Kranenburg, M. J., & van Ijzendoorn, M. H. (2011). Differential susceptibility to the environment: An evolutionary–neurodevelopmental theory. *Development and Psychopathology, 23*(01), 7–28.

Erhardt, D., & Hinshaw, S. P. (1994). Initial sociometric impressions of attention-deficit hyperactivity disorder and comparison boys: Predictions from social behaviors and from nonbehavioral variables. *Journal of Consulting and Clinical Psychology, 62*(4), 833–842.

Estell, D. B., Cairns, R. B., Farmer, T. W., & Cairns, B. D. (2002). Aggression in inner-city early elementary classroom: Individual and peer-group configurations. *Merrill-Palmer Quarterly, 48,* 52–76.

Farver, J. A. M. (1996). Aggressive behavior in preschoolers' social networks: Do birds of a feather flock together? *Early Childhood Research Quarterly, 11,* 333–350.

Feinberg, M. E., Button, T. M. M., Neiderhiser, J. M., Reiss, D., & Hetherington, E. M. (2007). Parenting and adolescent antisocial behavior and depression: Evidence of genotype x parenting environment interaction. *Archives of General Psychiatry, 64*(4), 457–465.

Fekkes, M., Pijpers, F. I. M., Fredriks, A. M., Vogels, T., & Verloove-Vanhorick, S. P. (2006). Do bullied children get Ill, or do Ill children get bullied? A prospective cohort study on the relationship between bullying and health-related symptoms. *Pediatrics, 117,* 1568–1574.

Fekkes, M., Pijpers, F. I. M., & Verloove-Vanhorick, S. P. (2005). Bullying: Who does what, when and where? Involvement of children, teachers and parents in bullying behavior. *Health Education Research, 20*(1), 81–91.

Fowler, T., Shelton, K., Lifford, K., Rice, F., McBride, A., Nikolov, I., et al. (2007). Genetic and environmental influences on the relationship between peer alcohol use and own alcohol use in adolescents. *Addiction, 102*(6), 894–903.

Gazelle, H., & Druhen, M. J. (2009). Anxious solitude and peer exclusion predict social helplessness, upset affect, and vagal regulation in response to behavioral rejection by a friend. *Developmental Psychology, 45,* 1077–1096. doi:10.1037/a0016165.

Giles, J. W., & Heyman, G. D. (2005). Preschoolers use trait-relevant information to evaluate the appropriateness of an aggressive response. *Aggressive Behavior, 31*(5), 498–509.

Gilson, M. S., Hunt, C. B., & Rowe, D. S. (2003). The friends of siblings. *Marriage and Family Review, 33*(2–3), 205–223.

Goldstein, D. B. (2009). Common genetic variation and human traits. *New England Journal of Medicine, 360*(17), 1696–1698.

Goldstein, S. E., Tisak, M. S., & Boxer, P. (2002). Preschoolers' normative and prescriptive judgments about relational and overt aggression. *Early Education and Development, 13*(1), 23–39.

Gottfredson, M. R., & Hirschi, T. (1990). *A general theory of crime.* Stanford, CA: Stanford University Press.

Guo, G. (2006). Genetic similarity shared by best friends among adolescents. *Twin Research and Human Genetics, 9*(1), 113–121.

Guo, G., Elder, G. H., Cai, T., & Hamilton, N. (2009). Gene-environment interactions: Peers' alcohol use moderates genetic contribution to adolescent drinking behavior. *Social Science Research, 38*(1), 213–224.

Hanish, L. D., & Guerra, N. (2000). Predictors of peer victimization among urban youth. *Social Development, 9*(4), 521–543.

Hanish, L. D., & Guerra, N. G. (2002). A longitudinal analysis of patterns of adjustment following peer victimization. *Development and Psychopathology, 14*(1), 69–89.

Harden, K. P., Hill, J. E., Turkheimer, E., & Emery, R. E. (2008). Gene-environment correlation and interaction in peer effects on adolescent alcohol and tobacco use. *Behavior Genetics, 38*(4), 339–347.

Hektner, J. M., August, G. J., & Realmuto, G. M. (2000). Patterns and temporal changes in peer affiliation among aggressive and nonaggressive children participating in a summer school program. *Journal of Clinical Child Psychology, 29,* 603–614.

Henry, D., Guerra, N., Huesmann, R., Tolan, P., Vanacker, R., & Eron, L. (2000). Normative influences on aggression in urban elementary school classrooms. *American Journal of Community Psychology, 28,* 59–81.

Hicks, B. M., South, S. C., DiRago, A. C., Iacono, W. G., & McGue, M. (2009). Environmental adversity and increasing genetic risk for externalizing disorders. *Archives of General Psychiatry, 66*(6), 640–648.

Hodges, E. V., Boivin, M., Vitaro, F., & Bukowski, W. M. (1999). The power of friendship: Protection against an escalating cycle of peer victimization. *Developmental Psychology, 35*(1), 94–101.

Hodges, E. V. E., & Perry, D. G. (1999). Personal and interpersonal antecedents and consequences of victimization by peers. *Journal of Personality and Social Psychology, 76*(4), 677–685.

Houndoumadi, A., & Pateraki, L. (2001). Bullying and bullies in Greek elementary schools: Pupils' attitudes and teachers'/parents' awareness. *Educational Review, 53*(1), 19–26.

Iervolino, A. C., Pike, A., Manke, B., Reiss, D., Hetherington, E. M., & Plomin, R. (2002). Genetic and environmental influences in adolescent peer socialization: Evidence from two genetically sensitive designs. *Child Development, 73*(1), 162–174.

Jaffee, S. R., Caspi, A., Moffitt, T. E., Dodge, K. A., Rutter, M., Taylor, A., et al. (2005). Nature × nurture: Genetic vulnerabilities interact with physical maltreatment to promote conduct problems. *Development and Psychopathology, 17*(1), 67–84.

Johnson, R. E., Marcos, A. C., & Bahr, S. (1987). The role of peers in the complex etiology of drug use. *Criminology, 25,* 323–340.

Keller, M. C., & Coventry, W. L. (2005). Quantifying and addressing parameter indeterminacy in the classical twin design. *Twin Research and Human Genetics, 8*, 201–213.

Kim, J. E., Hetherington, E. M., & Reiss, D. (1999). Associations among family relationships, antisocial peers, and adolescents' externalizing behaviors: Gender and family type differences. *Child Development, 70*, 1209–1230.

Kim-Cohen, J., Caspi, A., Taylor, A., Williams, B., Newcombe, R., Craig, I. W., & Moffitt, T. E. (2006). MAOA, maltreatment, and gene-environment interaction predicting children's mental health: New evidence and a meta-analysis. *Molecular Psychiatry, 11*(10), 903–913.

La Greca, A. M., & Lopez, N. (1998). Social anxiety among adolescents: Linkages with peer relations and friendships. *Journal of Abnormal Child Psychology, 26*, 83–94.

Ladd, G. W., Profilet, S. M., & Hart, C. H. (1992). Parents' management of children's peer relations: Facilitating and supervising children's activities in the peer culture. In R. D. Parke & G. W. Ladd (Eds.), *Family peer relationships: Modes of linkage* (pp. 215–253). Hillsdale, NJ: Erlbaum.

Lahey, B. B., Gordon, R. A., Loeber, R., Stouthamer-Loeber, M., & Farrington, D. P. (1999). Boys who join gangs: A prospective study of predictors of first gang entry. *Journal of Abnormal Child Psychology, 27*, 261–276.

Lamarche, V., Brendgen, M., Boivin, M., Vitaro, F., Dionne, G., & Pérusse, D. (2007). Do friends' characteristics moderate the prospective links between peer victimization and reactive and proactive aggression? *Journal of Abnormal Child Psychology, 35*, 665–680.

Lau, J. Y. F., & Eley, T. C. (2008). Disentangling gene-environment correlations and interactions on adolescent depressive symptoms. *Journal of Child Psychology and Psychiatry, 49*(2), 142–150.

Leve, L. D. L. (2001). Observation of externalizing behavior during a twin-friend discussion task. *Marriage and Family Review, 33*(2), 225–250.

Li, Q. (2006). Cyberbullying in schools. *School Psychology International, 27*(2), 157–170.

McDougall, P., Hymel, S., Vaillancourt, T., & Mercer, L. (2001). The consequences of childhood peer rejection. In M. Leary (Ed.), *Interpersonal rejection* (pp. 213–247). New York: Oxford University Press.

Meaney, M. J. (2010). Epigenetics and the biological definition of gene X environment interactions. *Child Development, 81*(1), 41–79.

Moffitt, T. E. (2005). The new look of behavioral genetics in developmental psychopathology: Gene-environment interplay in antisocial behaviors. *Psychological Bulletin, 131*(4), 533–554.

Moffitt, T. E., Caspi, A., & Rutter, M. (2005). Strategy for investigating interactions between measured genes and measured environments. *Archives of General Psychiatry, 62*, 473–481.

Monroe, S. M., & Simons, A. D. (1991). Diathesis–stress theories in the context of life stress research: Implications for depressive disorders. *Psychological Bulletin, 110*, 406–425.

Narusyte, J., Andershed, A.-K., Neiderhiser, J., & Lichtenstein, P. (2007). Aggression as a mediator of genetic contributions to the association between negative parent–child relationships and adolescent antisocial behavior. *European Child and Adolescent Psychiatry, 16*(2), 128–137.

Neale, M. C., & Cardon, L. R. (1992). Methodology for genetic studies of twins and families. Dortrecht, Netherlands: Kluwer Academic. Turkheimer, E. (2000). Three laws of behavior genetics and what they mean. *Current Directions in Psychological Science, 9*(5), 160–164.

Nishina, A., Juvonen, J., & Witkow, M. R. (2005). Sticks and stones may break my bones, but names will make me feel sick: The psychosocial, somatic, and scholastic consequences of peer harassment. *Journal of Clinical Child and Adolescent Psychology, 34*(1), 37–48.

Ouellet-Morin, I., Boivin, M., Dionne, G., Lupien, S. J., Arsenault, L., Barr, R. G., et al. (2008). Variations in heritability of cortisol reactivity to stress as a function of early familial adversity among 19-month-old twins. *Archives of General Psychiatry, 65*(2), 211–218.

Patterson, G. R., DeBaryshe, B. D., & Ramsey, E. (1989). A developmental perspective on antisocial behavior. *American Psychologist, 44*(2), 329–335.

Patterson, G. R., Dishion, T. J., & Yoerger, K. (2000). Adolescent growth in new forms of problem behavior: Macro- and micro-peer dynamics. *Prevention Science, 1*, 3–13.

Paul, J., & Cillessen, A. H. (2003). Dynamics of peer victimization in early adolescence: Results from a four-year longitudinal study. *Journal of Applied School Psychology, 19*(2), 25–43.

Pellegrini, A. D., & Long, J. D. (2002). A longitudinal study of bullying, dominance, and victimization during the transition from primary school through secondary school. *British Journal of Developmental Psychology, 20*(2), 259–280.

Piaget, J. (1932). *The moral judgment of the child.* New York: Harcourt.

Pike, A., McGuire, S., Hetherington, E. M., Reiss, D., & Plomin, R. (1996). Family environment and adolescent depressive symptoms and antisocial behavior: A multivariate genetic analysis. *Developmental Psychology, 32*, 590–603.

Plomin, R., DeFries, J. C., & Loehlin, J. C. (1977). Genotype–environment interaction and correlation in the analysis of human behavior. *Psychological Bulletin, 84*, 309–322.

Pope, A. W., & Bierman, K. I. (1999). Predicting adolescent peer problems and antisocial activities: The relative roles of aggression and dysregulation. *Developmental Psychology, 35*(2), 335–346.

Poulin, F., & Boivin, M. (2000). The role of proactive aggression and reactive aggression in the formation and development of boys' friendships. *Developmental Psychology, 36*, 233–240.

Purcell, S. (2002). Variance components models for gene-environment interaction in twin analysis. *Twin Research, 5*(6), 554–571.

Rhee, S., & Waldman, I. D. (2002). Genetic and environmental influences on antisocial behavior: A meta-analysis of twin and adoption studies. *Psychological Bulletin, 29*, 490–529.

Rice, F. (2009). The genetics of depression in childhood and adolescence. *Current Psychiatry Reports, 11*(2), 167–173.

Rice, F., Harold, G. T., & Thapar, A. (2003). Negative life events as an account of age-related differences in the genetic aetiology of depression in childhood and adolescence. *Journal of Child Psychology and Psychiatry, 44*(7), 977–987.

Rigby, K. (1999). Peer victimisation at school and the health of secondary school students. *The British Journal of Educational Psychology, 69*(1), 95–104.

Rose, R. J. (2002). How do adolescents select their friends? A behavior-genetic perspective. In L. Pulkkinen & A. Caspi (Eds.), *Paths to successful development: Personality in the life course* (pp. 106–125). New York: Cambridge University Press.

Rubin, K. H., Bukowski, W., & Parker, J. G. (1998). Peer interactions, relationships, and groups. In *Handbook of child psychology: Social, emotional, and personality development* (5th ed., Vol 3, pp. 619–700): Hoboken, NJ, US: Wiley.

Rubin, K. H., Bukowski, W., & Parker, J. G. (2006). Peer interactions, relationships, and groups In N. Eisenberg & W. Damon (Eds.), *Handbook of child psychology* (6 ed., Vol. 3, pp. 571–645). Hoboken, NJ, US: Wiley.

Rudolph, K. D., Hammen, C., & Burge, D. (1994). Interpersonal functioning and depressive symptoms in childhood: Addressing the issues of specificity and comorbidity. *Journal of Abnormal Child Psychology, 22*(3), 355–371.

Rutter, M., Moffitt, T. E., & Caspi, A. (2006). Gene-environment interplay and psychopathology: multiple varieties but real effects. *Journal of Child Psychology and Psychiatry, 47*(3/4), 226–261.

Rutter, M., Pickles, A., Murray, R., & Eaves, L. J. (2001). Testing hypotheses on specific environmental causal effects on behavior. *Psychological Bulletin, 127*(3), 291–324.

Scaramella, L. V., Conger, R. D., Spoth, R., & Simons, R. L. (2002). Evaluation of a social contextual model of delinquency: A cross-study replication. *Child Development, 73*(1), 175–195.

Scarr, S., & McCartney, K. (1983). How people make their own environments: A theory of genotype—environment effects. *Child Development, 54*(2), 424–435.

Schuster, B. (2001). Rejection and victimization by peers: Social perception and social behavior mechanisms. In J. Juvonen & S. Graham (Eds.), *Peer harassment in school: The plight of the vulnerable and victimized* (pp. 290–309). New York: Guilford Press.

Schwartz, D., Dodge, K. A., Coie, J. D., Hubbard, J. A., Cillessen, A. H. N., Lemerise, E. A., & Bateman, H. (1998a). Social-cognitive and behavioral correlates of aggression and victimization in boys' play groups. *Journal of Abnormal Child Psychology, 26*(6), 431–440.

Schwartz, D., McFadyen-Ketchum, S. A., Dodge, K. A., Pettit, G. S., & Bates, J. E. (1998b). Peer group victimization as a predictor of children's behavior problems at home and in school. *Development and Psychopathology, 10*(1), 87–99.

Schwartz, D., McFadyen-Ketchum, S., Dodge, K. A., Pettit, G. S., & Bates, J. E. (1999). Early behavior problems as a predictor of later peer group victimization: Moderators and mediators in the pathways of social risk. *Journal of Abnormal Child Psychology, 27*(3), 191–201.

Shanahan, M., & Hofer, S. (2005). Social context in gene-environment interactions: Retrospect and prospect. *Journal of Gerontology: Series B, 60B*, 65–76.

Silberg, J., Pickles, A., Rutter, M., Hewitt, J., Simonoff, E., Maes, H., & Eaves, L. (1999). The influence of genetic factors and life stress on depression among adolescent girls. *Archives of General Psychiatry, 5*, 225–232.

Simons, R. L., Chao, W., Conger, R. D., & Elder, G. H, Jr. (2001). Quality of parenting as mediator of the effect of childhood defiance on adolescent friendship choices and delinquency: A growth curve analysis. *Journal of Marriage and the Family, 63*, 63–79.

Slee, P. T. P. (1995). Bullying: health concerns of Australian secondary school students. *International Journal of Adolescence and Youth, 5*(4), 215–224.

Smith, P. K., Morita, Y., Junger-Tas, J., Olweus, D., Catalano, R., & Slee, P. (1999). *The nature of school bullying: A cross-national perspective*. New York: Routledge.

Smith, P. K., Shu, S., & Madsen, K. (2001). Characteristics of victims of school bullying: Developmental changes in coping strategies and skills. In J. Juvonen & S. Graham (Eds.), *Peer harassment in school: The plight of the vulnerable and victimized* (pp. 332–351). New York, NY, US: Guilford Press.

Snyder, J., Horsch, E., & Childs, J. (1997). Peer relationships of young children: Affiliative choices and the shaping of aggressive behavior. *Journal of Clinical Child Psychology, 26*, 145–156.

Snyder, J., Schrepferman, L., Oeser, J., Patterson, G., Stoolmiller, M., Johnson, K., et al. (2005). Deviancy training and association with deviant peers in young children: Occurrence and contribution to early-onset conduct problems. *Development and Psychopathology, 17*, 397–413.

Stormshak, E. A., Bierman, K. L., Bruschi, C. J., Dodge, K. A., Coie, J. D., & the Conduct Problems Prevention Research Group. (1999). The relation between behavior problems and peer preference in different classroom contexts. *Child Development, 70(1)*, 169–182.

Strauss, C. C., Smith, K., Frame, C., & Forehand, R. (1985). Personal and interpersonal characteristics associated with childhood obesity. *Journal of Pediatric Psychology, 10*(3), 337–343.

Sullivan, H. S. (1953). *An interpersonal theory of psychiatry*. New York: Norton.

Sutherland, E. (1947). *Principles of criminology* (3rd ed.). Philadelphia: Lippincott.

Sweeting, H., Wright, C., & Minnis, H. (2005). Psychosocial correlates of adolescent obesity, 'slimming down' and 'becoming obese'. *Journal of Adolescent Health, 37*, 409–417.

Teräsahjo, T., & Salmivalli, C. (2003). She is not actually bullied. The discourse of harassment in student groups. *Aggressive behavior, 29*(2), 134–154.

Thapar, A., Harold, G., Rice, F., Langley, K., & O'Donovan, M. (2007). The contribution of gene-environment interaction to psychopathology. *Development and Psychopathology, 19*(4), 989–1004.

Thorpe, K., & Gardner, K. (2006). Twins and their friendships: Differences between monozygotic, dizygotic same-sex and dizygotic mixed-sex pairs. *Twin Research and Human Genetics, 9*(1), 155–164.

Tremblay, R. E., Mâsse, L. C., Vitaro, F., & Dobkin, P. L. (1995). The impact of friends' deviant behavior on early onset of delinquency: Longitudinal data from 6 to 13 years of age. *Development and Psychopathology, 7*, 649–667.

Urberg, K. A., Degirmencioglu, S. M., & Tolson, J. M. (1998). Adolescent friendship selection and termination: The role of similarity. *Journal of Social and Personal Relationships, 15*, 703–710.

Van Der Sluis, S., Posthuma, D., & Dolan, C. V. (2012). A note on false positives and power in G × E modelling of twin data. *Behavior Genetics, 42*(1), 170–186.

Van Lier, P., Boivin, M., Dionne, G., Vitaro, F., Brendgen, M., Koot, H., et al. (2007). Kindergarten children's genetic vulnerabilities interact with friends' aggression to promote children's own aggression. *Journal of the American Academy of Child and Adolescent Psychiatry, 46*, 1080–1087.

Vernberg, E. M., Abwender, D. A., Ewell, K. K., & Beery, S. H. (1992). Social anxiety and peer relationships in early adolescence: A prospective analysis. *Journal of Clinical Child Psychology, 21*(2), 189–196.

Vitaro, F., Brendgen, M., & Tremblay, R. E. (2001). Preventive intervention: Assessing its effects on the trajectories of delinquency and testing for mediational processes. *Applied Developmental Science, 5*, 201–213.

Vitaro, F., Tremblay, R. E., Kerr, M., Pagani, L. S., & Bukowski, W. M. (1997). Disruptiveness, friends' characteristics, and delinquency: A test of two competing models of development. *Child Development, 68*, 676–689.

Vuijk, P., van Lier, P. A. C., Crijnen, A. A. M., & Huizink, A. C. (2007). Testing sex-specific pathways from peer victimization to anxiety and depression in early adolescents through a randomized intervention trial. *Journal of Affective Disorders, 100*(1–3), 221–226.

Waasdorp, T. E., Bagdi, A. & Bradshaw, C. P. (2009). Peer victimization among urban, predominantly African American youth: Coping with relational aggression between friends. *Journal of School Violence, 9*(1), 98–116.

Walden, B., McGue, M., Iacono, W. G., Burt, S. A., & Elkins, I. (2004). Identifying shared environmental contributions to early substance use: The respective roles of peers and parents. *Journal of Abnormal Psychology, 113*(3), 440–450.

Werner, N. E., & Crick, N. R. (2004). Maladaptive peer relationships and the development of relational and physical aggression during middle childhood. *Social Development, 13*(4), 495–514.

Williams, K., Chambers, M., Logan, S., & Robinson, D. (1996). Association of common health symptoms with bullying in primary school children. *British Medical Journal, 313*, 17–19.

Youniss, J. (1980). *Parents and peers and social development: A Sullivan-Piaget perspective.* Chicago: University of Chicago Press.

Zuckerman, M. (1999). *Vulnerability to psychopathology: A biosocial model.* Washington, DC: American Psychological Association.

Chapter 6
Gene-Environment Processes in Adolescent Family Relationships

Nan Chen and Kirby Deater-Deckard

6.1 Gene-Environment (GE) Processes in Adolescent Family Relationships

Each developing adolescent's thoughts, emotions and behaviors arise from a complex interplay of genetic and environmental influences. These work in combination to produce the individual differences in personality, psychopathology, and social cognitions that contribute to healthy or maladaptive development. In the current chapter, our goal was to examine and integrate research that has examined GE interplay as it pertains to a particular aspect of adolescents' social worlds: their family relationships, with emphasis on the parent–adolescent relationship, the adolescent sibling relationship, and the parents' marital relationship. In each case, these social relationships are powerful contexts in which the role of genetic factors must be considered, in order to better understand the roles of both biology and environment in adolescent individuality.

6.2 Family Relationships in Adolescence

The family is a critically important environment for adolescents to learn and implement social rules (e.g., altruism), to form self-identity, and to develop into responsible, competent, healthy adults. Support and caring relationships with family members are seen as protective factors against risks and stressors in the rapidly

N. Chen · K. Deater-Deckard (✉)
Department of Psychology, Virginia Polytechnic Institute and State University,
109 Williams Hall (0436), Blacksburg, VA 24061, USA
e-mail: kirbydd@vt.edu

© Springer Science+Business Media New York 2015
B.N. Horwitz and J.M. Neiderhiser (eds.), *Gene-Environment Interplay in Interpersonal Relationships across the Lifespan*,
Advances in Behavior Genetics 3, DOI 10.1007/978-1-4939-2923-8_6

changing developmental period of adolescence. In contrast, cold or hostile family relationships are assumed to serve to promote growth in social-emotional and academic difficulties including scholastic failure, internalizing problems (i.e., depression, anxiety) and externalizing problems (i.e., aggression, delinquency) (Barber 1992; Demo and Acock 1996).

Furthermore, as the basic functional unit in all societies, families exist in broader social settings, from communities to neighborhood districts, and all the way up to regions of states in nations. Because of this hierarchical structure of human physical and social environments, families play a critical role in connecting individual adolescents to a variety of contexts and as a result, family relationships may serve as mediators of more distal influences on adolescents' personal development and well-being (Bronfenbrenner and Evans 2000). For example, living in an area that has relatively few socioeconomic opportunities can influence adolescents' cognitive, emotional and behavioral development through the effects of poor access to resources on parenting and parent–adolescent relationship quality (Bornstein and Bradley 2003).

The important role that family relationships play in adolescent development makes it necessary to understand the underlying factors that contribute to individual difference in the quality of these family relationships. In the current chapter, we focus on three types of family relationships: parent–child relationships, sibling relationships, and adolescents' parents' marital relationships. We review studies that have employed behavioral genetic approaches, also with more studies using molecular biology, to gain better insight into GE interplay that gives rise to family relationships of varying quality that have implications for adolescent healthy and maladaptive development.

6.3 Which Personal Factors Influence Family Relationship Quality?

Family relationships are influenced, in part, by the individual difference attributes of each member of the dyadic relationship. For example, it is intuitively obvious that people who are agreeable and considerate toward others are easier to get along with when compared to those who are aggressive and irritable. This commonsense notion applies in family settings just as well as in other social settings. Studies show that parents' agreeableness, extroversion and neuroticism longitudinally predict parents' mood and sensitivity toward their children and adolescents (Belsky et al. 1995). For siblings, agreeableness is positively associated with warmth in the sibling relationship (Lanthier 2007). Similarly, in married couples the likelihood of conflict and divorce is higher for emotionally reactive individuals than for those with good emotion regulation (Jockin et al. 1996).

Besides the individual attributes of each member of a family relationship dyad, there are potentially many factors outside each individual that influence the relationship. In the literature, these external factors typically fall into two broad contrasting categories: environmental stressors and supports. In regard to stressors, families who live in poverty and in dangerous neighborhoods are more likely to have more chaotic home environments, and also to have parenting environments that are less warm and consistent, and more hostile and harsh (Deater-Deckard et al. 2009; Pinderhughes et al. 2001). In contrast, environmental supports show positive associations with positive qualities of parenting as well as parent–child interactions and relationships. Social support from family members and friends helps protect parents from depression, helps them to be more nurturing and con-sistent in their parenting, and is associated with less reliance on harsh punitive methods of managing adolescent behavior (McLoyd 1990, 1998; Mason et al. 1994), particularly when the broader context also includes neighbors who feel safe and connected (Ceballo and McLoyd 2002). Other examples can be seen in the results of interventions that target parent–adolescent relationships. Many of these programs offer training on appropriate use of adolescents' coping strategies in family interactions, in an effort to increase positive parenting and better adoles-cent outcomes (Compas et al. 2010). Social support within and between families in neighborhoods, when operating in conjunction with adolescents' own effective coping strategies, provide strong buffers against stressors when they arise.

In the current chapter, we consider as separate the potential effects of "internal" (e.g., genetic, personal) and "external" (e.g., family environment, social context) factors on family relationships in adolescence. However, this distinction is mostly a pragmatic one. The family is a complex and dynamic system, with internal and external factors operating interactively and continuously to shape family relation-ship qualities and adolescent outcomes. This is important to keep in mind when considering the multiple examples that follow of studies that have examined GE correlation and interaction in family relationships.

6.4 The Parent–Adolescent Relationship

Genetic and environmental factors may influence the parent–adolescent relation-ship in a variety of ways. Behavioral genetic studies on parent–child and parent–adolescent relationships have shown that both genes and environments give rise to warm, cold or antagonistic interactions through interpersonal and person-environment processes. Using studies of identical and fraternal twins, and of adoptive and nonadoptive parent–child and sibling dyads, researchers are able to estimate sibling and parent–child similarity as intra-class correlations for various measures and variables. Then, the patterns of sibling and parent–child similarities are examined, to see if they differ systematically as a function of the siblings' or parent–child dyads' genetic similarity, with genetic similarity quantified as 100 % for monozygotic/identical twins, 50 % (on average) for dizygotic/fraternal twins, full siblings, and biological parent–child dyads, 25 % for half-siblings, and 0 % for adoptive and step parent–child and sibling dyads. Regression methods are then used to estimate the portions of variance in a particular variable that accounts for family member similarity due to genetic factors (i.e., heritability) and nongenetic factors (i.e., shared environment), as well as the variance that is specific to each individual in the family (i.e., nonshared environment).

With regard to adolescents' perceptions of their home environments and relationships with their parents, the behavioral genetic evidence points to shared and nonshared environmental variance in youth reports of parental control, and genetic and non-shared environmental variance in reports of parental warmth and acceptance (Plomin 1994; Rowe 1983). However, because dyadic relationships involve two people, the qualities of each relationship are determined by bi-directional processes whereby the parent and adolescent alike are active agents in their interactions—as is seen in the impact on parenting of children's challenging conduct and emotional problems, from early in life through adolescence (Bell 1979). This becomes even clearer (in theory at least) as children develop into and through adolescence and gain more autonomy, exerting their own independent thoughts and actions in a desire to control their own lives and relationships with their parents and others (Maccoby 1992). In behavioral genetic research, theories have emphasized the role of *GE correlation*, to represent certain aspects of this bi-directionality in children's and adolescents' social relationships. In general, GE correlation refers to the covariation between genetic variation in the individual, and variation in that individual's environment (Plomin et al. 1977).

GE correlation typically is defined using three general classes that may operate differently over the course of development. These include passive, active, and evocative/reactive forms of GE correlation (Scarr and McCartney 1983). Passive GE correlation exists in biologically related family members who share genes as well as environments. For example, a parent and adolescent who both have warm, accepting personalities also may have a strong and warm relationship with each other because they have in common many of the genetic factors that contribute to personal warmth and positive emotion. Active GE correlation is exemplified in contexts in which the adolescent selects her or his own experiences, such as the choosing of particular peers and close friends. Using the same example just given, an adolescent who is warm and accepting of others may tend to gravitate toward others who are similarly warmth and accepting of others. Furthermore, their friendships may be very warm and supportive in part because they have in common some of the genetic factors that contribute to these aspects of personality. Evocative or reactive GE correlation is exemplified in the way in which people consistently react or respond to the adolescent's genetically influenced (behaviors or physical) characteristics. Returning to our current example, an adolescent who is warm and accepting of others is more likely to elicit in strangers and acquaintances alike a pattern of responses that are friendly and supportive. Furthermore, these interactions are likely to be positive and enjoyable, in part because the genetically influenced attributes of the adolescent consistently lead to positive responses and reactions from others.

In an effort to address the bidirectional nature of the parent–adolescent relationship, behavioral genetic studies have investigated the various ways in which this relationship can be shaped by the emotions, cognitions and behaviors of both partners.

In theory, "child effects" on the parent–adolescent relationship most likely operate through evocative GE correlation, whereby the parent responds in a particular way to their adolescent child's behaviors. For instance, researchers using adoption studies have shown that adolescents who are genetically at risk for antisocial behavior (i.e., had biological parents who were highly antisocial) are more likely to experience negative, harsh parenting from their adoptive parents (Ge et al. 1996; O'Connor et al. 1998). This likely reflects a process whereby genetically influenced adolescent conduct problems, as well as warm socially responsive behaviors, evoke or elicit higher levels of harsh parenting or warm caring parenting respectively, suggesting an important role of youth personal attributes in shaping parent–youth relationships and interactions (South et al. 2008).

In Scarr and McCartney's (1983) influential theory, evocative GE correlation processes become stronger as adolescents acquire more autonomy and influence in their relationships with their parents. In behavioral genetic research of adolescents' relationships and their developmental outcomes, researchers have found that adolescents' perceptions of warmth and conflict in the parent–adolescent relationship shift in the transition to puberty (i.e., from 10 to 15 years old) toward greater negativity and less positivity (Herndon et al. 2005; McGue et al. 2005). At the same time, the heritability of the variance in adolescents' perceptions of their relationships with their parents also increases—an effect found previously in cross-sectional data (Elkins et al. 1997). While not conclusive, such findings are consistent with the idea that GE correlation connecting adolescents' attributes and their relationships with their parents grows during this developmental period. These evocative GE correlation processes probably operate in tandem with passive GE correlation effects, as well as GE interaction effects, to account for parent–adolescent similarity in the negativity and positivity each expresses toward the other in their relationship (Neiderhiser et al. 2004).

Emotion is central to the parent–adolescent relationship and the frequency and qualities of the dyad's interactions. Within the same family, the levels of emotional warmth and closeness, as well as negativity and conflict, often differ markedly across each parent–adolescent dyad within the same family (Deater-Deckard 2009; Reiss 2005). From a behavioral genetic perspective, these within-family differences indicate that the emotional features of parent–adolescent relationship are due in part to individual difference of adolescent attributes that arise from heritable and nonshared environmental influences (see Dunn and Plomin 1990; Plomin 1994). For example, Neiderhiser et al. (2007) found that adolescents' reports of paternal positivity and negativity were distinct for each adolescent in the family. Furthermore, using behavioral genetic models, the variance in father-adolescent relationships was found to be attributable in part to evocative and passive GE correlation, indicating that the fathers' and the adolescents' genetically influenced attributes both were contributing to the experienced and expressed emotions in their relationship.

The vast majority of behavioral genetic studies of family relationships have utilized sibling designs, whereby different types of child or adolescent sibling pairs (e.g., identical and fraternal twins, adoptive siblings) are studied and compared in order to identify sources of genetic and nongenetic variance in relationship processes. However, it also is possible to examine these processes using a "parent" sibling design, whereby different types (e.g., identical and fraternal twins) of adult siblings who have children of their own are studied and compared. In one example of this approach, Neiderhiser et al. (2004) studied a sample of adult twins with children. They found that variation in mothers' reports of negativity and positivity in their relationships with their offspring was accounted for more or less equally by heritable and nonshared environmental variance; shared environmental variance was negligible. This pattern suggested a role of genetic and nongenetic influences on differential parenting of sibling children that is driven in part by parent attributes, and in part by adolescent attributes.

Other studies have examined adult twins as parents, focusing on identifying the specific parent personal attributes that might account for genetic and nongenetic variance in parenting behaviors within the parent–youth relationship—an important step in identifying potential passive GE correlation processes. Spinath and O'Connor (2003) investigated genetic and environmental factors underlying the association between the "Big Five" personality traits in adults and those same adults' parenting of their offspring. Results showed overlapping genetic influences that explained the covariation between personality and parenting behavior—for example, in the link between parental openness to experience and overprotection, and the link between parental neuroticism and rejection of the child. In another study, Horwitz and colleagues (Horwitz et al. 2010) examined the covariation between parental aggressive attributes and several aspects of family relationships, including negativity, marital conflict, and global family conflict. A substantial proportion of the genetic variance overlapped between parental aggressiveness, marital conflict, parent–adolescent negativity, and global family conflict. The remaining covariance between these aspects of individual and family negativity was accounted for by nonshared environmental factors, indicating that some of the family-wide hostility was specific to particular parent–adolescent dyads in the family. Thus, the results from these studies of adult twins as parents point to passive and evocative GE correlation and nonshared environment processes. Furthermore, these findings are consistent with previously conducted twin and adoption studies of children's and adolescents' parenting environments described earlier in the chapter.

To sum up, behavioral genetic studies have shown that the qualities of the parent–adolescent relationship arise from bidirectional influences of each partner on the other, with parents and adolescents alike playing active roles in shaping the dyad's daily interactions and broader relationship. These bidirectional influences can be discerned from offspring and parent sibling designs, which have converged to implicate genetic and nongenetic influences that work together to differentiate siblings in the positive and negative qualities of their relationships with the same parents. The main limitation of the work completed to date is that the results are inconclusive in regard to precisely which type of GE correlation process is operating. The promise of future behavioral genetics research, particularly with the inclusion of molecular genetic biomarkers of DNA variation (described later in the chapter), is for greater specificity in regard to these GE processes in parent–adolescent relationships.

6.5 The Sibling Relationship

Much of what we know in regard to GE processes in family relationships is based on studies of the parent–adolescent relationship. There has not been nearly as much theoretical and empirical research on the sibling relationship, but it is also important in its shaping of individual's development in childhood, adolescence and adulthood (Cicirelli 1995). Around the world (though with some notable exceptions such as the "single child policy" in China), the vast majority of adolescents have at least one sibling, and a sizable minority having multiple siblings (Dunn 1992). In addition, most siblings maintain their relationships over their lifetime, which means that the sibling relationship is by far the most enduring relationship that most of us experience (Kramer and Bank 2005).

The importance of the sibling relationship in adolescent development can be seen in a number of interpersonal processes. Various qualities of the sibling relationship are associated with healthy developmental outcomes (e.g., high self-esteem, good friendships) or maladjustment, including drug and alcohol use as well as psychological difficulties such as mood disturbance and delinquency (Yeh and Lempers 2004; Rende et al. 2005). The sibling relationship literature has applied distinct frameworks including social relations, social contagion, social learning, and evolutionary theory models, to distinguish the causes and consequences of individual differences in sibling relationship qualities and their effects on adolescent outcomes (Manke and Plomin 1997; Rende et al. 2005; McHale et al. 2009; Pollet 2007). As with the literature on parent–adolescent relationships and adolescent outcomes described above, behavioral genetic methods have provided a way to distinguish genetic and nongenetic sources of variation that give rise to individual difference in sibling relationship qualities and adolescents' developmental outcomes. However, the sibling behavioral genetic literature typically examines not only the genetic and nongenetic sources of links between sibling relationship processes and youth outcomes, but also examines connections between multiple family relationships, such as sibling and mother-adolescent relationships.

To date, the most comprehensive longitudinal behavioral genetic study of sibling and parent–adolescent relationships has been the Nonshared Environment and Adolescent Development (NEAD) project in which six different types of siblings spanning genetically identical twins to genetically unrelated step-siblings were compared. Genetic and environmental sources of variance were identified for a variety of sibling relationship dimensions including positivity (e.g., warmth, rapport, mutual involvement in enjoyable activities) and negativity (e.g., anger, conflict), using multiple informants including youth, parents, and teachers, as well as observers' ratings based on videotaped family member interactions in the home. Although the findings varied to some degree as a function of construct, informant and method, overall the results pointed to modest to moderate heritability and nonshared environmental variance in measures of the sibling relationship, but with more substantial shared environmental variance found compared to what is typically found for measures of parent–adolescent relationships (Reiss et al. 2000). In another study (Pike and Atzaba-Poria (2003), genetic and nongenetic sources of variance differed depending on the construct in question. For sibling affection, shared and nonshared environmental variance were substantial, for sibling rivalry there was a strong effect of genetic factors, and for sibling hostility the variance was distributed across genetic, shared environmental and nonshared environmental sources of variance.

A distinct innovation in the sibling relationships literature has involved integration of the behavioral genetic design with Kenny and La Voie's (1984) Social Relations Model or SRM. The SRM is used to distinguish actor, partner, and dyad-level statistical effects within systems of social relationships, such as those in families. This is done by partitioning the variance and covariance between each individual's measured variables within the family members that are studied. Manke and Plomin (1997) used behavioral observation and interviews as part of a "round robin" family interaction study design involving twin siblings and their mothers. Using the SRM, the researchers estimated actor and partner statistical effects (i.e., variance attributed to each individual) and sibling dyad or relationship statistical effects (i.e., variance specific to the sibling dyadic relationship that cannot be accounted for by variance attributable to either individual alone) in four aspects of family interaction including warmth, conflict, disclosure of positive things, and disclosure of negative things. They found distinct actor, partner and sibling relationship effects depending on the construct. For instance, there were significant actor effects for conflict and positive disclosures, and sibling dyadic relationship effects for warmth and disclosure of both positive and negative things. More importantly however, behavioral genetic analyses also revealed distinct genetic and environmental influences on variance in actor, partner and dyad effects depending on the construct in question. For the actor effect for conflicts, genetic and shared environmental effects were prominent. In contrast, for the actor effect for self-disclosure of positive things, only environmental factors (shared and nonshared) accounted for variance—a pattern found for other aspects of the sibling relationship as well (e.g., warmth). More recently, Rasbash et al. (2011) used the SRM to examine twin and non-twin siblings' observed interactions with each other and both parents. They found high levels of negativity and low levels of positivity for adolescent actor effects. Furthermore, like Manke and Plomin, the researchers found significant genetic and shared environmental actor effects, particularly for negativity in the relationship. Taken together, the results from these behavioral genetic SRM studies provide some of the most concrete evidence to date of the ways in which genetically influenced attributes of individuals—construed as actor and partner effects in the SRM—operate through social interactions within enduring dyadic relationships in families.

Behavioral genetic studies provide a detailed differentiation of sources of variance in sibling relationship processes. They also can inform the sources of overlap between, and uniqueness of, the qualities of relationships in different family dyads spanning offspring and adults. In one behavioral genetic study of the similarities between mother–adolescent and sibling relationships, investigators found positive correlations between mother–adolescent and sibling relationship positivity and negativity (Bussell et al. 1999). Nearly all of this covariance across relationships was attributable to shared environmental influences, with genetic and non-shared environmental factors accounting for small amounts of covariation across relationship dyads. This finding points to the importance of family-level factors that, although not identified in the study, contribute to similarities in parent–adolescent and adolescent sibling relationships within each family—perhaps via direct and indirect socialization processes that contribute to an overall family climate that influences the family members' dyadic relationships in similar ways (Jenkins et al. 2012).

Scholarly interest in the links between sibling social relationships and adolescents' other relationships extends beyond the family, to include links to peers and friendships. Pike and Atzaba-Poria (2003) investigated the covariance between siblings' relationships with each other and with their friends, with emphasis on the role of temperament. They found moderate positive correlations between sibling and friendship positivity and negativity. Furthermore, some of this similarity between the sibling and friend dyads was attributable to individual differences in adolescents' temperament—specifically, emotionality with relationship negativity, and sociability and activity level with relationship positivity. These links between relationships through temperament were accounted for largely by nonshared environmental sources of covariance, suggesting a critical role of siblings' distinct experiences within the same family in regard to connections between their sibling and friend relationships.

As with the literature on parent–adolescent relationships, the research on adolescents' sibling relationships will benefit from greater specificity of GE correlation and interaction processes in future research, something that will be aided by the inclusion of molecular genetic information. To summarize the behavioral genetic literature to this point, the evidence provides a fairly clear picture about the roles of various components of genetic and nongenetic variance in explaining individual differences in sibling relationship qualities and connections to other family and non-family relationships. It also extends the study of relationship processes to a more precise level, by distinguishing different sources of variance underlying each specific relationship component and process, such as actor/partner/dyad effects (in the case of the SRM) or the overlap between sibling, parent–adolescent, and adolescent peer relationships. The interconnectedness of family relationships is even more apparent when considering mothers' and fathers' relationships with each other—the topic to which we turn next.

6.6 The Marital Relationship

The marital relationship between mothers and fathers has long been the focus of theoretical and empirical inquiry. Between-family differences in qualities of the marital relationship are related not only to the well-being of each spouse but also to the relationship qualities of parent–adolescent and sibling relationships in the family. Furthermore, the variability in marital relationship qualities is associated with a variety of socioeconomic indicators and social functioning in the workplace and community (Cooke and Gash 2010; Harkonen and Dronkers 2006; Tzeng and Mare 1995).

The behavioral genetic literature has indicated that genetic and nonshared environmental factors contribute to the formation (i.e., partner selection through assortative mating—selecting partners who are like ourselves), maintenance (i.e., marital satisfaction), and dissolution (i.e., conflict and divorce) of the marital relationship (Ulbricht and Neiderhiser 2009). Furthermore, studies have shown that the genetic and nonshared environmental influences on marital relationship processes, including the propensity to get married and divorced, are statistically mediated in part by individual differences in each spouse's stable personality attributes (Jocklin et al. 1996; Johnson et al. 2004; Spotts et al. 2005). The behavioral genetic findings suggest that it is through their influence on individuals' personal attributes, that genetic and nongenetic factors alike exert their influence on marital relationships.

As mentioned earlier, the marital relationship is important for other types of relationships within families. Its critical role is illustrated by setting the tone for family atmosphere, influencing parent–child relationship and the adjustment of the children (Deater-Deckard et al. 1999; McCloskey and Lichter 2003). One of the key questions addressed in the marital relationship literature has been the nature of the link between marital relationship quality and the overall family atmosphere that is experienced by each family member in the household, including children and adolescents. Are parents' individual attributes and behaviors associated with marital relationship qualities, as well as levels of family warmth and conflict? Furthermore, are these associations attributable to genetic and non-genetic influences? These questions were addressed in a large study of adult same-sex twins and their adolescent offspring (Ganiban et al. 2007). The investigators replicated previous studies of non-twin families, showing moderate-sized statistical associations between greater marital satisfaction and lower levels of family conflict (Erel and Burman 1995; Feldman et al. 1997). Using behavioral genetic analyses, Ganiban et al. reported that these associations could be attributed to overlapping sources of nonshared environmental variance. In addition, there was a consistent link between marital relationship quality and parent–adolescent relationship quality that was attributable to overlapping genetic variance. Thus, within immediate and extended family networks, there are substantial differences in each adolescent's experience, and this differential experience is due in part to genetically influenced parent attributes as well as nonshared environmental processes (including systematic measurement error)—mechanisms that connect marital conflict to a poorer family climate and greater parent–adolescent negativity and conflict as well. Furthermore, like the studies of adolescent GE correlation effects reviewed above, this study also demonstrated that part of the genetic influence that connects marital relationship quality and parent–adolescent relationship quality is mediated by parental personality characteristics such as aggressiveness and anxiousness (Ganiban et al. 2009).

Just as with studies of parent–adolescent and sibling relationships, behavioral genetic studies of marital relationships and adolescent functioning can elucidate the interplay between genetic and environmental factors. Using a sibling design, Feinberg et al. (2005) examined genetic and environmental variance in adolescents' antisocial behavior and depressive symptoms in distinct family environments that varied on multiple dimensions including marital and co-parenting conflict, as well as parental and sibling negativity. They found that the heritability and shared environmental variance in adolescent antisocial behavioral problems—but not depressive symptoms—varied as a function of the degree of marital and co-parenting conflict in the home. Genetic variance was strongest and shared environmental variance weakest in high conflict homes; the opposite was true (weakest heritability, strongest shared environmental variance) in low conflict homes. This might represent qualitatively distinct GE and family processes in these two contexts, whereby in more cohesive, low-conflict environments, caregiving is more stable and systematic across sibling adolescents which results in greater similarity associated with socializing effects. In contrast, in high conflict homes, adolescents' genetically influenced attributes may be more important to their behavioral problems because caregiving is less consistent and systematic across parent–child relationships. However, behavioral genetic designs like the one used in this study are inconclusive. For example, the finding might also reflect complex patterns of GE correlations between genetic variance in individuals' and dyads' behaviors that are very distinct from each other at high versus low levels of family harmony and conflict that alter GE interaction effects.

In sum, there have been several behavioral genetic studies that have attempted to connect marital and parent–adolescent relationship processes with each other and with adolescent developmental outcomes. However, the literature in this area is based on just a few samples and there have been no published replications of effects. This remains as an important gap in the adolescent development and family relations literature. The field needs more behavioral genetic studies with designs spanning offspring sibling, adult parent sibling, and parent-offspring samples—and ideally also incorporating molecular genetic markers of DNA variation—in order to more precisely differentiate evocative, active, and passive GE correlation mechanisms and to identify replicable GE interaction effects (Narusyte et al. 2008). Only then will we be able to understand the intergenerational transmission of family relationship qualities during adolescence, when the foundations for future romantic and marital relationships are being formed (Hines and Saudino 2002).

6.7 The Family as a System

Although we have chosen to distinguish parent–adolescent, sibling, and co-parent marital relationships, all three types of relationship exist within a broader family system in most families. Furthermore, each relationship operates not in isolation but in combination with the other, forming a distinct if not truly unique family and household context that has strong effects on quality of life and mental health, and ultimately, adolescents' developmental outcomes in the transition to adulthood (Cox and Paley 1997). Indirectly, this can be seen in the patterns of GE correlation and interaction that connect qualities of each of these three types of relationships with adolescents' outcomes, as reviewed in the current chapter. These findings provide an essential complement to traditional socialization studies that have examined the broader family social environment (e.g., global measures of warmth and acceptance, as well as discord and conflict) and its link with adolescent adjustment and maladjustment (Parke 2004).

Related to this theoretical concept of the family system, most of the behavioral genetic research on the topic has focused on two aspects of the general family climate: conflict and acceptance or closeness—often, measured using the Family Environment Scale or FES (Moos and Moos 1981). Although effect sizes vary across samples, the general pattern indicates genetic and nonshared environmental variance in retrospective and concurrent reports of individuals' perceptions of these broad measures of family relationship closeness and conflict across relationships within the household (Rowe 1983; Bouchard and McGue 1990; Jacobson and Rowe 1999; Jang et al. 2001). This pattern of results generally is consistent with the findings reviewed above in regard to specific family dyadic relationships (i.e., marital, sibling, parent–adolescent), further supporting the broad conclusion that genetically influenced attributes of adolescents and their parents play an important role in the warmth and conflict that adolescents experience.

So what accounts for the genetic variance in measures of family environments and dyadic relationships? The behavioral genetic literature suggests that this arises from genetic influences on individuals' personal attributes, ranging from their personality traits to potential behavioral and emotional problems—behaviors that influence, and are influenced by, qualities of family relationships (Krueger et al. 2008; O'Connor et al. 1998). The literature also converges on the critical role of the environment, given that the effect sizes representing genetic and GE correlation statistical effects that connect family relationship and adolescent outcome variables often are far from substantial in magnitude.

GE interaction processes also are found in the links between adolescent outcomes and parent–adolescent relationship factors. For instance, Zimmermann et al. (2009) found that the security of adolescent attachment with parents moderated the statistical effect of adolescents' serotonin transporter gene (5HTT) on mother-adolescent hostility. Specifically, among insecurely attached adolescents, those with the short version of 5HTT were higher in hostility compared to those with other forms of 5HTT. In contrast, 5HTT structure had no statistical associations for securely attached adolescents. This finding parallels results from a longitudinal adoption study (O'Connor et al. 2003) in which adoptive parents' separation and divorce moderated genetic risks (measured as biological parents' negative emotionality prior to the child's birth and subsequent adoption into another family) on adolescents' social competence outcomes. Specifically, genetic risk for poorer social competence was evident only among adopted adolescents whose adoptive parents had separated or divorced. Findings like these suggest that strong, cohesive marital and parent–adolescent relationships can reduce or even completely mitigate genetic risks for maladjustment in adolescence. Troubled parent–adolescent relationships can also differentiate adolescents' maladjustment trajectories within the same family over time through nonshared environmental mechanisms, even with genetic factors statistically controlled, as shown in longitudinal research on identical twin differences (Burt et al. 2006).

Research also has pointed to GE interplay in the link between the sibling relationship and adolescent outcomes. Slomkowski et al. (2005) investigated sibling relationship qualities and their associations with on adolescent nicotine use. They found genetic and shared environmental influences on smoking behavior. Furthermore, the shared environmental effects on smoking were larger for adolescents with close, socially connected sibling relationships. This seems consistent with theory and empirical studies showing the sibling relationship to be an important conduit for transmission and reinforcement of antisocial behavior and substance use in adolescence (Bullock and Dishion 2002; Rowe 1986).

In sum, the behavioral genetic literature on family environments and relationships as networks or systems points to complex transactions between family-level factors and individual/dyad-level factors. But much remains to be discovered— it has been just over a decade since Reiss et al. (2000) highlighted the necessity of distinguishing GE processes in whole systems of family relationships. In that time, the field has become better positioned to identify specific nonshared environmental and GE correlation and interaction processes. If these can be specified and replicated, it not only informs basic research regarding GE interplay in adolescent development, but also informs approaches to individual and family prevention and intervention efforts that can improve the lives of adolescents and their parents and siblings (Leve et al. 2010).

6.8 Family Relationships Research in the Era of Genomics

The greatest opportunity and challenge to family scientists who wish to understand GE interplay is the effective incorporation of molecular biology methods into longstanding traditional population genetics statistical methods for estimating genetic and environmental effects, in order to achieve a more detailed and complete understanding of family processes and adolescent development (Deater-Deckard 2011). For reasonable costs, molecular genetic methods now permit precise measurement of structural variation in DNA, and the use of atheoretical or theoretical and empirically derived "candidate gene" biomarkers of behaviors of interest.

As a result, family researchers can measure specific genes and related biological processes (e.g., stress reactivity in cortisol or vagal tone measures) that pertain to individual differences in parent–adolescent, sibling, and marital relationships and how these operate in different family contexts. To this point, the molecular genetics of human behavior has been dominated by research on candidate genes involved in neuromodulation of dopamine and serotonin (e.g., 5HTT, DRD2, DRD4, DAT1, COMT). Results have pointed to some evidence—sometimes with mixed results, due in part to small effect sizes—for molecular genetic and molecular GE interaction effects for parental negativity, reactivity and responsiveness and child and adolescent outcomes (Bakermans-Kranenburg and van IJzendoorn 2006; Voelker et al. 2009; Lahey et al. 2011). A useful recent illustration of this burgeoning literature is a study by Berry and colleagues in which a 7-repeat version or allele of the dopamine receptor 4 gene was associated with longitudinal growth over middle childhood in children's attention problems. However, this statistical association was mitigated for children with the 7-repeat allele who had warm sensitive maternal care in early childhood (Berry et al. 2013).

Although these studies are informative in the sense of getting a better understand of the underlying biological process and opening up the opportunity for pharmaceutical intervention (e.g., by controlling gene expression), they are limited in several aspects. Most of the studies have focused on parents, and few have examined genes that influence adolescent reactivity to the environment. In addition, nearly all of the studies are cross-sectional in nature, and with this sort of design it is not possible to detect the stability and plasticity in the effects of genes over time in the whole period of adolescence.

Furthermore, the genes of greatest interest to behavioral scientists are those that are related to the secretion and modulation of neurotransmitters, but the neurocognitive mechanisms by which these genes and neurotransmitters come to influence family relationships and adolescents' outcomes are not specified or understood. Other approaches, such as genome wide association analysis (GWAS), may be useful for establishing evidence of links between heritability at the population level using traditional behavioral genetic methods, and heritability at the molecular level measured as genetic variation in genes, i.e., alleles and single nucleotide polymorphisms or "SNPs". However, the powerfulness of the GWAS approach is restricted in terms of the limitation of association analysis (e.g., unreliability due to population stratification) and the small effect sizes of SNPs on quantitative traits.

6.9 Conclusion

In closing, family relationship processes are difficult to study in adolescence, because they are shaped by multiple factors and over the second decade of life the adolescent is acquiring more autonomy in her or his relationships. However, it is crucial for researchers to examine family relationships because the qualities of these relationships are so strongly related to the health and well-being of all family members as well as the community and society in a broader sense. Behavioral genetic approaches in combination with modern molecular biology techniques provide powerful tools to disentangle the complexity of these relationships as well as offering suggestions to construct a warm and caring home environment for adolescents. With better designs involving more types of families and a broader range of contemporary methods (e.g., molecular genetics, neuroimaging), the coming decade is promising. We will gain insights into the ways in which genes and environments work together to produce the wide variety of social relationships in adolescence, and how these relationships influence development in adolescence and adulthood.

References

Bakermans-Kranenburg, M., & van IJzendoorn, M. (2006). Gene-environment interaction of the dopamine D4 receptor (DRD4) and observed maternal insensitivity predicting externalizing behavior in preschoolers. *Developmental Psychobiology, 48*, 406–409.

Barber, B. K. (1992). Family, personality, and adolescent problem behaviors. *Journal of Marriage and the Family, 54*, 69–79.

Bell, R. Q. (1979). Parent, child and reciprocal influences. *American Psychologist, 34*, 821–826.

Belsky, J., Crnic, K., & Woodworth, S. (1995). Personality and parenting: exploring the mediational role of transient mood and daily hassles. *Journal of Personality, 63*, 905–931.

Berry, D., Deater-Deckard, K., McCartney, K., Wang, Z., & Petrill, S. A. (2013). Gene-environment interaction between DRD4 7-repeat VNTR and early maternal sensitivity predicts inattention trajectories across middle childhood. *Development and Psychopathology, 25*, 295–306.

Bornstein, M. H., & Bradley, R. H. (Eds.). (2003). *Socioeconomic status parenting, and child development*. Mahwah, NJ: Erlbaum.

Bouchard, T. J., & McGue, M. (1990). Genetic and rearing environmental influences on adult personality: An analysis of adopted twins reared apart. *Journal of Personality, 58*, 263–292.

Bronfenbrenner, U., & Evans, G. W. (2000). Developmental science in the 21st century: Emerging theoretical models, research designs, and empirical findings. *Social Development, 9*, 115–125.

Bullock, B. M., & Dishion, T. J. (2002). Sibling collusion and problem behavior in early adolescence: Toward a process model for family mutuality. *Journal of Abnormal Child Psychology, 30*, 143–153.

Burt, S. A., McGue, M., Iacono, W. G. & Krueger, R. F. (2006). Differential parent-child relationships and adolescent externalizing symptoms: Cross-lagged analyses within a monozygotic twin differences design.*Developmental Psychology, 42*, 1289–1298.

Bussell, D. A., Neiderhiser, J. M., Pike, A., Plomin, R., Simmens, S., Howe, G. W., et al. (1999). Adolescents' relationships to siblings and mothers: A multivariate genetic analysis. *Developmental Psychology, 35*, 1248–1259.

Ceballo, R., & McLoyd, V. C. (2002). Social support and parenting in poor, dangerous neighborhoods. *Child Development, 73*, 1310–1321.

Cicirelli, V. G. (1995). *Sibling relationships across the lifespan*. New York: Plenum.

Compas, B. E., Champion, J., Forehand, R., Cole, D., Reeslund, K. L., Fear, J., & Roberts, L. (2010). Coping and parenting: Mediators of 12-month outcomes of a family group cognitive-behavioral preventive intervention with families of depressed parents. *Journal of Consulting and Clinical Psychology, 78*, 623–634.

Cooke, L. P., & Gash, V. (2010). Wives' part-time employment and marital stability in Great Britain, West Germany and the United States. *Sociology, 4*, 1091–1108.

Cox, M. J., & Paley, B. (1997). Families as systems. *Annual Review of Psychology, 48*(1), 243–267.

Deater-Deckard, K. (2009). Parenting the genotype. In K. McCartney & R. Weinberg (Eds.), *Experience and development: A festschrift in honor of sandra wood scarr* (pp. 141–161). New York: Taylor and Francis.

Deater-Deckard, K. (2011). Families and genomes: The next generation. *Journal of Marriage and Family, 73*, 822–826.

Deater-Deckard, K., Fulker, D. W., & Plomin, R. (1999). A genetic study of the family environment in the transition to early adolescence. *Journal of Child Psychology and Psychiatry, 40*, 769–775.

Deater-Deckard, K., Mullineaux, P. Y., Beekman, C., Petrill, S. A., Schatschneider, C., & Thompson, L. A. (2009). Conduct problems, IQ, and household chaos: A longitudinal multi-informant study. *Journal of Child Psychology and Psychiatry, 50*, 1301–1308.

Demo, D. H., & Acock, A. C. (1996). Family structure, family process, and adolescent well-being. *Journal of Research on Adolescence, 6*, 457–488.

Dunn, J. (1992). Siblings and development. *Current Directions in Psychological Science, 1*, 6–9. doi:10.1111/1467-8721.ep10767741.

Dunn, J., & Plomin, R. (1990). *Separate lives: Why siblings are so different*. New York: Basic Books.

Elkins, I. J., McGue, M., & Iacono, W. G. (1997). Genetic and environmental influences on parent–son relationships: Evidence for increasing genetic influence during adolescence. *Developmental Psychology, 33*, 351–363.

Erel, O., & Burman, B. (1995). Interrelatedness of marital relations and parent-child relations: A meta-analytic review. *Psychological Bulletin, 118*, 108–132.

Feinberg, M. E., Reiss, D., Neiderhiser, J. M., & Hetherington, E. M. (2005). Differential association of family subsystem negativity on siblings' maladjustment: using behavior genetic methods to test process theory. *Journal of Family Psychology, 19*, 601–610.

Feldman, S. S., Fisher, L., & Seitel, L. (1997). The effect of parents' marital satisfaction on young adults' adaptation: A longitudinal study. *Journal of Research on Adolescence, 7*, 55–80.

Ganiban, J. M., Spotts, E. L., Lichtenstein, P., Khera, G. S., Reiss, D., & Neiderhiser, J. M. (2007). Can genetic factors explain the spillover of warmth and negativity across family relationships? *Twin Research Human Genetics, 10*, 299–313. doi:10.1375/twin.10.2.299.

Ganiban, J. M., Ulbricht, J. A., Spotts, E. L., Lichtenstein, P., Reiss, D., Hansson, K., & Neiderhiser, J. M. (2009). Understanding the role of personality in explaining associations between marital quality and parenting. *Journal of Family Psychology, 23*, 646–660. doi:10.1037/a0016091.

Ge, X., Conger, R. D., Cadoret, R. J., Neiderhiser, J. M., Yates, W., Troughton, E., et al. (1996). The developmental interface between nature and nurture: A mutual influence model of child antisocial behavior and parent behaviors. *Developmental Psychology, 32*, 574–589.

Harkonen, J., & Dronkers, J. (2006). Stability and change in the educational gradient of divorce: A comparison of seventeen countries. *European Sociological Review, 22*, 501–517.

Herndon, R. W., McGue, M., Krueger, R. F., & Iacono, W. G. (2005). Genetic and environmental influences on adolescents' perceptions of current family environment. *Behavior Genetics, 35*, 373–380.

Hines, D. A., & Saudino, K. J. (2002). Intergenerational transmission of intimate partner violence—A behavioral genetic perspective. *Trauma, Violence, and Abuse, 3*, 210–225.

Horwitz, B. N., Ganiban, J. M., Spotts, E. L., Lichtenstein, P., Reiss, D., & Neiderhiser, J. M. (2010). The role of aggressive personality and family relationships in explaining family conflict. *Journal of Family Psychology, 25*, 174–183.

Jocklin, V., McGue, M., & Lykken, D. T. (1996). Personality and divorce: A genetic analysis. *Journal of Personality and Social Psychology, 71*, 288–299.

Jacobson, K. C., & Rowe, D. C. (1999). Genetic and environmental influences on the relationships between family connectedness, school connectedness, and adolescent depressed mood: Sex differences. *Developmental Psychology, 35*(4), 926–939.

Jang, K. L., Vernon, P. A., Livesley, W. J., Stein, M. B., & Wolf, H. (2001). Intra- and extra-familial influences on alcohol and drug misuse: A twin study of gene-environment correlation. *Addiction, 96*, 1307–1318.

Jenkins, J., Rasbash, J., Leckie, G., Gass, K., & Dunn, J. (2012). The role of maternal factors in sibling relationship quality: A multi-level study of multiple dyads per family. *Journal of Child Psychology and Psychiatry, 53*, 622–629.

Jockin, V., McGue, M., & Lykken, D. T. (1996). Personality and divorce: A genetic analysis. *Journal of Personality and Social Psychology, 71*, 288–299.

Johnson, W., McGue, M., Krueger, R. F., & Bouchard, T. J, Jr. (2004). Marriage and personality: A genetic analysis. *Journal of Personality and Social Psychology, 86*(2), 285–294.

Kenny, D. A., & La Voie, L. (1984). The social relations model. In L. Berkowitz (Ed.), *Advances in experimental social psychology* (Vol. 18, pp. 142–182). Orlando, FL: Academic Press.

Kramer, L., & Bank, L. (2005). Sibling relationship contributions to individual and family well-being: Introduction to the special issue. *Journal of Family Psychology, 19*, 483–485.

Krueger, R. F., South, S., Johnson, W., & Iacono, W. (2008). The heritability of personality is not always 50 %: Interactions and correlations between personality and parenting. *Journal of Personality, 76*, 1485–1522.

Lahey, B. B., Rathouz, P. J., Lee, S. S., Chronis-Tuscano, A., Pelham, W. E., et al. (2011). Interactions between early parenting and a polymorphism of the child's dopamine transporter gene in predicting future child conduct disorder symptoms. *Journal of Abnormal Psychology, 120*, 33–45.

Lanthier, R. P. (2007). Personality traits and sibling relationships in emerging adults. *Psychological Reports, 100*, 672–674.

Leve, L., Harold, G., Ge, X., Neiderhiser, J., & Patterson, G. (2010). Refining intervention targets in family-based research: Lessons from quantitative behavior genetics. *Perspectives on Psychological Science, 5*, 515–526.

Maccoby, E. E. (1992). The role of parents in socialization of children: An historical overview. *Developmental Psychology, 28*, 1006–1017.

Manke, B., & Plomin, R. (1997). Adolescent familial interactions: A genetic extension of the social relations model. *Journal of Social and Personal Relationships, 14*(4), 505–522.

Mason, C. A., Cauce, A. M., Gonzales, N., Hiraga, Y., & Grove, K. (1994). An ecological model of externalizing behaviors in African American adolescents: No family is an island. *Journal of Research on Adolescence, 4*, 639–655.

McCloskey, L. A., & Lichter, E. L. (2003). The contribution of marital violence to adolescent aggression acts across different relationships. *Journal of Interpersonal Violence, 18*, 390–412.

McGue, M., Elkins, I., Walden, B., & Iacono, W. (2005). Perceptions of the parent–adolescent relationship: A longitudinal investigation. *Developmental Psychology, 41*, 971–984.

McHale, S. M., Bissell, J., & Kim, J. (2009). Sibling relationship, family, and genetic factors in sibling similarity in sexual risk. *Journal of Family Psychology, 23*(4), 562–572.

McLoyd, V. C. (1990). The impact of economic hardship on black families and children: Psychological distress, parenting, and socioemotional development. *Child Development, 61*, 311–346.

McLoyd, V. C. (1998). Socioeconomic disadvantage and child development. *American Psychologist, 53*, 185–204.

Moos, R. H., & Moos, B. S. (1981). *Family environment scale*. Palo Alto, CA: Consulting Psychologists Press.

Narusyte, J., Neiderhiser, J. M., D'Onofrio, B. M., Reiss, D., Spotts, E. L., Ganiban, J., & Lichtenstein, P. (2008). Testing different types of genotype-environment correlation: an extended children-of-twins model. *Developmental Psychology, 44*, 1591–1603. doi:10.1037/a0013911.

Neiderhiser, J. M., Reiss, D., Lichtenstein, P., Spotts, E. L., & Ganiban, J. (2007). Father-adolescent relationships and the role of genotype-environment correlation. *Journal of Family Psychology, 21*, 560–571.

Neiderhiser, J. M., Reiss, D., Pedersen, N. L., Lichtenstein, P., Spotts, E. L., Hansson, K., et al. (2004). Genetic and environmental influences on mothering of adolescents: A comparison of two samples. *Developmental Psychology, 40*, 335–351.

O'Connor, T. G., Caspi, A., DeFries, J. C., & Plomin, R. (2003). Genotype-environment interactions in children's adjustment to parental separation. *Journal of Child Psychology and Psychiatry, 44*, 849–856.

O'Connor, T. G., Deater-Deckard, K., Fulker, D. W., Rutter, M., & Plomin, R. (1998). Gene-environment correlations in late childhood and early adolescence. *Developmental Psychology, 34*, 970–981.

Parke, R. D. (2004). Development in the family. *Annual Review of Psychology, 55*, 365–399.

Pike, A., & Atzaba-Poria, N. (2003). Do sibling and friend relationships share the same temperamental origins? A twin study. *Journal of Child Psychology and Psychiatry, 44*, 598–611.

Pinderhughes, E. E., Nix, R., Foster, E. M., Jones, D., & The Conduct Problems Prevention Research Group. (2001). Parenting in context: Impact of neighborhood poverty, residential stability, public services, social networks, and danger on parental behaviors. *Journal of Marriage and the Family, 63*, 941–953.

Plomin, R. (1994). *Genetic and experience: The interplay between nature and nurture*. Thousand Oaks, CA: Sage.

Plomin, R., DeFries, J. C., & Loehlin, J. C. (1977). Genotype-environment interaction and correlation in the analysis of human behavior. *Psychological Bulletin, 88*, 245–258.

Pollet, T. V. (2007). Genetic relatedness and sibling relationship characteristics in a modern society. *Evolution and Human Behaviour, 28*, 176–185.

Rasbash, J., Jenkins, J., O'Connor, T. G., Tackett, J., & Reiss, D. (2011). A social relations model of observed family negativity and positivity using a genetically informative sample. *Journal of Personality and Social Psychology, 100*, 474–491.

Reiss, D. (2005). The interplay between genotypes and family relationships. *Current Directions in Psychological Science, 14*, 139–143.

Reiss, D., Neiderhiser, J. M., Hertherington, E. M., & Plomin, R. (2000). *The relationship code: Deciphering genetic and social influences on adolescent development*. Cambridge, MA: Harvard University Press.

Rende, R., Slomkowski, C., Lloyd-Richardson, E., & Niaura, R. (2005). Sibling effects on substance use in adolescence: Social contagion and genetic relatedness. *Journal of Family Psychology, 19*, 611–618.

Rowe, D. C. (1983). A biometrical analysis of perceptions of family environment: A study of twins and singleton sibling kinships. *Child Development, 54*, 416–423.

Rowe, D. C. (1986). Genetic and environmental components of antisocial behavior: A study of 265 twin pairs. *Criminology, 24*, 513–532.

Scarr, S., & McCartney, K. (1983). How people make their own environments: A theory of genotype → environment effects. *Child Development, 54*, 424–435.

Slomkowski, C., Rende, R., Novak, S., Lloyd-Richardson, E., & Niaura, R. (2005). Sibling effects on smoking in adolescence: Evidence for social influence from a genetically informative design. *Addiction, 100*, 430–438.

South, S. C., Krueger, R. F., Johnson, W., & Iacono, W. G. (2008). Adolescent personality moderates genetic and environmental influences on relationships with parents. *Journal of Personality and Social Psychology, 94*, 899–912.

Spinath, F. M., & O'Connor, T. G. (2003). A behavioral genetic study of the overlap between personality and parenting. *Journal of Personality, 71*, 785–808.

Spotts, E. L., Lichtenstein, P., Pedersen, N., Neiderhiser, J. M., Hansson, K., Cederblad, M., et al. (2005). Personality and marital satisfaction: A behavioral genetic analysis. *European Journal of Personality, 19*, 205–227.

Tzeng, J. M., & Mare, R. D. (1995). Labor market and socioeconomic effects on marital stability. *Social Science Research, 24*, 329–351.

Ulbricht, J. A., & Neiderhiser, J. M. (2009). Genotype–environment correlation and family relationships. In Y.-K. Kim (Ed.), *Handbook of behavioral genetics* (pp. 209–221). New York, NY: Springer.

Voelker, P., Sheese, B. E., Rothbart, M. K., & Posner, M. I. (2009). Variations in catechol-o-methyltransferase gene interact with parenting to influence attention in early development. *Neuroscience, 164*, 121–130.

Yeh, H. C., & Lempers, J. D. (2004). Perceived sibling relationships and adolescent development. *Journal of Youth and Adolescence, 33*, 133–147.

Zimmermann, P., Mohr, C., & Spangler, G. (2009). Genetic and attachment influences on adolescents' regulation of autonomy and aggressiveness. *Journal of Child Psychology and Psychiatry, 50*, 1339–1347.

Chapter 7
Toward a Developmentally Sensitive and Genetically Informed Perspective on Popularity

S. Alexandra Burt and M. Brent Donnellan

7.1 Toward a Developmentally Sensitive and Genetically Informed Perspective on Popularity

Popularity is a multifaceted construct that captures likeability among one's peers, number of friends, and how well an individual gets along with others (for a review see Rubin et al. 2006). In developmental studies, popularity is often operationalized using sociometric procedures such as likeability ratings or the frequency with which an individual is chosen as a desired friend by peers (Brown 2004). In general, popular children are seen as sociable and cooperative and are described as the kinds of children who "have the social abilities to achieve interpersonal goals" (Newcomb et al. 1993). Popularity should also be conceptually distinguished from related constructs, namely social status. Social status is typically defined as one's prominence, respect, and influence within social groups. Social status thus has implications for access to group resources, but does not index likeability per se (Anderson et al. 2001). Despite these differences, popularity and social status are related constructs as they are moderately and positively correlated (e.g., rs in the 0.40–0.50 range), particularly in adolescent boys (Becker and Luthar 2007; Mayeux et al. 2008). This overlap is perhaps not surprising, as one might expect popular individuals to also have a higher social status. Likewise, it is often argued that many of the personal attributes that promote sociometric popularity such as sociability, social engagement, and cognitive ability also promote other peer-related variables, at least in children (e.g., Newcomb et al. 1993).

S.A. Burt (✉) · M.B. Donnellan
Department of Psychology, Michigan State University,
107D Psychology Building, East Lansing MI 48824, USA
e-mail: burts@msu.edu

© Springer Science+Business Media New York 2015
B.N. Horwitz and J.M. Neiderhiser (eds.), *Gene-Environment Interplay in Interpersonal Relationships across the Lifespan*,
Advances in Behavior Genetics 3, DOI 10.1007/978-1-4939-2923-8_7

The gist is that the personality attributes that facilitate mutually enjoyable social interactions are likely to contribute to a range of peer variables (Newcomb et al. 1993). Accordingly, although the current review focuses on popularity, research on social status will also be reviewed as needed.

A lack of popularity with one's peers (particularly during early adolescence) has been tied to the development of internalizing and externalizing symptoms/ behaviors (Dodge et al. 2003; Ellis and Zarbatany 2007; Lansford et al. 2007), and has been identified as a risk factor for academic problems (reviewed in Rubin et al. 2006). Popularity in these studies was indexed by peer nominations and peer reports of social networks. Moreover, this association may be causal to some extent, as experimentally manipulated peer rejection has been found to increase negative affect (Nesdale and Lambert 2007), a core component of both personality and internalizing and externalizing psychopathology (Krueger 1999). Indeed, experimental research in social psychology suggests that peer rejection is causally related to lower self-esteem, negative mood states, and in some paradigms, aggression (Gerber and Wheeler 2009).

Uncovering the origins of popularity may represent a step (and perhaps a key step) for understanding significant developmental influences on personality and psychopathology more generally. Genes represent one possible etiological source of popularity, particularly as genetic differences between people account for a significant portion of variance in virtually all human characteristics (e.g., Turkheimer 2000). It would thus be extremely surprising if genetic factors did not influence popularity, at least to some extent. Classic work by Dodge (1983; see also Coie and Kupersmidt 1983) provides tentative support for this possibility, in that pre-existing behavioral tendencies (particularly disruptiveness and antagonism, which are themselves heritable) were found to predict peer rejection in newly formed play groups. There is also compelling evidence that temperamental characteristics evident in early childhood predict adult interpersonal functioning (Newman et al. 1997). Thus, the connection between attributes of individuals and elements of peer relationships are likely to involve bi-directional processes (Rubin et al. 2006). In line with this emerging transactional perspective, the current chapter considers peer popularity from a genetically-informed developmental perspective.

But how exactly would one's genes influence his or her popularity? After all, an individual's genes cannot directly code for the reactions of others. As alluded to above, the impact of biology on peer popularity should be mediated by specific behavioral tendencies or attributes (e.g., attractiveness, personality) that evoke more or less consistent responses from others (i.e., genes → individual attributes/ behaviors → social consequences). This sort of process is referred to in theoretical discussions as an evocative gene-environment correlation (rGE) (Plomin et al. 1977; Scarr and McCartney 1983) and it is thought to represent a fundamental form of interplay between genes and environmental experiences (Jaffee and Price 2007). The essential idea is that biological factors have behavioral manifestations which themselves relate to popularity.

7.2 The Overarching Role of Development

Conceptualizing popular children as those with the ability to achieve developmentally appropriate interpersonal goals within the context of peer groups provides a useful starting point for thinking about the etiology of popularity at different stages of development. In particular, there are known developmental changes in the function and importance of peer relationships (Brown 2004; Rubin et al. 2006), and these changes occur alongside changes in interpersonal goals (e.g., Carstensen et al. 1999). Building on this point, Rubin et al. (2006) suggested that popularity in the peer group is "driven by conformity to or deviation from *behavioral* norms" (p. 612). In other words, because interpersonal goals and behavioral norms shift with development, there is thus the distinct possibility that the correlates of popularity might also shift with development.

A recent experimental study (Gardner and Steinberg 2005) nicely illustrates the role of developmentally different behavioral norms as they relate to peer groups. Gardner and Steinberg (2005) examined 306 participants in three age groups: 13–16 (i.e., adolescence), 18–22 (i.e., emerging adulthood), and 24 and older (i.e. adulthood). Participants completed behavioral assessments of risk-taking, risk preferences, and risky decision-making, and did so either alone or with two same-aged peers. Younger participants took more risks, made riskier decisions, and focused on the benefits of risky behavior (as opposed to the costs) as compared to adult participants ($r_{effect\ size} \geq 0.25$). More importantly, however, these age differences in risky behaviors were far more pronounced when completing the tasks in peer groups than when alone (i.e., peer effects were largely specific to participants in adolescence and emerging adulthood, and were far less pronounced for those ages 24 or older). Such results are consistent with the notion that behavioral norms and peer effects change with development. In short, we would argue that it is essential to conceptualize associations between popularity and specific personality and behavioral correlates within a developmental framework. By implication, specific genetic contributions to popularity might change with development. We now turn to a brief review of what is known about the biological underpinnings of popularity/social status.

7.3 Biological Underpinnings of Popularity

Previous experimental work in primates has suggested that the serotonin system plays a causal role in social status.[1] Raleigh et al. (1991) created settings of uncertain social status by removing dominant males from troops of vervet monkeys

---

[1] Of note, we refer to social status literature for animal studies as likeability cannot, to our knowledge, be studied in non-human animals.

(Cercopithecus aethiops sabaeus) (Raleigh et al. 1991). They then treated one of the two remaining males with either serotonergic-enhancing or serotonergic-reducing drugs. In every case, the male treated with the serotonergic-enhancing drug became dominant, and did so by increasing a constellation of socially affiliative behaviors that in turn led to higher social status. Such results are clearly consistent with the premise that social status has some biological origins.

Importantly, these results have since been paralleled in humans (Knutson et al. 1998). Knutson et al. (1998) administered a selective serotonin reuptake inhibitor (SSRI; a common antidepressant) or a placebo to normal adult volunteers in a randomized, double-blind design for 4 weeks. They found that socially affiliative behavior, as measured by observer-rated behavioral indices in a task-controlled setting, increased with SSRI administration and was correlated 0.65 with plasma SSRI levels. Other pharmacological studies have since replicated and extended these results (Moskowitz et al. 2001; Tse and Bond 2002). Given that socially affiliative behaviors should be linked to likeability, these results circumstantially imply that popularity may be, at least in part, be related to the serotonin system.

Consistent with the above speculation, a polymorphism within the $5HT_{2A}$ serotonin receptor gene, $-G1438A$, has since been associated with experimentally-derived estimates of popularity in late-adolescent males, an effect that replicated across two independent samples (Burt 2008). Analyses revealed that those with the G-allele were significantly more popular than those with the A-allele, and that $-$ G1438A accounted for up to 8 % of the variance in popularity. These findings thus imply that particular polymorphisms lined with the serotonin system predispose individuals to particular social experiences.

7.4 Current Issues

There are two key caveats to this conclusion, however. First, all of the above studies (including the primate study) examined social outcomes in adolescent/adult samples. Because individuals exert an increasingly greater influence on their choice of peers and activities as part of normal human development (Scarr and McCartney 1983), it remains unclear the extent to which these conclusions also apply prior to adolescence. Indeed, we know of only one (Brendgen et al. 2009) genetically-informed sociometric study of peer acceptance/rejection in children (as defined via peer nominations of those they most liked to play with and those they least liked to play with). Brendgen et al. (2009) collected peer nominations in 336 twin pairs. Univariate analyses of peer acceptance and rejection revealed that 54 % of the variance was genetic in origin, results that are consistent with those obtained in adolescent/adult samples. Nevertheless, more work is needed before any firm conclusions can be drawn about the heritability of sociometric status.

Second, although conclusions regarding genetic influences on popularity are certainly provocative, they must remain speculative until the full pathway is illuminated.

Put differently, there is a critical need to identify interpersonal correlates of popularity, and to evaluate whether any of these correlates mediate the association between −G1438A and popularity. Each of these fundamental issues will be examined in turn, using new empirical data when necessary. We begin with an empirical examination of the etiology of popular peer affiliation, a key component of popularity, in a large sample of child twins.

7.5 Part 1: Additional Evidence of Genetic Influences on Popular Peer Affiliation in Children

7.5.1 Method

7.5.1.1 Participants

The Michigan State University Twin Registry (MSUTR) includes several independent twin projects (Klump and Burt 2006). The 587 families examined here (containing 1171 twins) were assessed as part of the on-going Twin Study of Behavioral and Emotional Development in Children (TBED-C) within the MSUTR. The TBED-C includes both a population-based sample (current $n = 442$ families) and an independent at-risk sample for which inclusion criteria also specified that participating twin families lived in moderately to severely disadvantaged neighborhoods (current $n = 145$ families). Conclusions were the same with and without the at-risk sample and thus these families were retained for analysis. Recruitment procedures, response rates, and participation rates are detailed elsewhere (Burt and Klump 2013; Klump and Burt, 2006).

Children gave informed assent, while parents gave informed consent for themselves and their children. The twins were 48 % female and ranged from in age 6 to 10 years [mean age (SD) = 8.3 years (1.49)], although some had turned 11 by the time they actually participated. Zygosity was established using a physical similarity questionnaire administered to the twins' primary caregiver (Peeters et al. 1998). This questionnaire has accuracy rates of 95 % or better (Peeters et al. 1998).

7.5.1.2 Popular Peer Affiliation

Mothers provided ratings for each of their twins' entire peer groups using the Friends Inventory (Burt 2009b; Burt and Klump in press; Burt et al 2009; Walden et al. 2004), with items scored using a 4-choice response format (1 = *none of my child's friends are like that,* 2 = *just a few of my child's friends are like that,* 3 = *most of my child's friends are like that,* and 4 = *all of my child's friends are like that*). Items were summed to index affiliation with popular peers, a key component of popularity (i.e., 4 items, including "My child's friends are popular with other kids" and "Other kids look up to my child's friends"; $\alpha = 0.724$). Data were

available for 98.9 % of twins. As expected, maternal ratings of popular peer affiliation were correlated with individual items on the CBCL tapping specific social
outcomes for the twins themselves, including "number of friends" ($r = 0.27$,
$p < 0.001$), "not liked by other kids" ($r = -0.20$, $p < 0.001$), "gets teased a lot"
($r = -0.16$, $p < 0.001$), and "gets along with peers" ($r = 0.16$, $p < 0.001$).

7.5.1.3 Analyses

Twin studies make use of the difference in the proportion of genes shared between
monozygotic (MZ; share 100 % of their segregating genetic material) and dizygotic (DZ; share an average of 50 % of their segregating genetic material) twins to
estimate the relative contributions of additive genetic (a^2), shared environmental
(i.e., factors that make twins similar to each other; c^2), and non-shared environmental effects plus measurement error (i.e., factors that make twins different from
each other; e^2) to the variance within observed behaviors or characteristics (phenotypes). To evaluate the possibility of genetic influences on popular peer affiliation,
we ran a simply univariate model, decomposing the phenotypic variance in popular peer affiliation into a^2, c^2, and e^2 variance components. More information on
twin studies is provided elsewhere (Plomin et al. 2008).

7.5.2 Results

7.5.2.1 Descriptives

The affiliation with popular peers variable was approximately normally-distributed
in these data (range = 4–16, mean = 11.01, SD = 1.84). Moreover, mean levels of
affiliation with popular peers did not vary across sex (mean (SD) = 11.0 (1.9) and
11.1 (1.8) for boys and girls, respectively). The MZ intraclass correlation was 0.75
and the DZ intraclass correlation was 0.58, results which are indicative of both
genetic and shred environmental influences on popular peer affiliation.

   ACE model fit results. There was no evidence that ACE estimates varied by sex
(the no-sex-differences model provided a better fit to the data (AIC = 2050.26)
than did the sex-differences model (AIC = 2051.39). Shared and non-shared environmental influences both made significant contributions to popular peer affiliation (accounting for 32 and 22 % of the variance, respectively). However, genetic
influences accounted for the largest proportion of variance in popular peer affiliation (46 % of the variance). Such findings are at least consistent with the argument
that genetic influences on popularity extend to childhood, thereby suggesting that
the biological underpinnings of popularity precede the niche-picking thought to
characterize adolescence and emerging adulthood.

7.6 Part 2: What Types of Genetically-Influences Traits/Behaviors Might Be Related to Popularity?

As we noted, popularity is *conferred upon* the individual by others. In keeping with this proposal, our underlying model is that an individual's attributes and stable patterns of behavior evoke relatively consistent reactions from others (i.e., genes → behaviors → social consequences). Accordingly, there is a need to evaluate the specific interpersonal correlates of popularity, and to test whether any of these correlates mediate the association between −G1438A and popularity.

To date, popularity has been associated with a range of attributes and behaviors such as physical attractiveness (Becker and Luther 2007), perceived athletic ability (Becker and Luthar 2007), and general social skills (Newcomb et al. 1993). In adolescent males, risk-taking or rule-breaking behaviors (RB) also appear to confer increased likeability (Allen et al. 2005; Luthar and D'Avanzo 1999). RB is defined here to include "covert" antisocial/illegal behaviors such as licit and illicit substance use, risky driving behaviors, vandalism, and stealing. Some readers may be initially skeptical that behaviors typically conceptualized as antisocial could lead to such a positive social outcome given that antisocial behaviors are presumed to typically harm others. However, it is important to recognize that antisocial behaviors cover a broad range of activities. Early sociometric studies indicating that antisocial behavior was related to low levels of popularity primarily examined behaviors that seemed to "interfere with the actions and goals of others" (Rubin et al., 2006, p. 609). Consistent with this caveat, RB is explicitly defined to exclude the more serious "overt" antisocial behaviors like physical aggression, intimidation, or violence against others (Burt 2012); in other words, RB does not include antisocial actions aimed at harming peers. Moreover, RB is thought to be a preferred social activity, particularly among adolescent males (Mayeux et al. 2008; Moffitt 1993). This more nuanced conceptualization of adolescent RB is also supported by the ubiquity of RB among adolescent males; only 7 % of 18-year-old boys deny all forms of RB (as discussed in Moffitt 1993). In short, RB may be better conceptualized as a normal part of teenage life, such that it is in fact statistically aberrant for adolescent boys to completely *refrain* from RB. Nonetheless, RB activities are antisocial to the extent that they often violate societal norms and ideals.

In her seminal theory on the development of antisocial behavior, Moffitt (1993) made sense of this normalization of otherwise pathological behavior. Essentially, she theorized that there were two primary subtypes of antisocial individuals that could be distinguished by age-of-onset and the presence or absence of physical aggression (Moffitt 1993). Child-onset or life-course-persistent antisocial behavior was postulated to represent a severe and often violent condition that begins early in life and persists throughout the lifespan. These individuals often engage in high levels of physical aggression with little regard for the welfare of others (a pattern of behavior likely to alienate peers, as noted above), and are often incarcerated or addicted to substances as adults. Adolescent-onset or adolescence-limited antisocial behavior, by contrast, was postulated to represent a normative increase in non-violent, rule-breaking delinquency and substance use (i.e., RB as defined in the current chapter) that largely dissipates as they incorporate themselves into adult roles. Moreover, unlike life-course-persistent antisocial behavior, adolescence-limited antisocial behavior was thought to result (at least in part) from "peer mimicry" or the modeling of peer RB (Bot et al. 2007; Moffitt 1993).

Moffitt (1993) further argued that adolescent-onset RB served an adaptive purpose from the perspective of contemporary teens: it provides them access to some of the freedoms and privileges of adult life. As an example, one study found that, among high school students, 61 % of marijuana users were sexually experienced (a highly valued outcome by many adolescent males) as compared to only 18 % of non-users (Jessor 1991). Thus, adolescent RB can be reframed as a circumscribed response to the particular developmental challenges of adolescence given that adolescents are reproductively mature but do not fulfill adult roles in society (i.e., they experience a so-called maturity gap; Moffitt 1993).

Within this broader developmental conceptualization, the proposed association between RB and increased popularity makes a good deal more sense. Delinquency, substance use, and rebellion against authority are sometimes admired and rewarded among adolescent peer groups and, as a result, are generally linked to increased popularity (Allen et al. 2005; Luthar and D'Avanzo 1999; Moffitt 1993; Sussman et al. 2007; Valenteet al. 2005), though not always (see Becker and Luthar 2007). For example, Allen et al. (2005) found that alcohol/substance use and "minor" delinquency (generally akin to the current definition of RB) were respectively correlated 0.22 and 0.25 with sociometric popularity (i.e., adolescents were asked to nominate 10 peers with whom they would most like to spend a Saturday night). In contrast, those who follow adult rules are typically viewed as socially undesirable by their peers (Allen et al. 1989; Bukowski et al. 2000). In short, we suspect that some amount of rule-breaking might be related to popularity in adolescence and emerging adulthood, especially for males.

More recent work (Burt 2009a) confirmed these impressions. Burt (2009a) examined associations between RB and popularity, as measured after interactions with previously unacquainted peers, in two independent college-aged samples of males. RB was moderately and positively associated with rated popularity in both samples ($r = 0.22$ and 0.25 in samples 1 and 2, respectively; both $ps < 0.05$), such that the higher an individual's rule-breaking, the more likely he was to be popular with his peers. To better understand the association between RB and popularity, participants were divided into popularity quartiles (bottom 25 % = least popular, top 25 % = most popular) and then mean levels of RB were computed for each group. As seen in Fig. 7.1, the most popular participants reported the highest levels of RB, whereas the least popular participants reported the lowest levels of RB. For those with less extreme popularity estimates (i.e., the middle 50 % of participants), however, there appeared to be little association between popularity and RB (both the "less popular" and the "more popular" participants exhibited average levels of RB). In sum, prior research convincingly demonstrates that, at least in adolescent males, RB appears to constitute an important and socially advantageous set of behaviors.

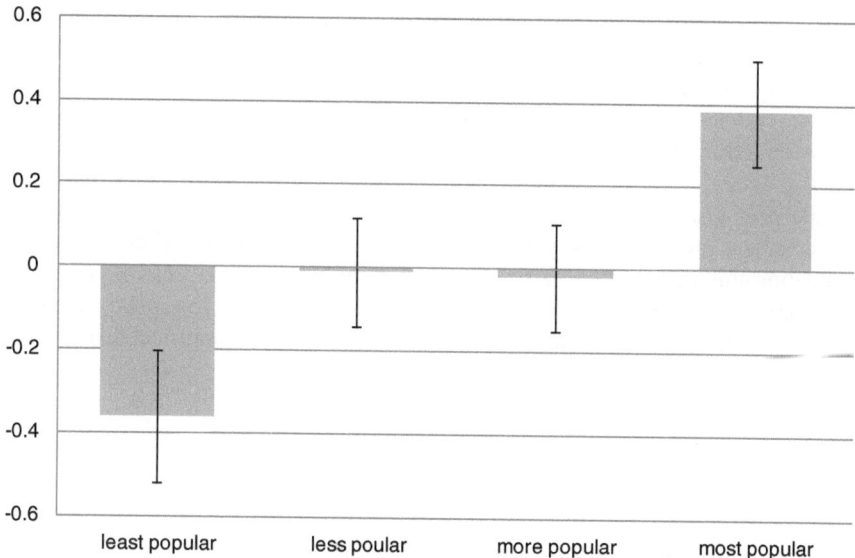

Fig. 7.1 The association between rule-breaking and popularity, collapsed across samples. *Note* Figure reproduced from Burt (2009a). Popularity was divided into quartiles. Mean levels (and standard errors) of RB (standardized to have a mean of zero and a standard deviation of 1.0) were then plotted. The most popular males thus nearly 0.4 standard deviations above the mean in their RB in these data

A subsequent set of analyses further evaluated whether RB mediated the afore-mentioned link between −G1438A and popularity. Mediation analyses suggested that roughly 20 % of this association could indeed be accounted for by a common association with RB (Burt 2009a). Such findings are fully consistent with the concept of evocative rGE, implying that particular genes predispose individuals to exhibit particular behaviors and to reap the social consequences of those behaviors.

7.7 Do the Above Associations in Males also Extend to Females?

An intriguing finding from studies (Gardner and Steinberg 2005) examining peer groups and risky decision making may involve gender differences. Namely, males gave significantly more weight to the likely benefits of risky decisions than did females ($r_{effect\ size} \geq 0.26$; although this association appeared to vary with age, such that it was specifically driven by participants in adolescence/emerging adulthood) (Gardner and Steinberg 2005). Such findings make sense in light of the fact that males typically engage in far more risk-taking than do like-aged females (Burt et al. 2007; Moffitt 2003).

In short, interpersonal goals and behavioral norms may vary as function of the gender composition of a given peer group. Thus, it is important to consider the possibility of gender differences in the correlates of popularity, especially when studying adolescent/emerging adult samples. For example, issues of dominance and even intrasexual competition might play a larger role in all-male peer groups whereas issues of mutual cooperation and closeness might play a larger role in all-female peer groups. It seems reasonable to hypothesize that risk taking may be a less salient predictor of popularity for females. As we know of no studies yet evaluating this possibility, we sought to examine whether the links between RB and popularity in male adolescents/emerging adults extended to females using as-yet unpublished data. The method and results of these new empirical analyses are described below.

7.7.1 Method

7.7.1.1 Participants

The sample of female-female peer groups included 86 undergraduate females assessed in 21 groups. The sample of male-male peer groups included 132 undergraduate males assessed in 32 groups, assessed using identical recruitment procedures (the male sample was examined in Burt (2008), (2009a)[2]; the second sample examined in Burtstudies was ultimately not examined here, as its eligibility criteria differed from those of the female-female groups). Participants were representative of the ethnic composition of the area; 79 % were Caucasian, 7 % were African–American, 5 % were Asian, and 9 % were of other ethnicities. Participants were in the developmental period of late adolescence/emerging adulthood (average age = 19 years).

7.7.1.2 Procedure

Up to ten individuals participated in each experimental session, separated into two, roughly equivalent, groups of up to five individuals. Prior to assigning groups, participants were explicitly asked whether they were acquainted with the other participants in any way. If so, those individuals were placed in separate groups. All groups were composed of participants previously unknown to each other.

---

[2]As we were preparing for this chapter, we conducted additional analyses on the SRM data collected as part of Burt (2008, 2009a, b), adding a small sample of adolescent females to identify possible gender differences. In preparing these files, the first author found that a single case was mistakenly omitted from the replication sample in her previously published analyses. Fortunately, the results are virtually identical with this additional case. The correlation between 5HT2A and −G1438A and popularity is now 0.20 ($p < 0.05$, one-tailed) rather than 0.21, and RB accounts for 22 % of this association rather than 21 %. The prior conclusions are thus entirely unchanged.

Participants in group "A" were then placed in individual rooms to complete a series of questionnaires. They then returned to the group room, where they collectively planned two parties and completed a brief series of brain teasers (i.e., anagrams, puzzles, etc.). They were told that the purpose of their interaction was to study "group dynamics", and were instructed to "be creative and have fun". They were not told they would later complete a sociometric questionnaire.

After interacting for 45–60 min, participants were returned to their individual rooms, where they immediately completed a sociometric scale in which they both ranked (on a 1–4 scale, with 1 meaning "I liked this person best") and rated (on a 1–10 scale, with 1 meaning "strongly dislike relative to other people I know") each of the other group members. Rankings were reverse-scored prior to analysis to maintain consistency with the ratings. Group "B" simultaneously completed the same activities, but did so in "reverse" order (i.e., they completed the group interaction and then completed the questionnaires). There were no differences in sociometric or questionnaire data based on the ordering of the tasks.

7.7.1.3 Popularity

The Social Relations Model (Kenny 1994) is a two-way random-effects model that partitions the variance in sociometric/dyadic data into individual-level and dyad-level components to allow researchers to unearth the source(s) of variation in interpersonal perception (e.g., how much does Kelly like Susan?). The first of these components is the actor effect, which indexes the general tendency of a person to like other group members. For example, how much Kelly likes Susan may be, in part, a function of how much Kelly tends to like people in general. By contrast, the partner effect indexes the general tendency of each person to be liked by other group members. Thus, how much Kelly likes Susan may also be a function of how likeable Susan is generally perceived to be. The final source of variance is the relationship effect, which is that part of the dyadic score unique to each dyad after the actor and partner effects are removed. Because this study involved only a single assessment of dyadic data, measurement error is also contained within the relationship effect.

Both actor and partner effects are individual-level variables, and therefore individual estimates of these effects (with the group mean removed) can be computed for use in additional analyses. These partner effect estimates served as the focal point of the current analyses (hereafter referred to as "popularity"), as they explicitly capture the evocative processes involved in peer interactions (i.e., how likeable was each participant generally perceived to be?).

Rankings and ratings contained statistically significant partner effects in both samples (accounting for 14–47 % of the total variance in popularity), indicating there was consensus within groups concerning each member's overall likeability (Burt 2008). Partner effect estimates for the sociometric ratings and rankings were highly correlated ($r > 0.80$ in males and females, $p < 0.001$), and were therefore averaged to create a composite popularity variable. All analyses were conducted using this popularity composite.

7.7.1.4 Personality

Personality was assessed using the 155-item Multidimensional Personality Questionnaire—Brief Form (MPQ-BF) (Patrick et al. 2002) administered via computer. Six participants ($n = 4$ female) were missing data due to computer malfunction. The MPQ-BF is comprised of 10 primary scales that coalesce into three higher-order factors: Positive Emotionality (PEM; the dispositional tendency to experience positive affect/emotions), Negative Emotionality (NEM; the dispositional tendency to experience negative affect/emotions), and Constraint (CON; reverse-scored impulsivity and lack of behavioral restraint). The PEM primary scales include Well-being (e.g., optimistic, happy disposition), Social Potency (e.g., likes being in charge), Achievement (e.g., ambitious, persistent), and Social Closeness (e.g., sociable, affectionate). The NEM primary scales include Stress Reaction (e.g., unaccountable mood changes, easily upset), Aggression (e.g., physically violent), and Alienation (e.g., estrangement). Lastly, the CON scales include Control (e.g., cautious, plans ahead), Harm Avoidance (e.g., avoids risk), and Traditionalism (e.g., conventionality). Consistent with the concept of evocative rGE, all of these personality traits are quite heritable (Krueger 2000; McGue et al. 1993; Tellegen et al. 1988).

7.7.1.5 Analyses

We tested the possibility of gender-specific predictors of popularity using two interrelated sets of analyses. We first examined and compared correlations between popularity and personality across gender. We next regressed popularity onto the relevant personality factors, gender, and the interaction between personality and gender. A significant interaction would suggest that the association of personality with popularity varies by gender. The latter analyses were conducted using multi-level modeling (MLM) software to account for the non-independence of popularity observations within groups while maximizing statistical power. In particular, because participants are nested within groups, our data have a two-level structure with the individual as the lower-level unit and the group as the upper-level unit.

7.7.2 Results and Discussion

7.7.2.1 Descriptives

Means are presented separately by gender in Table 7.1. As seen there, males reported higher levels of impulsive, risk-taking behaviors and lower levels of conventionality (as seen in their significantly lower levels of CON, Harm Avoidance, and Traditionalism). Males also reported higher levels of Aggression. By contrast, PEM and its primary scales did not differ across gender.

7.7.2.2 Correlations

The correlations of popularity with each of the personality dimensions are also presented in Table 7.1, separately by gender. In women, popularity was moderately and positively associated with PEM and several of its primary scales, specifically, Well-being, Social Potency, and to a lesser extent, Achievement. In short, those women with an optimistic, happy disposition and who liked being in charge were more popular with their peers than women lower on those traits. There was also a marginally-significant and negative association with two of the NEM primary scales, Stress Reaction and Alienation. However, this association did not extend to NEM more broadly or to the third primary scale, Aggression. Similarly, there was no association between popularity and CON in women.

In men, the associations with popularity were somewhat different. Popularity was associated with low levels of CON and all three of its primary sub-scales, such that those men who were more impulsive, risk-taking, and unconventional were more popular than men with lower levels of these traits. However, there was little to no association between popularity and either PEM or NEM for the men, results that clearly contrast those in the female sample. Statistical comparisons of the association of popularity with PEM and CON across gender offered additional support for these gender differences. The association of popularity with PEM was somewhat stronger in women than in men ($X^2 = 3.73$ on 1 df, $p = 0.053$). Conversely, the association of popularity with CON was somewhat stronger in men than in women ($X^2 = 3.43$ on 1 df, $p = 0.064$).

7.7.2.3 Regression Results

Table 7.2 presents the MLM regression results for PEM and CON. NEM was not included as it was not associated with popularity for either gender (however, all parameters that are statistically significant in Table 7.2 remain significant when NEM and NEM*gender are added to the model). CON and gender both evidenced significant main effects on popularity (note that the latter emerged only with the interaction terms were added to the model; gender alone had no predictive value for popularity). Importantly, moderation results confirmed our impressions of the correlations: gender significantly moderated the association between PEM and popularity, as well as the association between CON and popularity.

7.7.2.4 Discussion

These results are suggestive of gender differences in the prediction of experimentally-derived estimates of popularity from self-reported personality traits during late adolescence/emerging adulthood. In women, popularity was moderately and positively associated with Positive Emotionality and several of its primary scales, such that those women with an optimistic, happy disposition and who liked being

Table 7.1 Descriptive statistics and correlations among personality scales and popularity in males and females

Popularity	WB	SP	ACH	SC	SR	AL	AG	CT	HA	TR	PEM	NEM	CON+
Males (n = 132)													
Mean	9.07	7.48	6.65	9.29	5.31	2.75	3.95	7.62	6.55	5.95	75.58	42.02	71.87
Standard deviation	2.78	2.81	3.22	2.62	3.79	2.69	3.24	2.81	2.73	2.65	14.50	17.07	13.93
Correlations with popularity	0.16~	0.11	−0.03	0.08	−0.01	−0.08	0.15~	−0.17*	−0.15~	−0.20*	0.10	0.03	−0.25**
Female (n = 82)													
Mean	9.10	7.38	7.34	9.62	6.17	2.61	2.13	8.02	7.59	7.21	77.20	39.11	78.12
Standard deviation	3.04	3.40	3.22	2.40	2.41	2.43	2.39	3.17	3.02	2.54	15.14	14.57	14.42
Correlation with popularity	0.35**	0.32**	0.20~	0.14	−0.19~	−0.20~	−0.03	−0.03	−0.02	0.07	0.36**	−0.17	0.01

Note WB, SP, ACH, SC, SR, AL, AG, CT, HA, TR, and ABS represent the personality scales of Well-being, Social Potency, Achievement, Social Closeness, Stress Reaction, Alienation, Aggression, Control, Harm Avoidance, Traditionalism, and Absorption, respectively. PEM, NEM, and CON represent the higher-order factors of Positive Emotionality, Negative Emotionality, and Constraint, respectively. + indicates that means differ significantly across gender at $p < 0.05$. ** and * indicate that the correlation with popularity is significantly greater than zero at $p \leq 0.01$ or $p \leq 0.05$, respectively

Table 7.2 Predictors of Adolescent Popularity: MLM Results

Model	b (SE)	p-value
Gender	−2.345 (0.821)	0.005
PEM	0.005 (0.006)	0.246
CON	−0.013 (0.005)	0.007
PEM*gender	0.015 (0.007)	0.034
CON*gender	0.017 (0.007)	0.022

Note PEM and CON represent the higher-order personality factors of Positive Emotionality and Constraint, respectively. Estimates are unstandardized. Gender was dummy-coded such that $0 =$ women and $1 =$ men

in charge were found to be particularly popular with their peers. By contrast, popularity in males was associated with low levels of Constraint and all three of its primary sub-scales, such that those men who were described themselves as impulsive, risk-taking, and unconventional were especially well-liked by their group members. The latter finding was fully expected given prior associations with rule-breaking (behaviors robustly linked to impulsivity) in these data (Burt 2009a, b).

One important caveat to the conclusions of the present study is that results are specific to popularity in initial encounters and not to more lasting peer relationships. That said, the associations with "initial" popularity observed in males are consistent those in studies examining longer and more multi-layered peer sociometrics (Allen et al. 2005; Luthar and D'Avanzo 1999; Moffitt 1993; Sussman et al. 2007; Valente et al. 2005), implying that the dynamics of male popularity may be similar over time. However, prior work has suggested that status hierarchies take longer to emerge and stabilize in female peer groups. Given this, future work is clearly needed to evaluate whether popularity also takes longer to stabilize in female peer groups, and whether PEM would continue to predict popularity in longer-term acquaintanceships.

Despite this limitation, the current set of analyses are provocative and suggest that popularity with one's same-sex peers is associated with different personality traits in males and females during late adolescence/cmerging adulthood. Impulsivity and rebelliousness are key elements of popularity in male-male peer groups, but have no association with popularity in female-female peer groups. Instead, popularity in female-female groups is uniquely predicted by the tendency to experience positive mood states (i.e., having an optimistic, happy disposition and enjoying being in charge). Such findings likely reflect the fact that, among adolescent males, rule-breaking and risk-taking behaviors constitute an important and socially advantageous set of behaviors that allow them to access some of the freedoms and privileges of adult life (see Burt 2009a, b and Moffitt 1993 for more discussion on this point). Indeed, perhaps the only surprising aspect of the association of behavioral disinhibition with popularity is that it did not extend to females. That said, males typically engage in far more rule-breaking and risk-taking behaviors than do females (Burt et al. 2007; Moffitt 2003). Likewise, males appear to be particularly susceptible to peer effects on risky decision-making (Gardner and Steinberg 2005), highlighting possible differences in the social salience of risk-taking and RB across gender, especially during late adolescence and emerging adulthood.

7.8 Conclusion

A classic finding in the peer literature is that pre-existing behavioral tendencies are associated with later peer rejection (e.g., Dodge 1983). Such findings indicate that individual attributes matter for peer relationships. A critical lesson of behavioral genetic research is that genetic factors matter for understanding individual differences (Burt 2009b; Turkheimer 2000); thus, it seemed reasonable to speculate that genetic factors matter for peer relationships. The objective of this chapter was to sketch recent attempts to bring a genetically-informed and explicitly developmental perspective to the study of peer popularity and likeability. Our core genetic insight is that evocative rGE appears to underlie certain aspects of popularity: genetic effects are likely to influence particular patterns of behavior or individual attributes, which in turn, evoke reactions from peers. Further, we drew attention to a core developmental insight: Individual attributes that predict popularity might change with age and differ for males versus females.

In many ways the biggest challenges we faced in preparing this chapter stemmed from gaps in the existing literature. For example, Rubin et al. (2006) note that the "relevant data base for examining sex differences in the correlates of peer acceptance and rejection is sparse" (p. 611). Likewise, genetically-informed studies of peer outcomes are extremely rare, particularly when considering studies that focus on observed peer interactions as opposed to self-report measures. There are thus many unknowns in this area. On the other hand, these gaps in the literature could be construed as a sign of good things to come, in that they highlight the important and exciting questions to be answered by future researchers.

Building on that final possibility, we believe that the framework proposed here will be a useful one when planning future studies of popularity and peer rejection. We would argue that future such research should begin by considering the "tasks" of a given developmental periods in conjunction with the function of the peer group. That is, what goals are individuals likely to pursue and what role do peer groups play in facilitating these goals? A key issue is to identify the behavioral norms of a given period. This approach allows researchers to make developmentally-sensitive predictions about the kinds of individual attributes that will be tied to popularity. As we described, attributes linked with risk taking and RB appear to be related to popularity in the late adolescent/emerging adult males. By contrast, it seems unlikely that such attributes will predict popularity in childhood or later adulthood, as rule breaking is far less normative during these developmental periods, and it seems likely to interfere with the mastery of common developmental tasks such as doing well in elementary school and succeeding in the workplace (contexts that demand conformity for success and actively discourage rule-breaking).

We should also point out that a genetically-informed developmental perspective can accommodate a causal role for social relationships in human outcomes. Indeed, experiences of peer rejection and acceptance are likely contributors to the development and expression of psychopathology. The potential advantage of the approach outlined herein is that it helps to identify individual difference factors that underlie popularity/rejection, and as such, potential targets for intervention. In this way, the genetically-informed developmental perspective could actually prove useful for therapists, parents, and teachers who want to help children to function more effectively in peer groups. Research in this developmentally-informed perspective also points to the fact that peer popularity at certain points in the life span might even be related to less desirable behaviors. Future work should explore all of these possibilities.

References

Allen, J. P., Porter, M. R., McFarland, F. C., Marsh, P., & McElhaney, K. B. (2005). The two faces of adolescents' success with peers: Adolescent popularity, social adaptation, and deviant behavior. *Child Development, 76*, 747–760.

Allen, J. P., Weissberg, R. P., & Hawkins, J. A. (1989). The relation between values and social competence in early adolescence. *Developmental Psychology, 25*, 458–464.

Anderson, C., John, O. J., Keltner, D., & Kring, A. M. (2001). Who attains social status? Effects of personality and physical attractiveness in social groups. *Journal of Personality and Social Psychology, 81*, 116–132.

Becker, B. E., & Luthar, S. S. (2007). Peer-perceived admiration and social preference: Contextual correlates of positive peer regard among suburban and urban adolescents. *Journal of Research on Adolescence, 17*, 117–144.

Bot, S. M., Rutger, C. M. E., Engels, M. E., Knibbe, R. A., & Meeus, W. H. J. (2007). Sociometric status and social drinking: Observations of modelling and persuasion in young adult peer groups. *Journal of Abnormal Child Psychology, 35*, 929–941.

Brendgen, M., Vitaro, F., Boivin, M., Girard, S., Bukowski, W. M., Dionne, G., & Perusse, D. (2009). Gene-environment interplay between peer rejection and depressive behavior in children. *Journal of Child Psychology and Psychiatry, 50*(8), 1009–1017.

Brown, B. B. (2004). Adolescents' relationships with peers. In R. M. Lerner & L. Steinberg (Eds.), *Handbook of Adolescent Pscyhology* (2nd ed., pp. 363–394). New York: Wiley.

Bukowski, W. M., Sippola, L. K., & Newcomb, A. F. (2000). Variations in patterns of attraction to same- and other-sex peers during early adolescence. *Developmental Psychology, 36*, 147–154.

Burt, S. A. (2008). Genes and popularity: Evidence of an evocative gene-environment correlation. *Psychological Science, 19*, 112–113.

Burt, S. A. (2009a). A mechanistic explanation of popularity: Genes, rule-breaking, and evocative gene-environment correlations. *Journal of Personality and Social Psychology, 96*, 783–794.

Burt, S. A. (2009b). Rethinking environmental contributions to child and adolescent psychopathology: A meta-analysis of shared environmental influences. *Psychological Bulletin, 135*, 608–637.

Burt, S. A. (2012). *How do we optimally conceptualize the heterogeneity within antisocial behavior?*. An argument for aggressive versus non-aggressive behavioral dimensions: Clinical Psychology Review. doi:10.1016/j.cpr.2012.02.006.

Burt, S. A., Carter, L. A., McGue, M., & Iacono, W. G. (2007). The different origins of stability and change in Antisocial Personality Disorder symptoms. *Psychological Medicine, 37*, 27–38.

Burt, S. A., & Klump, K. L. (2013). The Michigan State University Twin Registry (MSUTR): An update. *Twin Research and Human Genetics, 16*, 344–350.

Burt, S.A., & Klump, K.L. (in press). Delinquent peer affiliation as an etiological moderator of childhood delinquency *Psychological Medicine*.

Burt, S. A., McGue, M., & Iacono, W. G. (2009). Non-shared environmental mediation of the association between deviant peer affiliation and adolescent externalizing behaviors over time: Results from a cross-lagged monozygotic twin differences design. *Developmental Psychology, 45*, 1752–1760.

Carstensen, L. L., Isaacowtiz, D. M., & Charles, S. T. (1999). Taking time seriously: A theory of socioemotional selectivity. *American Psychologist, 54*, 165–181.

Coie, J. D., & Kupersmidt, J. B. (1983). A behavioral analysis of emerging social status in boys' groups. *Child Development, 54*, 1400–1416.

Dodge, K. A. (1983). Behavioral antecedents of peer social status. *Child Development, 54*, 1386–1399.

Dodge, K. A., Lansford, J. E., Burks, V. S., Bates, J. E., Pettit, G. S., Fontaine, R., & Price, J. M. (2003). Peer rejection and social information-processing factors in the development of aggressive behavior problems in children. *Child Development, 74*, 374–393.

Ellis, W. E., & Zarbatany, L. (2007). Peer group status as a moderator of group influence on children's deviant, aggressive, and prosocial behavior. *Child Development, 78*, 1240–1254.

Gardner, M., & Steinberg, L. (2005). Peer influence on risk taking, risk preference, and risky decision making in adolescence and adulthood: An experimental study. *Developmental Psychology, 41*, 625–635.

Gerber, J., & Wheeler, L. (2009). On being rejected: A meta-analysis of experimental research on rejection. *Perspectives on Psychological Science, 4*, 468–488.

Jaffee, S. R., & Price, T. S. (2007). Gene-environment correlations: A review of the evidence and implications for prevention of mental illness. *Molecular Psychiatry, 12*, 432–442.

Jessor, R. (1991). Risk behavior in adolescence: A psychosocial framework for understanding and action. *Journal of Adolescent Health, 12*, 597–605.

Kenny, D.A. (1994). *Interpersonal perception: A social relations analysis*. New York: Guilford Press.

Klump, K. L., & Burt, S. A. (2006). The Michigan State University Twin Registry (MSUTR): Genetic, environmental, and neurobiological influences on behavior across development. *Twin Research and Human Genetics, 9*, 971–977.

Knutson, B., Wolkowitz, O.M., Cole, S.W., Chan, T., Moore, E.A., et al. (1998). Selective alteration of personality and social behavior by serotonergic intervention. *American Journal of Psychiatry, 155*, 373–379.

Krueger, R. F. (1999). Personality traits in late adolescence predict mental disorders in early adulthood: A prospective-epidemiological study. *Journal of Personality, 67*, 39–65.

Krueger, R. F. (2000). Phenotypic, genetic, and nonshared environmental parallels in the structure of personality: A view from the Multidimensional Personality Questionnaire. *Journal of Personality and Social Psychology, 79*, 1057–1067.

Lansford, J. E., Capanna, C., Dodge, K. A., Caprara, G. V., Bates, J. E., Pettit, G. S., & Pastorelli, C. (2007). Peer social preference and depressive symptoms of children in Italy and the United States. *International Journal of Behavioral Development, 31*, 274–283.

Luthar, S. S., & D'Avanzo, K. (1999). Contextual factors in substance use: A study of suburban and inner-city adolescents. *Development and Psychopathology, 11*, 845–867.

Mayeux, L., Sandstrom, M. J., & Cillessen, A. H. N. (2008). Is being popular a risky proposition? *Journal of Research on Adolescence, 18*, 49–74.

McGue, M., Bacon, S., & Lykken, D. T. (1993). Personality stability and change in early adulthood: A behavior genetic analysis. *Developmental Psychology, 29*, 96–109.

Moffitt, T. E. (1993). Adolescence-limited and life-course-persistent antisocial behavior: A developmental taxonomy. *Psychological Review, 100*, 674–701.

Moffitt, T. E. (2003). Life-course persistent and adolescence-limited antisocial behavior: A research review and a research agenda. In B. Lahey, T. E. Moffitt, & A. Caspi (Eds.), *The causes of conduct disorder and serious juvenile delinquency* (pp. 49–75). New York: Guilford.

Moskowitz, D. S., Pinard, G., Zuroff, D. C., Annable, L., & Young, S. N. (2001). The effect of tryptophan on social interaction in everyday life: A placebo-controlled study. *Neuropsychopharmocology, 25*, 277–289.

Nesdale, D., & Lambert, A. (2007). Effects of experimentally manipulated peer rejection on children's negative affect, self-esteem, and maladaptive social behavior. *International Journal of Behavioral Development, 31*, 115–122.

Newcomb, A. F., Bukowski, W. M., & Pattee, L. (1993). Children's peer relations: A meta-analytic review of popular, rejected, neglected, controversial, and average sociometric status. *Psychological Bulletin, 113*, 99–128.

Newman, D. L., Caspi, A., Moffitt, T. E., & Silva, P. A. (1997). Antecedants of adult interpersonal functioning: Effects of individual differences in age 3 temperament. *Developmental Psychology, 3*, 206–217.

Patrick, C. J., Curtin, J. J., & Tellegen, A. (2002). Development and validation of a brief form of the Multidimensional Personality Questionnaire. *Psychological Assessment, 14*, 150–163.

Peeters, H., Van Gestel, S., Vlietinck, R., Derom, C., & Derom, R. (1998). Validation of a telephone zygosity questionnaire in twins of known zygosity. *Behavior Genetics, 28*(3), 159–161.

Plomin, R., DeFries, J. C., & Loehlin, J. C. (1977). Genotype-environment interaction and correlation in the analysis of human behavior. *Psychological Bulletin, 84*, 309–322.

Plomin, R., DeFries, J. C., McClearn, G. E., & McGruffin, P. (2008). *Behavioral Genetics* (5th ed.). New York: Worth Publishers.

Raleigh, M. J., McGuire, M. T., Brammer, G. L., Pollack, D. B., & Yuwiler, A. (1991). Serotonergic mechanisms promote dominance acquisition in adult male vervet monkeys. *Brain Research, 559*, 181–190.

Rubin, K.H., Bukowski, W.M., & Parker, J.G. (2006). Peer interactions, relationships, and groups. In W. Damon & R. M. Lerner (Eds.), *Handbook of child psychology: Vol. 3. Social, emotional, and personality development* (6th ed., pp. 571–645). New York: Wiley.

Scarr, S., & McCartney, K. (1983). How people make their own environments: A theory of genotype-environment effects. *Child Development, 54*, 424–435.

Sussman, S., Pokhrel, P., Ashmore, R. D., & Brown, B. B. (2007). Adolescent peer group identification and characteristics: A review of the literature. *Addictive Behaviors, 32*, 1602–1627.

Tellegen, A., Lykken, D. T., Bouchard, T. J, Jr, Wilcox, K. J., Segal, N. L., & Rich, S. (1988). Personality similarity in twins reared apart and together. *Journal of Personality and Social Psychology, 54*, 1031–1039.

Tse, W. S., & Bond, A. J. (2002). Serotonergic intervention affects both social dominance and affiliative behaviors. *Psychopharmacology (Berl), 161*, 324–330.

Turkheimer, E. (2000). Three laws of behavior genetics and what they mean. *Current Directions in Psychological Science, 13*, 160–164.

Valente, T. W., Unger, J. B., & Johnson, C. A. (2005). Do popular students smoke? The associations between popularity and smoking in middle school students. *Journal of Adolescent Health, 37*, 323–329.

Walden, S. B., McGue, M., Iacono, W. G., Burt, S. A., & Elkins, I. (2004). Identifying shared environmental contributions to early substance use: The importance of peers and parents. *Journal of Abnormal Psychology, 113*, 440–450.

Chapter 8
Spouse, Parent, and Co-workers: Relationships and Roles During Adulthood

Erica L. Spotts and Jody M. Ganiban

Middle adulthood includes adults aged 40–60 years, with some researchers extending this definition by a decade on either end. Relationships and roles change across middle adulthood as most adults become spouses, parents, and workers. Within Erikson's lifespan developmental model (1968), this time period is demarcated by the establishment of families and careers, but also defined by a growing care for the next generation, as expressed by the nurturance of children and leadership in the workplace. One's readiness for these new roles and the emotional quality of interpersonal relationships are determined by many factors, including one's social context and stable personal characteristics. The goal of this chapter is to explore the utility of a behavioral genetic approach in developing a comprehensive understanding of the determinants of relationships during middle adulthood. In the first section we provide a brief overview of theories and phenotypic research focusing on the determinants of relationships during middle adulthood. This section is followed by a review of pertinent behavioral genetic research, and a discussion of the importance of the environment and genetic factors in relationships during middle adulthood. Last, we highlight gaps in the research literature and outline future research directions.

E.L. Spotts (✉)
Office of Behavioral and Social Sciences Research, National Institutes of Health, 31 Center Drive, Building 31, Room B1C19, Bethesda, MD 20892, USA
e-mail: Erica.Spotts@nih.gov

J.M. Ganiban
Department of Psychology, The George Washington University, Washington, USA

© Springer Science+Business Media New York 2015
B.N. Horwitz and J.M. Neiderhiser (eds.), *Gene-Environment Interplay in Interpersonal Relationships across the Lifespan*,
Advances in Behavior Genetics 3, DOI 10.1007/978-1-4939-2923-8_8

8.1 Determinants of Adult Relationships

Numerous theories emphasize the impact of social experiences on social competence throughout the lifespan. It is frequently expected that one's social behaviors
and the quality of relationships are shaped by specific experiences with partners and
by families and culture. For example, in his seminal trilogy on attachment theory,
Bowlby (1973) proposed that individuals build generalized templates or "internal
working models" of relationships during childhood that are shaped by children's
actual experiences with their primary caregivers. Bowlby (1973) further argued that
working models created during childhood influence adults' behaviors and perceptions of partners as they develop new attachment relationships with romantic partners and their own children. Support for this contention is primarily derived from
studies that rely upon adults' recollections of their relationships with their parents
during childhood and perceptions of their current relationships with romantic partners or children (Cohn et al. 1992; Cowan et al. 1996; Van IJzendoorn 1992). One
longitudinal study that followed participants from infancy through young adulthood
also suggests that early experiences predict the quality of relationships over time
(Simpson et al. 2007). However, this study also indicates that the impact of early
experiences on adult relationships is indirect, rather than direct. One's attachment
history may initially affect social opportunities and friendships from childhood
through adolescence, but the latter have a direct impact on adult relationships.

Others have emphasized the impact of adults' contemporary social experiences on their social behaviors and the quality of their relationships. Attachment
research indicates that adults' current romantic relationships can cause internal
working models to be reformulated (Treboux et al. 2004). Obviously, spouses'
expressions of support and affection versus hostility and conflict also have a substantial impact on marital quality (Mehta et al. 2009). Likewise, children play
important roles in the quality of the parent-child relationship (Sanson et al. 2004)
and can affect parents' behaviors (e.g., Ganiban et al. 2011; Ge et al. 1994). Last,
previous research has shown that different family relationships influence each
other (for comprehensive reviews, see Cox and Paley 1997; Erel and Burman
1995). Thus, there is ample evidence that social experiences affect relationships
during middle adulthood. Early experiences may affect one's perception of relationships and partner selection, while contemporary experiences with partners also
have a strong influence on social behaviors and relationship quality.

A second body of research has emphasized the influence of stable, genetically-
influenced characteristics on social behaviors and relationships during adulthood.
Trait theories of personality and temperament generally propose that a core set of
stable, genetically influenced tendencies guide one's perceptions, reactivity and
behavior across situations (Boyle et al. 2008). Although specific theories differ on
which personality characteristics reflect traits, there is general consensus that neu-
roticism/anxiousness, aggressiveness/hostility or disagreeableness, extraversion, sensation seeking, and conscientiousness/self-regulation are genetically influenced
characteristics that are relatively stable by adulthood (e.g., Cloninger et al. 1993;

Gray and McNaughton 2003; McCrae et al. 2000; Rothbart et al. 2000; Zuckerman 2005). These characteristics could influence how individuals perceive relationships and guide their behaviors within different relationships (Karney et al. 1994; Rusting 1998), and contribute to similarities across different relationships (Ganiban et al. 2009; Horwitz et al. 2011; Robins et al. 2003). Consistent with this proposal, several studies have reported significant links between adults' personality and the quality of their marital relationship (Karney and Bradbury 1995), likelihood to divorce (Jocklin et al. 1996; Kelly and Conley 1987), and to parenting (Clark et al. 1997; Kochanska et al. 2004; Metsapelto and Pulkkinen 2003).

In reality, it is more likely that relationships and social behaviors are the products of social experiences *and* genetically influenced characteristics such as personality traits. As a consequence, interest has shifted to estimating the relative importance of both factors and understanding how they work together to affect social functioning during adulthood. Researchers, however, are faced with a number of conceptual and methodological obstacles in answering these questions. At first glance it seems sensible to categorize all social experiences as environmental determinants of relationships, while stable characteristics represent the effects of genetic influences. Yet, the underlying processes may be less straightforward. For example, the expression of personality characteristics could be regulated by social contexts and social cognitive processes, rather than genes (Mischel 2004). Conversely, genetically influenced personality characteristics may elicit specific reactions from social partners or shape social niches throughout the lifespan (Caspi 1998). Most studies cannot determine if a "social influence" is truly environmentally based, or if associations between personality and relationships reflect genetic effects. Perhaps even more importantly, how both factors work together to shape social development is poorly understood at all ages.

8.2 Behavioral Genetic Research Designs

Behavioral genetic methods provide powerful tools for addressing these issues, and providing further insight into the determinants of relationships during adulthood. These designs include pairs of individuals who vary in regard to genetic relatedness, such as monozygotic and dizygotic twin pairs, full siblings, half-siblings or unrelated siblings. As such, variance in any phenotype within the population can be partitioned into genetic and environmental components (Fig. 8.1). Estimates of genetic variance capture the sum of genetic factors that contribute to similarities between pair members. While most studies focus on additive genetic influences on phenotypes, larger scale studies have the power to estimate dominant genetic effects. Environmental influences are broken down into two parts: Shared and nonshared environmental variance. Shared environmental variance relates to experiences that two pair members have in common that make them similar to each other. For adult twins who were raised together, shared environmental effects could include experiences that stretch back to their childhood, when they lived

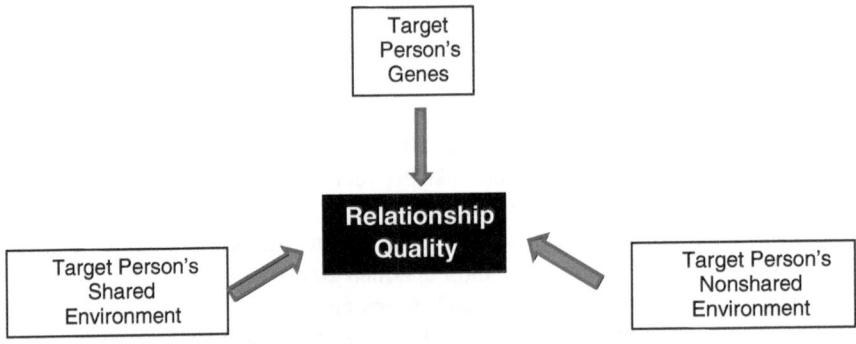

Fig. 1 Genetic and environmental contributions to relationship quality

together or might represent current experiences. Nonshared environmental variance captures the impact of unique experiences that make pair members different from each other from childhood through adulthood.

Figure 8.2 expands this model to include potential mediators or "conduits" of genetic and environmental influences on relationships and social behaviors. The right side of Fig. 8.2 depicts the impact of the target person's genetic makeup and environments on their relationships. Attachment theory would anticipate that shared and nonshared environmental effects on relationships are explained by one's relationship histories or other social experiences. However, if these mediators actually function as conduits of genetic influences on relationships, then a different interpretation is warranted. This pattern of findings would be more consistent with a person-based model of relationships, in which genetic predispositions influence the social environments individuals create for themselves. This process is sometimes referred to as active gene-environment correlation: genetically influenced tendencies lead one to select compatible environments or elicit similar reactions

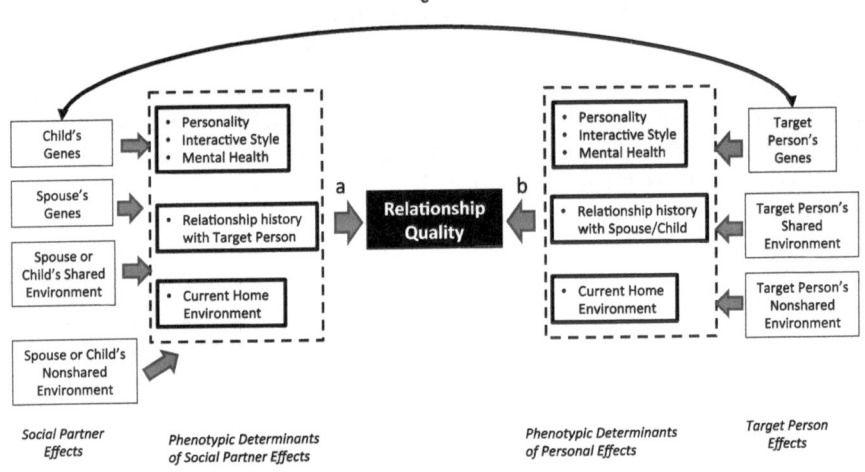

Fig. 2 Mediators of Genetic and environmental contributions to relationship quality

from social partners (Scarr and McCartney 1983). However, this effect may also reflect passive gene-environment correlation if the target person and social partner are genetically related, as in the case of parents and children. In this scenario, genetic influences on the parent-child relationship may be caused by genetically influenced tendencies that are shared by parents and children (path a). The left side of Fig. 8.2 includes the impact of social partners on the target person's relationships. If the target person and social partner are not genetically related, the social partner's personalities, behaviors, or own relationship history could act as environmental influences on the target person's behavior and relationship quality.

Behavior genetic research designs can also be instrumental in understanding gene-environment interplay. Numerous reviews have already characterized the ways in which genes and environment interact to shape phenotypes (e.g. Belsky and Pluess 2009; Ottman 1996; Shanahan and Hofer 2005). For example, the expression of genetic-based risks for emotional and behavioral problems may be mitigated by favorable social environments or enhanced by stressful, chaotic, or hostile environments (e.g., the diathesis-stress model). On the other hand, genetically-based tendencies may affect one's overall responsiveness to both supportive and nonsupportive environments (e.g., Differential Susceptibility; Belsky and Pluess 2009). It is also possible that environments exert strong control over social behavior, essentially canalizing specific behavioral and emotional outcomes (Shanahan and Hofer 2005). Alternatively, genetic factors may also override environmental effects in the development of some phenotypes (Shanahan and Hofer 2005). There is some empirical support for each type of gene-environment interaction in the development of psychiatric disorders and symptoms, drug use, and overall mental health, but there is less research pertaining to social relationships or social development within normative populations. This line of research is particularly important for intervention and prevention sciences as it can elucidate pathways through which genetic and/or environmental risk work together to shape individual differences in specific behaviors and social relationships throughout the lifespan.

In subsequent sections of this review, behavioral genetic research pertinent to marital quality, parenting, and work place relationships will be presented to address three points: (1) relative contributions of genetic and environmental factors to adult relationships; and (2) phenotypic mediators of genetic and environmental contributions; (3) interplay between genes and environment in shaping relationships during adulthood.

8.3 Marriage

Marriage is a key relationship of adulthood for most people. There are a number of theories about how an individual functions within their marriage. The marital relationship might be an extension and function of attachment bonds formed in childhood. Children might learn early on, from their own parents, how to interact with a spouse, and this might form the foundation for their own marriage years later.

Yet other theories put marriage into an evolutionary context and consider it along-side the pairbonding of other species. Regardless of the framing, it is of great scientific interest because of its robust link with mental and physical health, lon-gevity, and mortality. In this section, we will discuss the genetically informative research that has been conducted on marriage and discuss how it has furthered our understanding of this important relationship. Marriage phenotypes (e.g. status, quality) have been linked with health and mortality; however, it's not clear if these associations are causal or due to selection effects. Genetically informative studies can help to lend support for one or the other explanations.

8.4 Marital Status

Most genetically-informed studies of marriage focus on marital *status* rather than marital *quality*. The first study to examine a component of marital status was a study of the genetic and environmental influences on divorce, using a sample drawn from the Minnesota Twin Registry (McGue and Lykken 1992). Participants were between the ages of 34 and 53 years. Both male and female MZ twins were more similar for divorce status than DZ twins, indicating genetic influences. Having a divorced MZ co-twin increased one's odds of getting a divorce nearly sixfold, whereas having a divorced parent or DZ co-twin only increased risk by less than twofold. Heritability estimates for divorce were nearly equally divided between genetic and nonshared environmental influences, for both men and women.

The propensity to marry seems to be a different story. Another study from the Minnesota Twin Registry examined the genetic and environmental influ-ences on getting married in a sample with a mean age of approximately 40 years of age (Johnson et al. 2004). The total heritability, including additive and domi-nant effects, was 0.70. Gender differences were examined, and it was found that the heritability for men and women were similar, but the genetic components of variance were not the same for men and women, suggesting that different genetic factors influence men's and women's propensity to marry.

The question has also been asked about whether people who have never married and people who have divorced can both be categorized as having "failed to marry suc-cessfully." If they are two sides of the same coin, MZ cotwins of those who have never married should have higher rates of divorce than DZ co-twins, and the MZ co-twins of those who have divorced should have a higher rate of never marrying than DZ co-twins. This question was addressed using the National Academy of Sciences–National Research Council (NAS–NRC) sample of World War II veterans (Trumbetta and Gottesman 1997). These males were born between 1917 and 1927 and were surveyed in 1972 and 1985, when the men were 45–55 and 58–68 years old, respectively. Never marrying and divorcing were not co-heritable, though there was a nonsignificant trend for more co-twins of those who had divorced to have never married. The authors sug-gest that this indicates the possibility of a continuum of pair bondedness. The same authors later explored marital status in the evolutionary terms of pair bondedness and mate diversification, saying that being married resembles the former and divorce maps onto the latter (Trumbetta and Gottesman 2000). Again using the sample of World War II veterans, they found that pair bonding seemed to be more genetically influenced than mate diversification, though the heritability declined over time.

Noticing that there were age differences in heritability in other studies, Trumbetta et al. (2007) examined whether or not the relative contributions of genetic and environmental factors to phenotypic variance in current marriage change over time. They used data on individuals aged 20–70 years of age collected for the Duke Dementia Study (a followed subset of the World War II Veteran Twin Registry). Environmental influences appeared to accumulate with time in that influences from earlier years continued to exert influence in later years. Proportional influences of genetic and environmental factors changed, but the common variance factor was constant throughout adult life. Notably, there were some additional sources of genetic influences at age 30, which the authors speculate is likely due to post WWII stability. They also offer further discussion about the likely role of other historical phenomena (e.g. relaxation of divorce laws) on the fluctuation of genetic and environmental influences on marital status across time.

Jerskey and colleagues (2010) explored the genetic and environmental influences on marriage and divorce and the nature of the association between the two constructs. They used the Vietnam Era Twin Registry (VETR) to do this. The subjects were asked if they had ever married, were still married, and how the marriage ended (if it did). Because one variable is dependent on the other (i.e. being divorced is dependent on ever being married), the authors used a Causal Contingent Common (CCC) model, which is often used in substance use research, but not often used for social constructs such as marriage. Notably, while both getting married and getting divorced were found to be heritable, the genetic influences on ever marrying and divorce were found to be independent of each other. In other words, different heritable characteristics influence the likelihood of someone getting married than influence the likelihood of divorce.

8.5 Attachment Relationships

The genetic literature on adult romantic attachment is somewhat sparse. This is probably largely due to the fact that the typical in-person measures of attachment are too cumbersome to use in adequately powered sample sizes. The advent of self-report attachment measures has initiated more genetic inquiry into adult attachment styles. That said, there are limitations to the work that has been conducted, which will be discussed after a review of the literature.

The first genetically informed study of adult attachment was conducted on a twin sample of just over 200 pairs (age range 16–79 years) using items from the Relationship Scales Questionnaire (RSQ) (Brussoni et al. 2000). Substantial genetic influences (25–43 %) were found on secure, fearful, and preoccupied attachment styles, with the remaining variance explained by nonshared environmental influences.

The dismissive attachment style was found to be influenced by shared and nonshared environmental influences. A later study, using what appears to be the same data set, created anxious and avoidant attachment scales from the RSQ and examined the underlying genetic and environmental links between attachment and personality disorders (Crawford et al. 2007). They found that anxious attachment was genetically influenced, but avoidant attachment was entirely influenced by environmental effects. Covariation between avoidant attachment and personality problems was attributed to nonshared environmental effects, while links between anxious attachment and personality disorder were predominantly explained by genetic effects.

A study of over 250 twin pairs (average age ~21 years) examined the links between anxious and avoidant attachment (as assessed by the Adult Attachment Scale) with the Big Five personality dimensions (Donnellan et al. 2008). In contrast to the first studies, these findings showed genetic influences on both dimensions of adult attachment. They also found that much of the overlap between attachment dimensions and Big Five personality traits could be explained by shared genetic factors. The authors are clear in pointing out that they do not see this is evidence of attachment and personality being two sides of the same coin, but rather that personality influences the development and maintenance of internal working models throughout the life course.

An Italian study of nearly 250 twin pairs aged 23–24 years used the Experiences in Close Relationships Inventory (ECRI) and reported heritiabilties of 45 and 36 % for anxious and avoidant attachment respectively (Picardi et al. 2011). The covariation among the two types of attachment was explained predominantly by genetic factors.

The largest study to date of adult attachment ($n = 1237$ pairs) was conducted on a sample of middle-aged male veterans also using the ECRI (Franz et al. 2011). Both dimensions of attachment were found to be heritable, but no evidence was found for any shared environmental influences. In addition to examining the constructs individually, they also examined the covariance between the dimensions of attachment, with findings suggesting that the genetic correlation between anxious and avoidant attachment was 0.41.

The only study using interview based attachment measures showed a higher concordance for securely versus nonsecurely attached twin pairs (Torgersen et al. 2007). This study also found shared environmental influences on nonsecure attachment. Unfortunately, the small sample size ($n = 41$ pairs) leaves us questioning the reliability of these results, though they are somewhat consistent with the other studies presented here.

There are limitations to the literature on adult romantic attachments. First, unlike the assessment of attachment relationships in childhood, assessments of attachment in adulthood often focus on a general idea of romantic relationships rather than a specific romantic relationship. Many of the studies cited above did not even require the subjects to be in a romantic relationship at the time of the study. In contrast, childhood attachment studies focus on a specified relationship (i.e. the parent-child relationship). As a result, it is difficult to know who the attachment figure is that is being reported on, which likely dilutes the strength of study findings. Corollary to this first limitation is the fact that adult attachment work relies on self-report, rather than the behavioral assessment of child-based studies, making comparisons between the two parts of the lifespan difficult at best.

Second, with the exception of the VETSA findings reported here, the samples sizes of the twin studies used to examine adult attachment are quite small, leading to broad confidence intervals (when they are reported). Third, in spite of stemming from a very developmentally conscious theory, the genetically informed adult attachment literature lacks a developmental and/or lifespan perspective. There are no longitudinal studies, and most of the studies presented here do not take into account the age of the subjects or where they are in terms of relationship status.

Finally, there is very little overlap in the attachment measures that are used in the extant studies. This calls into question whether or not the same construct is being assessed across samples, and how reliably the construct is being assessed. Without some sort of cross walk or calibration between the different attachment measures, we cannot know if differing findings of heritability are the result of different ages, populations, measures, or other factors. Future research in this area would do well to use adequate sample sizes, conduct their studies within a developmental framework, and address the issue of the comparability of different adult attachment measures.

8.6 Marital Quality

Only recently have the genetic and environmental influences on marital quality been explored. This was first done in a study of Swedish twin mothers, their husbands, and their adolescent children (the Twin Moms study; Spotts et al. 2004). In this study, the Dyadic Adjustment Scale (DAS; Spanier 1976) and the Marital Adjustment Test (Locke and Wallace 1987), both often-used measures of marital quality, were administered to both the twin women and their spouses. Most measures of wife's marital quality were at least moderately influenced by genetic factors (standardized parameter estimates of 0.24–0.33) and primarily influenced by nonshared environmental influences (standardized parameter estimates of 0.67–0.85; Spotts et al. 2004). One exception was DAS Affectional Expression, which showed negligible amounts of genetic influence. Shared environmental influences played no role in the twin women's reports of their marital quality. Genetic influences on husbands' reports of their marital quality were also examined, with results similar to those of wives' reports. It needs to be noted that influences on husband reports were those of the wife's genetically and environmentally influenced characteristics, so these findings of genetic influences represent the effects of wife's genetically influenced characteristics on their husband's perceptions of their marriage. This could be the result of either active GE correlations, whereby the wife actively selects a husband that fits with her genetically influenced characteristics, or evocative GE correlation, whereby the wife's genetically influenced characteristics elicit a particular response from her husband.

These findings also held for a slightly younger (mean age = 35.6 years) American sample (Spotts et al. 2006). This sample allowed for testing gender differences in genetic and environmental contributions to marital quality. The best-fitting model for both marital warmth and conflict showed that the same genetic influences operate for men and women and allowed for quantitative differences in genetic and environmental influences between the genders, lending support to previous research suggesting that any differences that exist between men and women are of degree (quantitative) rather than of kind (qualitative; Rhyne 1981). It is intriguing that, to the extent that there were differences, these differences fell along gender-congruent lines; women showed higher heritabilities for warmth, whereas men showed higher heritabilities for conflict.

We can speculate on the findings of nonshared environmental influences. As depicted in Fig. 8.2, one's social partners contribute to relationship quality. Since adult twins and siblings have different spouses, the spouse is a potential source of nonshared environmental effects. For each of the analyses described above, husbands' reports of marital quality were substituted for the wives' reports, and in all cases, nonshared environmental influences were shared among marital quality, social support, and the measure of mental health (Spotts et al. 2005, 2005). These findings suggest that the husband is an important source of nonshared environmental influence on his wife's feelings about her interpersonal relationships and her mental health. Kessler et al. (1992) used the female twin sample drawn from the Virginia Twin Registry to examine the processes by which perceived social support and adjustment to stress, as indicated by onset of major depression (MD), were linked. Among the models tested was one hypothesizing that the link could be explained by a common genetic cause; the model was not supported. At first glance, these results seem to contradict the finding from the Twin Moms study reported above. However, the two studies used very different outcome measures: depressive symptomatology versus acute onset of MD. Also, the nature of the variables used in the analyses resulted in very different sample sizes, with the Virginia sample being reduced to 22 cases. Wade and Kendler (2000) tested several possibilities explaining associations between lifetime MD (as indicated by an onset of MD within the past 12 months at either Time 1 or Time 2) and social support using the Virginia female twin sample. Three hypotheses were tested, but the one discussed here postulates that social support and the risk for MD may be linked by common, genetically influenced traits. To test this hypothesis, the level of social support in one twin was predicted by MD in her cotwin. A significant association was found, and it was stronger for MZ than DZ twins, indicating genetic influences on the association.

Candidate genes have also been directly associated with familial behaviors. Since there are many conserved regions of the genome across species, work with specific genes can help bridge human and non-human work. An example comes from our growing understanding of vasopressin, a peptide involved in the processing of social information. The work of Larry Young and others on voles has shown that vasopressin receptor genes play a large role in the regulation of social behavior in these animals (see Donaldson et al. 2008 for a review). Specifically, vasopressin facilitates social contact, partner preferences, parental behaviors, and variation in sexual and social fidelity in voles (for example, Insel et al. 1994; Ophir et al. 2008). There is not a direct genetic homolog in humans, but there are three repetitive sequences in the AVPR1A region that are polymorphic. Previous work suggested associations between AVPR1A polymorphisms and social phenomena such as autism (Kim et al. 2002), age at first sexual intercourse (Pritchard et al. 2007), and altruism (Knafo et al. 2007). This work sparked interest in finding out if AVPR1A was associated with relationship quality and pair bonding in human males. The Twin/Offspring Study in Sweden (TOSS) sample is an extension of the previously described Twin Moms study that added a cohort of twin fathers to complement the twin mothers. This study was used to look at the association between pair bonding and AVPR1A (Walum et al. 2008). Findings indicated pair bonding in men, but not women, was significantly associated with the RS3 alleles of the vasopressin receptor gene. Additionally, pair bonding varied by what allele the men were carrying with bonding being lower if men carried the 334 allele. This allele had a dose dependent effect depending on how many copies the man was carrying. Men carrying the 334 allele were more likely to have experienced a marital crisis or threat of divorce in the past year and were more likely to be in a cohabiting relationship without being married than carriers of other alleles. There was also a dyadic effect, in that the wives of male carriers of the 334 allele were less satisfied in their marriages than wives of non-carriers. While hardly conclusive, this study sets the stage for future work examining the effects of specific genes on social behavior, and suggests biological pathways that influence human interpersonal interactions.

A corollary study was conducted looking at the oxytocin receptor gene (OXTR) (Walum et al. 2012) in a developmental context using the TOSS sample as well as two studies of younger-aged individuals. Oxytocin is a neuropeptide with typically sexual dimorphic effects. It is known for its involvement in uterine contractions during labor, lactation, and on a wide range of social behaviors such as social memory and the regulation of aggression, sexual behavior and maternal care (Neumann 2008; a further review is in Walum et al. 2012). Walum et al. found that the OXTR SNP rs7632287 was associated with pairbonding behavior in adult and adolescent women. It was also associated with childhood social problems, which in other studies is predictive of behavior in romantic relationships later in life. Finally, this SNP was linked with autistic traits in childhood. What is particularly interesting about this study is the developmental nature of the findings, suggesting that these SNPs have an early and perhaps continuous impact on behaviors associated with pairbonding.

What accounts for the genetic influences on aspects of marriage? As depicted in Fig. 8.2, a host of phenotypic mediators may function as conduits of genetic effects. Amongst these mediators, it appears that personality plays a large role. A number of studies have used multivariate analyses to examine covariance between personality and marriage. Using a sample drawn from the Minnesota Twin Registry, Jockin and colleagues extended McGue and Lykken's (1992) study of divorce to examine the extent to which personality accounted for the genetic influences on divorce (Jockin et al. 1996). They found that 30 and 42 % of the genetic influences on divorce could be accounted for by personality for women and men, respectively. They also found that the environmental factors influencing personality and divorce were almost entirely independent of each other.

Another study using the same sample examined the extent to which personality characteristics accounted for the genetic influences on the propensity to marry (Johnson et al. 2004). Using composite scales of the Multidimensional Personality Questionnaire, they found that genetic influences accounted for 68 and 83 % of the covariance between personality and the propensity to marry for men and women, respectively. Another study using the Twin Moms sample tried to account for the genetic influences on marital quality and found that the personality characteristics aggression and optimism accounted for all of the genetic influences on wives' reports of marital quality (Spotts et al. 2005). It had also been hypothesized that husbands' personality characteristics would account for a substantial portion of the nonshared environmental influences on wives' reports of marital quality, but this was not supported.

8.7 Marital Quality Moderates the Impact of Genetics on Various Outcomes

It is beyond the scope of this chapter to explore the mechanisms underlying the robust links between marital quality and mental health (e.g. see review by Beach et al. 1998). Suffice it to say that there is evidence for genetically influenced overlap (e.g. Spotts et al. 2004, 2005; South et al. 2008). What will be discussed in this section is the possibility that the quality of the marital relationship can moderate the impact of genetic and environmental influences on a range of outcomes. An early study (Heath et al. 1998) found that the heritability of depression was decreased in married vs unmarried women, suggesting that marriage was protective against depression, at least the genetic liability thereof. It wasn't until recently that this question has been revisited using more sophisticated modeling techniques. In this study, South and colleagues (2008) found that genetic influences were more important for internalizing disorders in the context of low marital quality. What this suggests is that people reporting low quality marriages are more susceptible to any genetic predilection for internalizing problems that they might have.

Heath and colleagues (1989) also found that marital status moderated the heritability of alcohol use. Unmarried respondents had a substantially higher heritability of alcohol consumption than married individuals, and the heritabilities varied by age. The heritability of alcohol consumption for unmarried versus married groups aged 30 years or less was 31 versus 60 %; for those 31 or older, the heritabilities were 76 % to about 50 %. Trumbetta used the NAS-NRC World War II Veteran Twin Registry to examine causal models of marriage buffering against health risk behaviors (e.g. exercise, eating fruits and vegetables, smoking, alcohol consumption) in men (Trumbetta 2004). In this way the authors could control for any selection effects in terms of familial predilections toward certain health behaviors would also be associated with the likelihood of marriage. Findings indicated that being married reinforced health behaviors, either for better or worse. This was somewhat dependent on the heritability of the health behavior in question. For example, in this study, being married amplified the phenotypic expression of the genetic predisposition for being a smoker/nonsmoker. These findings only applied to men, as there were no women in this study. They were also unable to account for positive assortment for the health behaviors in question, which could contribute substantially to the findings.

A Norwegian study of twins in their early to mid-40s examined the effects of a marriage-like relationship on the heritability of subjective well-being (Nes et al. 2010).

Findings suggest that genetic influences on subjective well-being were greater in unmarried than married individuals. They also found that, in part, different sets of genes influence well-being across marital status, but only in females. The discussion had by these authors about the reasons for the difference in heritability across marital status is applicable the other findings reported in this session, and we recommend this paper for additional reading. In short, there is a small body of work that shows that heritability estimates are attenuated when there are higher levels of social control (e.g. Boomsma et al. 1999; Heath et al. 1998). The authors argue that marriage is a small, well-defined social network that has clear social cues about behavior. In this context, heritability is constrained. The findings of differing sets of genes operating on subjective well-being across marital status suggest that different social environments trigger different genetic responses.

In sum, the addition of a genetic approach has furthered our understanding of marriage and the dyadic relationships within this important relationship. Not only has this work placed the marital relationship within an evolutionary context, it has also permitted a better understanding of how spouses influence each other and health outcomes. That said, there are a number of limitations to the current body of genetically informed research on marriage. Primarily, there has yet to be a truly developmental focus driving any of this work. Genetic studies could be very informative to questions about the timing of marriage and divorce. For the most part, genetically informed studies thus far have only considered marriage as a static phenomenon, which it most certainly is not. There is much to be gained by looking at trajectories of marital quality over time, particularly in relation to health outcomes. In this way, scientists would get better traction on issues of selection and the directionality of associations between marriage and health. Future work will need to account for societal changes that impact marriage and marital quality, such as the increased prevalence of cohabitation, same-sex and multiple marriages, and stepfamilies. The increased prevalence of dual-worker families is another issue that should be addressed in the coming years. Finally, many of the studies reviewed here were not designed to address questions about marriage, and therefore, aggregate a broad age range within a set of analyses. It will be important in the future to consider the chronological age of the spouses in conjunction with the length of marriage.

8.8 Determinants of Parenting

Most behavioral genetic research on parenting has explored the extent to which genetic and environmental factors contribute to specific dimensions of parenting behavior, such as negativity, warmth and monitoring. Only a handful of studies have explored determinants of parent-child attachment quality. Furthermore, while some studies have focused on the impact of parent-based genetic and environmental contributions to parenting, others have examined the impact of children on the parenting they receive. In the next sections of this review, the overall rationales of parent- and child-based research designs are described, followed by a summary of research from both perspectives.

As depicted in the right side of Fig. 8.2, parent-based effects describe the extent to which parents' own genetic makeup and experiences influence the parenting they *provide*. Within the context of the parent-child relationship, shared environment captures experiences that adult siblings have in common and contribute to similarities in their parenting styles. For example adult cotwins' parenting behaviors may be modeled after their own parents, or shaped by a shared religion or contemporary culture. Nonshared environmental effects encompass siblings' unique experiences that contribute to differences in their parenting styles. Since marital conflict and satisfaction are related to parenting (Erel and Burman 1995), differences in marital quality between adult twins could contribute to differences in their parenting styles. It is important to note that only research designs that include *adult* sibling pairs who are concordant for children can be used to assess parent-based effects.

In contrast as depicted in the left side of Fig. 8.2, child-based studies have examined children's influences on parenting. Variance in parenting across children can be attributed to children's genes, or to qualities of the shared and nonshared environments. As described previously, child-based genetic effects are frequently interpreted as evidence of children's influence on parenting or evocative gene-environment correlation: parents alter their behavior in response to their children's genetically influenced characteristics (Fig. 8.2, path b). However, genetic contributions to parenting in child designs could also be explained by genes that are shared by parents and children (Fig. 8.2, path a).

Child-based designs also estimate environmental-based contributions to parenting, and differentiate between shared environmental versus nonshared environmental effects. The presence of significant shared environmental effects indicates that parents treat siblings similarly, regardless of their genetic relatedness. For example, parents' mental health, parenting philosophy and culture could all prompt them to treat all of their children the same way. Conversely, parents may also treat siblings very differently, making parental warmth and negativity unique or nonshared experiences for children. For example, parents may have very different expectations based upon previous experiences with each child. These expectations may be translated into different parenting strategies for each child. Generally, child-based effects can be estimated if the study design includes *child* sibling pairs that vary in regard to their genetic relatedness, such as twin or stepfamily designs. Adoption studies that include siblings who differ in regard to genetic relatedness or permit comparisons of children's similarities to biological and adoptive parents can also be used to assess child-based effects.

8.9 Findings from Parent-Based Designs

Previous research using parent-based designs has focused on parents of infants (Boivin et al. 2005), preschool to school age children (Losoya et al. 1997), and adolescents (Neiderhiser et al. 2004, 2007), with little consideration to parents' ages.

However, for studies that reported parental age, most parents tended to be in their 30s to 40s. Other studies have included middle-aged parents, without consideration of their children's ages (e.g., Kendler 1996; Spinath and O'Connor 2003; Wade and Kendler 2000). One additional study included adult twin retirees, and collected their retrospective reports of their parenting styles during the first 16 years of their children's lives (Perusse et al. 1994). Across most studies, parents' genetic makeup accounts for about one-fifth to one-third of variance in parenting dimensions (Kendler and Baker 2007). Heritability estimates tend to be highest when self-reports of parenting are used and lowest when behavioral observations are used (Neiderhiser et al. 2004, 2007). Studies that include child reports of parenting tend to yield heritability estimates that are similar to parent self-reports, suggesting that children and parents are sensitive to variations in parenting that relate to parents' genetic makeup. Lower estimates of genetic contributions within observational studies may be due to several factors, including differences in the time period over which parenting is assessed and participants' reactivity to the testing situation. While questionnaires typically ask respondents to consider behavior over extended periods of time and across multiple situations, observations are based upon parents' behaviors during a brief period of time, within the context of a research study. As such, questionnaires may be better equipped to capture stable patterns of behavior, while observational measures may be more sensitive to situational effects.

Most studies include assessments of how much parents' express negativity and warmth to their children, and exert control over their children. Previous reviews have concluded that heritability estimates are higher for parent positivity and negativity than for indices of parental control (Kendler and Baker 2007). For example, studies that have examined genetic contributions to parental warmth or support using a variety of assessment methods, have yielded heritability estimates that range from 0.04 to 0.60, with most heritability estimates falling between 0.30 and 0.40 (Kendler 1996; Losoya et al. 1997; Neiderhiser et al. 2004, 2007; Perusse et al. 1994; Spinath and O'Connor 2003). There is a greater range in heritability estimates for parental negativity, and in some studies heritability estimates for parental negativity are lower than for warmth (Kendler and Baker 2007). However, heritability estimates for self-reported parental negativity also seem to cluster in the 0.30–0.40 range (Bovian et al. 2005; Losoya et al. 1997; Neiderhiser et al. 2004, 2007). In regard to parental control, estimates of genetic contributions tend to be lower than for positivity and negativity, with most studies reporting heritability estimates for control between 0 and 0.20 (Bovian et al. 2005; Kendler 1996; Neiderhiser et al. 2004, 2007). Nevertheless, there are some exceptions. Higher heritability estimates for control have been found when parenting is assessed via retrospective reports and when specific dimensions of control (e.g., physical discipline and limit setting) are included (Perusse et al. 1994; Wade and Kendler 2000).

In summary, parent-based twin studies indicate that parents' genetic makeup contributes to their parenting styles. Recent studies have started to explore the degree to which genes related to the production and regulation of dopamine (COMT, DRD4, DAT1) and serotonin (5-HTT), predict maternal sensitivity or negativity (Bakersmans-Kranenburg and van IJzendoorn 2008; Lee et al. 2010;

van IJzendoorn et al. 2008). These studies indicate that associations between genes and parenting behavior are neither direct nor additive. Rather, some genes may have dominant effects, or interact with each other and with environmental stressors (e.g., daily hassles, children's disruptive behaviors) to predict parenting. Consequently, more complex models of genetic effects on parenting that go beyond simple associations are needed. In addition, all of these studies have critical limitations. First, findings were based on small, homogenous samples, increasing the possibility of false positives. Further replications with larger and more diverse samples are needed. Second, the studies targeted single loci on each gene, and may have missed other relevant loci. Furthermore, these studies did not take into account the possibility of passive-gene correlation. Specifically, if parents and children share the same "risky" genes, then it would be difficult to determine if the observed genetic effects are driven by the parents' reactivity or by the child's behaviors.

Environmental factors also make substantial contributions to parenting. Within parent-based designs, nonshared rather than shared environmental factors play important roles in parent negativity and warmth. Nonshared environmental contributions to parenting tend to be high, but also vary across studies. For example, estimates for parental warmth range from 0.40 to 0.77, and estimates for negativity range from 0.16 to 0.95 across self-, child- and observational ratings of parenting. Across all adult twin studies, shared environmental contributions to parental negativity and warmth tend to be negligible (Bovian et al. 2005; Kendler 1996; Losoya et al. 1997; Neiderhiser et al. 2004, 2007; Perusse et al. 1994; Spinath and O'Connor 2003). This pattern suggests that there are important differences in adult co-twins or siblings' parenting styles. In contrast, measures related to parental control and overprotection demonstrate significant shared environmental effects, and accordingly, lower nonshared environmental contributions than parental warmth and negativity (Bovian et al. 2005; Kendler 1996; Neiderhiser et al. 2004, 2007). Therefore, co-twins are somewhat similar in their beliefs and behaviors related to controlling and monitoring their children's behaviors.

In summary, most parent-based studies identify moderate genetic and nonshared environmental contributions to parental negativity and warmth. In contrast, measures related to parental control and protections display significant shared environmental effects. These findings give credence to nurture and nature perspectives on parenting. To a large degree, parenting is influenced by experiences. However, adults' genetically influenced characteristics play an important role in parents' warmth and negativity towards their children.

As the field moves beyond recognizing that both nature and nurture are important contributors to parenting, greater attention needs to be paid to potential mediators of genetic and environmental effects. In other words, which phenotypes account for these effects. As illustrated in Fig. 8.2, the impact of genetic or environmental factors on parenting may be explained by more proximal determinants, such as parents' personality or the quality of their relationships with their spouses and children. Identification of proximal determinants of parenting would provide a more complete picture of the paths through which genes and the environment affect parenting, and, potentially yield specific targets for intervention efforts. Very few studies, however, have focused on mediators of genetic and environmental effects.

Kendler and colleagues first reported that parents' stable characteristics (personality and mental health) account for differences in parenting styles across families, but did not distinguish between genetic and shared environmental effects on parenting. In a subsequent twin study, the associations between parents' personality and parenting were explained by environmental rather than genetic mechanisms (Spinath and O'Connor 2003). However, this relatively small study of 98 twin pairs may have been underpowered to determine whether "between family" differences are related to genetic or shared environmental factors. Associations between personality and parenting were also explored within the Twin-Offspring Study in Sweden, which included 909 adult same-sex twin pairs (TOSS; Neiderhiser et al. 2007). Biometric analyses indicated that overall, parents' personality characteristics explained about one-third of the genetic contributions to parent negativity and warmth (Ganiban et al. 2009), suggesting that personality does function as a mediator of genetic contributions to parenting. However, it is not the only mediator and it is likely that other genetically influenced characteristics such as mental health also account for parent-based genetic contributions to parenting.

Likewise, what are the mediators of nonshared or shared environmental effects on parenting? Phenotypic research suggests that marital quality affects parental negativity and warmth (Erel and Burman 1995), and may contribute to nonshared environmental effects. As such, marital quality may be a key mediator of nonshared environmental influences on parenting. Ganiban et al. (2009) tested this possibility with the TOSS sample. Marital quality and parenting were moderately correlated for mothers and fathers, and this covariance was explained by genetic and nonshared environmental factors. However, when the effects of the parents' personality were statistically controlled, the residual association between marital quality and parenting was primarily accounted for by nonshared environmental factors (Ganiban et al. 2009). This pattern of findings suggests that marital quality is an important contributor to nonshared environmental influences on parenting. Studies have not examined sources of shared environmental effects on parental control or monitoring. These effects could represent current cultural norms, living in similar neighborhoods, or tendencies that were shaped during their childhood. However, these hypotheses remain to be tested.

Future research should also attend to factors that may moderate the relative contributions of genes and environments to parenting. For example, parent gender and the broader family context could be important moderators of genetic and environmental effects. Although some studies have found evidence of parent gender effects on the contributions of genetic and environmental factors to mothers' versus fathers' parenting styles, most studies have not explored this issue. Last, the degree to which adults' environments moderate genetic and environmental contributions to parenting is not understood. Parenting does not occur within a vacuum: other family relationships, as well as the social and cultural context of families could regulate gene expression, and accordingly, the magnitude of genetic and environmental contributions to parenting. Consistent with this possibility, Ulbricht et al. (in press) found that marital conflict moderated genetic and environmental contributions to parental negativity. Although phenotypic research suggests that there are significant cultural variations in parenting, there is no extant research examining the role of culture on genetic and environmental contributions to parenting.

8.10 Child-Based Research

Child-based research has also explored genetic and environmental contributions to the parent-child relationship and to specific parenting dimensions. Again, however, little attention has been paid to the parents' age: most studies report children's ages, but not parental age. In addition, within child-based research, the focus has also been on specific dimensions of parenting behavior, and only a few attachment studies have examined child-based contributions to the quality of the child-parent bond.

In regard to relationship quality, a few twin studies have examined the degree to which genetic and environmental factors contribute to young children's observed attachment security. Collectively, they suggest that shared and nonshared environmental factors account for most variance in attachment security (Bokhorst et al. 2003; Finkel and Mathen 2000; O'Connor and Croft 2001). Small sample sizes, however, are a critical problem for each of these studies: it is likely that they were underpowered to differentiate between genetic and shared environmental contributions. Nevertheless, a larger scaled twin study supported these findings without detecting genetic contributions to attachment security, and further noted that associations between parenting and attachment security were explained by environmental factors (Roisman and Fraley 2008). Similarly, many molecular genetic studies have failed to find associations between specific candidate genes and children's attachment security (e.g., Luikik et al. 2011), and when significant associations are found, they have not been replicated consistently (Gervai 2009). Other studies have reported significant interactions between specific gene variants related to catecholamine functioning (COMT, DRD4, 5-HTTLPR; e.g. Luijk et al. 2011) or between genes and parenting quality (Barry 2008; Gervai et al. 2007) in predicting attachment quality. However, these findings require replication before strong conclusions can be drawn.

Within the behavioral genetic studies cited, child-based genetic influences on attachment quality are minimal. However, assessments of attachment security were based on the organization of children's emotional and behavioral responses to their caregivers during times of stress, and designed to reflect children's working models of their parents' responsiveness and protection. These types of assessments were not intended to examine how parents have adapted to their children's characteristics over time, and are not well-suited to estimate a child's role in the evolution of the attachment relationship. In contrast, parenting studies tend to focus on specific behavioral dimensions, and it is likely that at this level of assessment, the impact of children on parents' behaviors is more salient.

When parent report measures of negativity are used, there are substantial child-based genetic contributions to parenting. Amongst a small sample of 3-year old twins from the UK, child-based genetic factors accounted for most variance in parents' reports of negativity ($h^2 = 0.55$; Deater-Deckard 2000). Similar estimates were reported for a larger sample of U.K. twins at 4 ($h^2 = 0.48$) and 7 ($h^2 = 0.42$) years. During adolescence, child-based genetic factors also account for significant variance in parents' self-reported negativity, with heritability estimates ranging from 0.30 to 0.55 (Moberg et al. 2011; Neiderhiser et al. 2004, 2007). Additional analyses have relied upon child reports or composites of parent and child reports (Burt et al. 2005; Elkins et al. 1997; Neiderhiser et al. 2004, 2007; Plomin et al. 1994). These studies yield comparable estimates of child-based genetic influences on parenting, and range from 0.17 to 0.60. However, there is a trend for heritability estimates to be slightly lower when child-reports are used, and lowest when observational ratings are used. Studies conducted with young children and with adolescents report very low genetic effects (Deater-Deckard 2000; Neiderhiser et al. 2004, 2007).

Heritability estimates for parental warmth are significant, but vary by assessment method as well. Studies that have used parent reports are indicative of moderate child-based genetic effects during early childhood ($h^2 = 0.46$; Deater-Deckard 2000), and adolescence ($h^2 = 0.46$ for maternal positivity, and 0.30 for paternal positivity). Child reports of parental warmth also yield evidence of moderate child-based genetic effects. However, these estimates are slightly higher for father positivity [h^2 ranging from 0.24 to 0.37 for fathers, and from 0.19 to 0.27 for mothers (Elkins et al. 1997; Neiderhiser et al. 2004, 2007)]. Slightly higher estimates are obtained for adults' retrospective accounts of the warmth they received when they were children (Kendler 1996). Last, observational measures of parent positivity have suggested very little genetic influences on parental warmth. Given the relatively few studies that have reported estimates for mother versus father positivity, it is unclear if this trend reflects different influences on mother and father behaviors.

A number of studies have also focused on parental control. Some studies have used general measures of parental control (Neiderhiser et al. 2004, 2007), while others have focused on specific forms of control, including physical discipline and punishment (Jaffee et al. 2004; Wade and Kendler 2000), limit setting (Wade and Kendler 2000), harsh discipline and negative control (Deater-Deckard 2000). Additional studies have examined child-based contributions to indices of overinvolvement (Moberg et al. 2011), and over protection (Kendler 1996). Across studies, child-based genetic effects range from nonsignificant to moderate, but are generally within the 0.10–0.30 range (Deater-Deckard 2000; Jaffee et al. 2004; Neiderhiser et al. 2004, 2007; Wade and Kendler 2000). Last, a series of papers have examined genetic contributions to parents' emotional over involvement in their children's lives and tendency to be overprotective. Moberg et al. (2011) reports higher child-based genetic effects for child reports of parental over involvement for girls (0.32–0.44) than for boys (0.26–0.29) from 16–20 years. Retrospective reports of parental overprotectiveness yield somewhat similar estimates, with adults' ratings of their mothers and fathers' overprotectiveness estimated as 0.29.

Shared and nonshared environmental factors also account for variance in parenting within most child-based studies. However, there is a wide range of estimates. For parent negativity, estimates of shared environmental contributions range from 0 to 0.61, while estimates of nonshared environmental contributions range from 0.08 to 0.78. When parent reports are used, moderate to high estimates of shared environmental effects to parent negativity are usually obtained (c^2 ranges from 0.23 to 0.48), while nonshared environmental effects tend to be more modest (e^2 ranges from 0.02 to 0.22; Deater-Deckard 2000; Larsson et al. 2008; Neiderhiser et al. 2004, 2007). The opposite pattern is observed for child-reported parent negativity: nonshared environmental estimates are moderate, while shared environmental estimates are very low (Neiderhiser et al. 2004, 2007). Similar variations in estimates of shared and nonshared environmental contributions are found for parent positivity and control. This general pattern suggests that parents tend to perceive themselves as treating their children similarly, leading to higher estimates of shared environmental effects on parenting. Children, however, are attuned to differences in the way parents treat themselves and their siblings. This tendency is expressed via higher estimates of nonshared environmental contributions to parental control.

Collectively, research has yielded considerable evidence that children's genetic makeup influences parenting. Which child behaviors or phenotypes account for genetic or environmental influences on parenting? Furthermore, when there is evidence of child-based effects on parenting, do we always understand the mechanisms that account for these effects? As illustrated in Fig. 8.2, these effects may reflect passive gene-environment correlation (path a) or evocative gene-environment correlation (child genotype→child behavior→parenting). A number of studies have focused on the first question, and examined the degree to which children's antisocial or aggressive behaviors act as mediators of child-based genetic contributions to parent negativity. For example, within the E-Risk twin sample, Jaffee et al. (2004) reported that young children's antisocial behavior accounts for significant genetic variance in mothers' negativity. These findings could reflect evocative GE correlation: children's negative behaviors elicit more negative parenting. It is also possible that genes that are shared by mothers and their biological children underlie the concurrent association between child behavior and mother negativity (i.e., passive GE correlation).

Several methods can be used to assess the presence of evocative GE correlation. Moderation of genetic contributions to parenting by children's characteristics would be consistent with an evocative effects model. In this instance the magnitude of child based genetic effects upon parenting *depends* on the child's behaviors. Two studies have examined whether child temperament moderates genetic and environmental contributions to parenting (Ganiban et al. 2011; South et al. 2008). Both studies have provided evidence of moderation, consistent with evocative GE correlation. Temperament moderates genetic and environmental contributions to the quality of the parent child relationship (South et al. 2008), and to specific parenting dimensions (Ganiban et al. 2011).

Recent studies have also utilized cross-lagged designs to examine the degree to which children's behavioral characteristics elicit changes in parenting over time. These studies include parent and child assessments at two time points, permitting estimation of their influence on each other over time, while controlling for their initial association and stability over time. Demonstrating that children's characteristics account for changes in parenting would be consistent with evocative GE correlation, especially if this effect is explained by child-based genetic factors. Neiderhiser and colleagues (1999) first examined the mutual influence of parenting and child adjustment across adolescence within the NEAD project. Findings were consistent with child-based evocative processes: adolescents' externalizing and depressive symptoms predicted mothers' and fathers' negativity over time, and genetic factors primarily accounted for these effects. Burt et al. (2005) identified small, but significant cross-lagged associations between children antisocial behavior and parent-child conflict across early adolescence: children's antisocial behavior at Time 1 predicted more conflict at Time 2, while more conflict at Time 1 independently predicted more antisocial behavior at Time 2. Importantly, children's impact on parenting over time was primarily explained by genetic factors. In contrast, cross-lagged parent effects on children's behaviors were primarily related to environmental factors. Similar findings are reported by Larsson et al. (2008) within a different set of preschool, same-sex twins. A recent cross-lagged study conducted by Moberg et al. (2011), suggests that there may be gender differences in child-based effects on parenting during adolescence. Within this latter study, girls' internalizing symptoms at Time 1 predicted changes in parents' emotional over involvement, and these effects were primarily explained by child-based genetic factors. Boy's internalizing problems did not predict changes in parenting. Therefore, the balance of evidence suggests that evocative GE correlation does contribute to the relationship between children's challenging behaviors and parental negativity. However, different effects may be present for boys and girls.

Adoption studies also represent a powerful method for identifying evocative child effects on parenting through eliminating the possibility of passive GE correlation since adoptive parents and children are genetically unrelated. These studies have provided evidence for evocative child effects on parenting during adolescence. For example, Ge et al. (1994) found that adopted children whose biological parents were antisocial tended to be the recipients of less warmth and harsher discipline from their adoptive parents. Consistent with an evocative effects mechanism, this association was mediated by the adoptees' own antisocial behavior. These findings were partially replicated by O'Connor et al. (1998) with an independent sample of adopted children and their families. However, a key limitation of these studies is that they did not account for prenatal influences on children's behavior. As a result, associations between biological parents' characteristics and the adopted children's behavior may have been influenced by prenatal factors such as mothers' drug use during pregnancy or emotional state. In other words, the observed evocative influences on parenting may represent biological effects, rather than genetic effects.

A final set of studies has started to disentangle evocative and passive GE correlation through combining child- and parent-based studies into a single model. In one series of studies, Neiderhiser et al. (2004, 2007) tested whether estimates of genetic and environmental contributions to parental warmth, negativity and control significantly differ across a child-based study (NEAD project) and a parent-based twin study (TOSS). Patterns of differences were then used to infer the presence of evocative GE correlation versus passive GE correlation. Analyses based upon mothers' parenting provided evidence of evocative GE correlation for negativity and control, and passive GE correlation for negativity and control (Neiderhiser et al. 2004). Similar results were reported for fathers, with one notable difference: for fathers only parental positivity appeared to be influenced by evocative processes (Neiderhiser et al. 2007). These results, thus, indicate that children may affect parenting through evocative and passive processes. However, the specific mechanism depends upon the dimension of parenting explored as well as parent gender. Although this approach highlighted the complexities of interpreting child-based genetic influences on parenting, a key limitation was that the passive and evocative GE paths to parenting behavior could not be assessed directly. This limitation is addressed within the Extended Children of Twins (ECOT) model (Narusyte et al. 2008). The ECOT design includes two twin cohorts within a single model: (1) a cohort of adult twins who are concordant for children; and (2) a child twin cohort. The inclusion of both cohorts permits simultaneous estimation of child- and parent-based genetic contributions to phenotypes, as well as the direct impact of the environment on phenotypes. Narusyte and colleagues have used ECOT to examine child and parent-based contributions to specific dimensions of negative parenting, emotional overinvolvement and criticism (Narusyte et al. 2008, 2011). Results suggest that mothers' criticism and emotional over involvement are influenced by child-based evocative processes. The same effect was not present for fathers: children's externalizing behavior was predicted by fathers' criticism. Thus, again, mothers' and fathers' parenting may be influenced by different mechanisms.

In summary, recent longitudinal and cross-generational studies have increased knowledge of the mechanisms that underlie child-based genetic influences on parenting. Evocative and passive GE correlation contribute to child-based effects, but current research also presents a complex picture of gene environment interplay. The relative importance of each mechanism may vary by parenting dimension, parent gender, and child gender. Results derived from TOSS and NEAD have started to identify different mechanisms for mothers and fathers related to parental positivity (Neiderhiser et al. 2007); Narusyte et al. (2011) work with the ECOT model is also suggestive of parent-based gender differences in child effects. Additional research is needed to further understand the potentially different effects of children on mothers and fathers, and the developmental importance of these differences. Last, children's gender may also influence their impact on parents (Moberg et al. 2011). This latter possibility has been explored rarely within current research, and represents a critical gap in the research literature. Such investigations would provide important insight into the dynamics of parents' relationships with their sons versus daughters.

Additional factors could also impact which mechanism is operative. For example, child age, parent personality, and family stress could all influence the degree to which passive versus evocative GE correlation occurs. Many have proposed that passive GE correlation is more likely to occur during infancy, while evocative GE correlation predominates as children become more mobile and capable of influencing their environments. However, although this developmental model makes intuitive sense, to date there is little empirical evidence of changes in the relative importance of passive and evocative GE correlation.

8.11 Work

Work relationships also form a key facet of adults' lives. Most adults spend more of their waking hours at their job than anywhere else. Concepts related to work and the work environment have been examined within a behavior genetic framework, primarily to understand the origins of organizational behavior. Ilies and colleagues have recently reviewed the behavior genetic literature on work and work-related phenotypes and placed organizational behavior within an evolutionary context (Illes et al. 2006). Their paper goes into a depth and breadth that is not possible in this chapter, so we recommend their paper for further reading.

The focus of genetic examination on work thus far has been on traits and attitudes leading to behaviors within organizations. For example, the following have been examined: job satisfaction, work values, perceptions of organizational climate, vocational interests, job and occupational switching, leadership, performance behaviors, and entrepreneurship. Reports of job satisfaction have been found to be heritable in several studies (Arvey et al. 1989, 1994). The study that did not find genetic influences on job satisfaction assessed the construct in ways that may not have truly been tapping into one's attitude towards the job (Hershberger et al. 1994).

A very small study of twins reared apart ($n = 43$ pairs) examined a range of work values that included achievement, comfort, status, altruism, safety and autonomy. The study found a range of heritabilities from 18 to 56 % (Keller et al. 1992). These findings were replicated in a later, larger ($n = 2401$ pairs) study (Arvey et al. 1994) that used somewhat different assessments of work values. The average heritability in this study was 35 %. There are a number of studies that have looked at the heritability of vocational interests. Lykken et al. (1993) found fairly high heritabilities for vocational interests—around 50 %. Examined the genetic and environmental influences on an individual's likelihood of leaving his/her job, and the frequency with which an individual changes his/her job. As with most other traits, these were partially heritable (36 and 26 %, respectively).

The origins of leadership and related traits are often debated: are leaders born or made? Johnson et al. (1998) looked at leadership abilities and behaviors and found significant genetic influences. Arvey, Zhang et al, put a developmental lens on their research and looked at genetic and environmental influences on leadership in high school. High school leadership was highly heritable, and the heritability increased with the number of leadership roles taken on in high school. This finding suggesting the possibility that early genetically influenced leadership characteristics set the stage for later leadership roles and behavior. Another study (Olson et al. 2001) showed that attitudes towards being a leader were heritable. In a primarily female sample, Nicolaou et al. 2008 found moderate heritability of entrepreneurial behaviors, and that sensation-seeking partially mediated the genetic effects on entrepreneurship. Zhang and colleagues expanded on this study using a sample from the Swedish Twin Registry that had a more equal number of men and women. They found high heritability for women (approximately 60 %), but none for men. Entrepreneurship in men was influenced primarily by shared environmental influences. The findings of Nicolaou et al. (2008) were replicated in that extraversion and neuroticism mediated the genetic influences for women. Extraversion partially mediated the shared environmental influences for men. Finally, Nicolaou and Shane (2009) used a small twin sample from MIDUS (National Survey of Midlife Development in the United States; $n = 650$ pairs) to examine the tendency to be an entrepreneur across gender, as well as factors related to self-employment. They found that being self-employed is substantially heritable for men and women, as is the intention to be self-employed in the future.

As with most other behavioral phenomenon, those of organizational psychology are also influenced by genetic factors. Ilies has an extensive discussion of what this means for organizational psychology and how it can move the field forward, which the current authors highly recommend. In looking towards the future, there are large gaps in the genetically informed literature of work. No studies were found about relationships within the workplace, or how work and work relationships are linked with interpersonal relationships outside the workplace. There should also be a focus on how work changes across the lifespan and what factors influence stability and change over time.

8.12 Conclusion

Behavioral genetic research has been crucial for better understanding relationships of midlife—parent-child, marital, and work. This work has enhanced understanding of the joint contributions of genetic and environmental factors to adult relationships, and the pathways through which both factors affect relationships during adulthood. In all cases, genetic influences have been found to play a substantial role in individual variability in these concepts. Personality seems to play an important role in mediating these influences. As discussed previously, these characteristics could elicit similar reactions from different social partners, leading to similarities across different relationships or lead adults to create social niches that support these characteristics (Caspi 1998; Robins et al. 2002).

These findings, however, do not negate developmental theories that focus on the importance of one's environment. Behavior genetic research informed us about the genetic component of midlife, but just as importantly, this research has reaffirmed the critical role of environmental influences that play a role in midlife relationships. During midlife, adults' unique experiences play key roles in determining relationship quality, while the effects of their shared experiences are minimal.

As behavior genetic research on social relationships moves forward, more attention should be paid to the mechanisms through which genetic and environmental factors affect relationships and the roles of gender, ethnicity and race as moderators of genetic and environmental influences. To start, mediators of genetic and environmental effects are poorly understood. This is important because mediators will contribute to our understanding of mechanisms, and from a mental health perspective, aid in promoting healthy relationships, potentially through therapeutic relationships. The goal of many types of therapy is often to produce changes in individuals' social environments. These therapies may be all the more effective when based on the knowledge that a particular psychosocial intervention has the potential to alter the mechanisms of genetic expression for a particular trait. Work is in progress on the best ways to use genetically informative studies to inform intervention efforts (Leve et al. 2010). Awareness of the different genetic mechanisms underlying both individual traits and relationships will also help to inform clinicians as to why certain forms of intervention are effective in treating some problems but not others.

The impact of broader environmental factors on social relationships also requires additional attention in behavioral genetic research. Culture, religion, and socioeconomic class contribute to one's social context, and may regulate the extent to which genetically-influenced tendencies are permitted free expression as opposed to being controlled. For example, Shanahan and Hofer (2005) propose that some environments may suppress genetic expression through placing stringent restrictions upon behavior and relationships or limiting one's social circle. On the other hand, non-restrictive environments could facilitate genetic expression through providing a greater variety of social opportunities and acceptance of a wider range of behaviors. As illustration, there is some evidence that genetic contributions to alcohol use and age at first intercourse are affected by one's social context (Dunne et al. 1997; Koopmans et al. 1999). It is likely that estimates of heritability or environmental contributions to social relationships could also depend upon social context.

There is also a great need to focus on ethnicities other than white. The bulk of behavioral genetic, and even medical genetic research, is conducted on white individuals, which is an issue both from a genetic and an environmental point of view. From the context standpoint, there are obviously different cultures, social mores, and levels of social stimulus (e.g. discrimination) that are linked with race and ethnicity. These important changes are being missed in a large part of the genetic research conducted today. From a genetic standpoint, there are known differences in allele frequencies and effects based on race and ethnic category. For example, APOE4 is a known risk factor for Alzheimer's disease (AD) in white populations. However, APOE4 has a higher frequency in blacks than whites, yet there is a weak to nonexistent association with AD in black populations. This is only one known example, and there are likely many more. Such differences might be the result of different epistatic effects, or different gene by environment effects. Focusing research efforts on these differences can only further our understanding of mechanisms and outcomes for all people.

Likewise, we need to focus greater attention on determining whether or not the same models of genetic influence and gene-environment interplay apply equally to males and females. A common reason for not examining gender effects is issues of power, which is a valid concern. Future studies need to work towards rectifying this, either by conducting larger studies, or by pooling data across a number of smaller studies. There are current efforts working on harmonizing across twin studies; iGEMS (The Consortium on Interplay of Genes and Environment across Multiple Studies; Pedersen et al. 2012) is an excellent example. This group is harmonizing across nine twin studies in a number of areas including depression, social support, well-being, and loneliness. By doing so, they will have extensive power to look at G × E interplay, as well as gender effects. They can also serve as a model for future harmonization and pooling efforts.

Finally, in order to further our understanding of family and social relationships, and the role that genetics and other biological processes play in these relationships, we need to foster integrative research combining all relevant domains. Science is currently moving at such rapid pace that we cannot expect scientists to be fully versed in both the biological and the behavioral/social sciences. Just as we encourage good communication in families, we need to encourage and develop a common language that will facilitate communication across various areas of science in order to make this possible.

References

Arvey, R. D., Bouchard, T. J., Segal, N. L., & Abraham, L. M. (1989). Job satisfaction: Environmental and genetic components. *Journal of Applied Psychology, 74*, 187–192.

Arvey, R. D., McCall, B. P., Bouchard, T. J., Taubman, P., & Cavanaugh, M. A. (1994). Genetic influences on job satisfaction and work values. *Personality and Individual Differences, 17*, 21–33.

Bakersmans-Kranenburg, M.J., & van IJzendoorn, M.H. (2008). Oxytocin receptor (OXTR) and serotonin transporter (5-HTT) genes associated with observed parenting. *Social Cognitive Affective Neuroscience, 3*, 128–134.

Barry, R. A., Kochanska, G., & Philibert, R. A. (2008). GxE interaction in the organization of attachment: Mothers' responsiveness as a moderator of children's genotypes. *Journal of Child Psychology and Psychiatry, 49*, 1313–1320.

Beach, S. R., Fincham, F. D., & Katz, J. (1998). Marital therapy in the treatment of depression: Toward a third generation of outcome research. *Clinical Psychology Review, 18*, 635–661.

Belsky, J., & Pluess, M. (2009). Beyond diathesis stress: Differential susceptibility to environmental influences. *Psychological Bulletin, 135*, 885–908.

Boyle, G. J., Matthews, G., & Saklofske, D. H. (2008). Personality theories and models: An overview. In G. J. Boyle, G. Matthews, & D. H. Saklofske (Eds.), *Personality theory and assessment* (pp. 1–30). Los Angeles: Sage.

Blum, J.S., Mehrabian, A. (1999). Personality and temperament correlates of marital satisfaction. *Journal of Personality, 67,* 93–125.

Boivin, M., Perusse, D., Dionne, G., Saysset, V., Zoccolillo, M., & Tarabulsy, G.M. (2005). The genetic-environmental etiology of parents' perceptions and self-assessed behaviours toward their 5-month-old infants in a large twin and singleton sample. *Journal of Child Psychology and Psychiatry, 46,* 612–630.

Bokhorst, C.L., Bakersmans-Kranenburg, M.J., Fearon, R.M.P., van IJzendoorn, M.H., Fonagy, P., & Schuengel, C. (2003). The importance of shared environment in mother-infant attachment security: A behavioral genetic study. *Child Development, 74,* 1769–1782.

Boomsma, E.I., de Geus, E., van Baal, G.C.M. & Koopmans, J.R. (1999). A religious upbringing reduces the influence of genetic factors in disinhibition: Evidence for interaction between genotype and environment on personality. *Twin Research, 2,* 115–125.

Bowlby, J. (1973). *Separation: Anxiety and anger attachment and loss* (Vol. II). New York: Basic Books.

Brussoni, M. J., Jang, K. L., Livesley, W. J., & Macbeth, T. M. (2000). Genetic and environmental influences on adult attachment styles. *Personal Relationships, 7,* 283–289.

Burt, S.A., McGue, M., Krueger, R.F., & Iacono, W.G. (2005). How are parent-child conflict and childhood externalizing symptoms related over time? Results from a genetically informative cross-lagged study. *Development and Psychopathology, 17,* 145–165.

Caspi, A. (1998). Personality development across the lifespan. In N. Eisenberg (Ed.), *Handbook of Developmental Psychology Vol. III: Social, emotional, and personality development* (5th ed., pp. 311–388). New York: Wiley.

Cloninger, C. R., Svrakic, D. M., & Przybeck, T. R. (1993). A Psychobiological model of temperament and character. *Archives of General Psychology, 50,* 975–990.

Cohn,D., Cowan, P.A., Cowan, C, & Pearson, J. (1992). Working models of childhood attachment and couples' relationships. *Journal of Family issues, 13,* 432–449.

Cowan, P.A., Cohn, D.A., Cowan, C.P., & Pearson, J.L. (1996). Parents' attachment histories and children's externalizing and internalizing behaviors: Exploring family systems models of linkage. *Journal of Consulting and Clinical Psychology, 64,* 53–63.

Cox, M.J., & Paley, B. (1997). Families as systems. *Annual Review of Psychology, 48,* 243–267.

Crawford, T. N., Livesley, W. J., Jang, K. L., Shaver, P. R., Cohen, P., & Ganiban, J. (2007). Insecure attachment and personality disorder: A twin study of adults. *European Journal of Personality, 21,* 191–208.

Deater-Deckard, K. (2000). Parenting and child behavioral adjustment in childhood: A quantitative genetic approach to studying family processes. *Child Development, 71,* 468–484.

Donnellan, M. B., Burt, S. A., Levendosky, A. A., & Klump, K. L. (2008). Genes, personality, and attachment in adults: A multivariate behavioral genetic analysis. *Personality and Social Psychology Bulletin, 34,* 3–16.

Dunne, M. P., Martin, N. G., Statham, D. J., Slutske, W. J., Dinwiddie, S. H., & Bucholz, K. K. (1997). Genetic and environmental contributions to variance in age at first sexual intercourse. *Psychologyical Science, 8,* 211–216.

Elkins, I.J., McGue, M., & Iacono, W.G. (1997). Genetic and environmental influences on parent-son relationships: Evidence for increasing genetic influence during adolescence. *Developmental Psychology, 33,* 351–363.

Erel, O., & Burman, B. (1995). Interrelatedness of marital relations and parent–child relations: A meta-analytic review. *Psychological Bulletin, 118,* 108–132.

Erikson, E. H. (1968). Life cycle. In D. L. Sills (Ed.), *International encyclopedia of the social sciences* (Vol. 9, pp. 286–292). New York: Crowell, Collier and MacMillan Inc.

Finkel, D., & Mathen, A. P. (2000). Genetic and environmental influences on a measure of infant attachment security. *Twin Research, 3,* 242–250.

Franz, C. E., York, T. P., Eaves, L. J., Prom-Wormley, E., Jacobson, K. C., Lyons, M. J., et al. (2011). Adult romantic attachment, negative emotionality, and depressive symptoms in middle aged men: A multivariate genetic analysis. *Behavior Genetics, 41*, 488–498.

Ganiban, J.M., Ulbricht, J., Spotts, E.L., Lichtenstein, P., Reiss, D., & Neiderhiser, J.M. (2009). Can parent personality explain links between marital satisfaction and parenting? *Journal of Family Psychology 23*(5), 646–660. NIHMSID 174245.

Ganiban, J.M., Ulbricht, J., Saudino, K.J., Reiss, D., & Neiderhiser (2011). Understanding child-based effects on parenting: Temperament as a moderator of genetic and environmental contributions to parenting. *Developmental Psychology 47*(3), 676–692.

Ge, X., Conger, R. D., Cadoret, R. J., Neiderhiser, J. M., Yates, W., Troughton, E., & Stewart, M. A. (1996). The developmental interface between nature and nurture: A mutual influence model of child antisocial behavior and parent behaviors. *Developmental Psychology, 32*, 574–589.

Gervai, J. (2009). Environmental and genetic influences on early attachment. *Child and Adolescent Psychiatry and Mental Health, 3*, 25–37.

Gervai, J., Novak, A., Lakatos, K., Toth, I., Danis, I., Ronai, Z., et al. (2007). Infant genotype may moderate sensitivity to maternal affective communications: Attachment disorganization, quality of care, and the DRD4 polymorphism. *Social Neuroscience, 2*, 307–391.

Gray, J. A., & McNaughton, N. (2003). *The Neuropsychology of anxiety* (2nd ed.). New York: Oxford University Press.

Heath, A. C., Eaves, L. J., & Martin, N. G. (1998). Interaction of marital status and genetic risk for symptoms of depression. *Twin Research I,* 119–122.

Heath, A.C., Jardine, R., & Martin, N.G. (1989). Interactive effects of genotype and social environment on alcohol consumption in female twins. *Journal of Studies on Alcohol, 50*(1), 38–48.

Hershberger, S.L.; Lichtenstein, Paul; Knox, Sarah S. (1994). Genetic and environmental influences on perceptions of organizational climate. *Journal of Applied Psychology,* Vol 79(1), 24-33.

Horwitz, B. N., Ganiban, J. M., Spotts, E. L., Lichtenstein, P., Reiss, D., & Neiderhiser, J. M. (2011). The role of aggressive personality and family relationships in explaining family conflict. *Journal of Family Psychology, 25*, 174–183.

Ilies, R., Arvey, R. D., & Bouchard, T. J, Jr. (2006). Darwinism, behavioral genetics, and organizational behavior: A review and agenda for future research. *Journal of Organizational Behavior, 27*, 121–141.

Insel, T. R., Wang, Z. X., & Ferris, C. F. (1994). Patterns of brain vasopressin receptor distribution associated with social organization in microtine rodents. *Journal of Neuroscience, 14*, 5381–5392.

Jaffe, S.R., Caspi, A., Moffitt, T.E., Polo-Tomas, M., & Price, T.S. (2004). The limits of child effects: Evidence for genetically mediated child effects on corporal punishment but not on physical maltreatment (2005). *Developmental Psychology, 40*, 1047–1058.

Jerskey, B.A., Panizzon, M.S., Jacobson, K.C., Neale, M.C., Grant, M.D., Schultz, M., Eisen, S.A., Tsuang, M.T. & Lyons, M.J. (2010). Marriage and divorce: A genetic perspective. *Personality and Individual Differences, 49*(5), 473–478.

Jockin, V., McGue, M., & Lykken, D. T. (1996). Personality and divorce: A genetic analysis. *Journal of Personality and Social Psychology, 71*, 288–299.

Johnson, W., McGue, M., Krueger, R.F., & Bouchard, T.J. Jr. (2004). Marriage and personality: A genetic analysis. *Journal of Personality and Social Psychology, 86*(2), 809–833.

Johnson, A.M., Vernon, P.A., McCarthy, J.M., Molson, M., Harris, J.A., & Jang, K.L. (1998). Nature vs. nurture: Are leaders born or made? A behavior genetic investigation of leadership style. *Twin Research, 1*(4), 216–223.

Karney, B.R., Bradbury, T.N., Fincham, F.D., & Sullivan, K.T. (1994). The role of negative affectivity in the association between attributions and marital satisfaction. *Journal of Personality and Social Psychology, 46*, 413–424.

Karney, B.R., & Bradbury, T.N. (1995). The longitudinal course of marital quality and stability: A review of theory, method, and research. *Psychological Bulletin, 118*, 3–34.

Keller, L., Bouchard, T., Arvey, R., Segal, N., & Dawes, R. (1992). Work values: Genetic and environmental influences. *Journal of Applied Psychology, 77*, 79–88.

Kelly, E. L., & Conley, J. J. (1987). Personality and compatibility: A prospective analysis of marital stability and marital dissatisfaction. *Journal of Personality and Social Psychology, 52*, 27–40.

Kendler, K.S. (1996). Parenting: A genetic-epidemiologic perspective. *American Journal of Psychiatry, 153*, 11–20.

Kendler, K.S., & Baker, J.S. (2007). Genetic influences on measures of the environment: A systematic review. *Psychological Medicine, 37*, 615–626.

Kessler, R.C., Kendler, K.S., Heath, A., Neale, M.C., Eaves, L.J., (1992) Social support, depressed mood, and adjustment to stress: A genetic epidemiologic investigation.*Journal of Personality and Social Psychology*, Vol 62(2), 257–272.

Kim, S. J., et al. (2002). Transmission disequilibrium testing of arginine vasopressin receptor 1A (AVPR1A) polymorphisms in autism. *Molecular Psychiatry, 7*, 503–507.

Knafo, A., et al. (2007). Individual differences in allocation of funds in the dictator game associated with length of the arginine vasopressin 1a receptor RS3 promoter region and correlation between RS3 length and hippocampal mRNA. *Genes, Brain and Behavior, 7*, 266–275.

Koopmans, J. R., Slutske, W. S., van Baal, G. C. M., & Boomsma, D. I. (1999). The influence of religion on alcohol use initiation: Evidence for genotype x environment interaction. *Behavior Genetics, 29*, 445–453.

Larsson, H., Viding, E., Rijsdijk, F. V., & Plomin, R. (2008). Relationships between parental negativity and childhood antisocial behavior over time: A bidirectional effects model in a longitudinal genetically informative design. *Journal of Abnormal Child Psychology, 36*, 633–645.

Lee, S.S., Chronis-Tuscano, A., Keenan, K., Pelham, W.E., Loney, J., & H., van Hulle (2010). Association of maternal dopamine transporter genotype with negative parenting: Evidence for gene x environment interaction with child disruptive behavior. *Molecular Psychiatry, 15*, 548–488.

Leve, L. D., Harold, G. T., Ge, X., Neiderhiser, J. M., & Patterson, G. (2010). Refining intervention targets in family-based research: Lessons from quantitative behavioral genetics. *Perspectives on Psychological Science, 5*, 516–526.

Locke, H., & Wallace, K. (1987). Marital Adjustment Test. In N. Fredman & R. Sherman (Eds.), *Handbook of measurements for marriage and family therapy* (pp. 46–50). New York: Brunner/Mazel.

Losoya, S.H., Callor, S., Rowe, D.C., & Goldsmith, H.H. (1997). Origins of familial similarity in parenting: A study of twins and adoptive siblings. *Developmental Psychology, 33*, 1012–1023.

Luijk, M.P.C.M., Roisman, G.I., Haltigan, J.D., Henning, T., Both-LaForce, C., & van IJzendoorn, M.H. (2011). Dopaminergic, serotonergic, and oxytonergic candidate genes associated with infant attachment security and disorganization? In search of main and interaction effects. *Journal of Child Psychology and Psychiatry, 52*, 1295–1307.

Lykken, D. T., Bouchard, T. J, Jr, McGue, M., & Tellegen, A. (1993). Heritability of interests: A twin study. *Journal of Applied Psychology, 78*, 649–661.

McCrae, R. R., Costa, P. T., Ostendorf, D., Angleitner, A., Hrebickiova, M., Avia, M. D., et al. (2000). Nature over nurture: Temperament, personality, and life span development. *Journal of Personality and Social Psychology, 78*, 173–186.

McGue, M., & Lykken, D.T. (1992). Personality and divorce: A genetic analysis. *Journal of Personality and Social Psychology, 71*(2), 288–299.

Mehta, N., Cowan, P.A., & Cowan, C.P. (2009). Working models of attachment to parents and partners: Implications for emotional behavior between partners. *Journal of Family Psychology, 23*, 895–899.

Metsapelto, R., & Lukkinen, L. (2003). Personality traits and parenting: Neuroticism, extraversion, and openness to experience as discriminative factors. *European Journal of Personality, 17*, 59–78.

Mischel, W. (2004). Toward an integrative science of the person. *Annual Review of Psychology, 55*, 1–22.

Moberg, T., Lichtenstein, P., Forsman, M., & Larsson, H. (2011). Internalizing behavior in adolescent girls affects parental overinvolvement: A cross-lagged twin study. *Behavior Genetics, 41*, 223–233.

Narusyte, J., Neiderhiser, J.M, D'Onofrio, B.M., Reiss, D., Spotts, E.L, Ganiban J.M., & Lichtenstein, P. (2008). Testing different types of genotype-environment correlation: An extended children-of-twins model. *Developmental Psychology, 44,*1591–1603.

Narusyte, J., Neiderhiser, J. M., Andershed, A. K., D'Onofrio, B., Reiss, D., & Spotts, E. (2011). Parental criticism and externalizing behavior problems in adolescents—the role of environment and genotype-environment correlation. *Journal of Abnormal Psychology, 120*(2), 365–376.

Neiderhiser, J.M, Reiss, D., Hetherington, E.M., & Plomin, R. (1999). Relationships between parenting and adolescent adjustment over time: Genetic and environmental contributions.*Developmental Psychology, 35*, 680–692.

Neiderhiser, J.M., Reiss, D., Lichtenstein, P., Spotts, E.L., & Ganiban, J. (2007). Father-adolescent relationships and the role of genotype-environment correlation.*Journal of Family Psychology, 21*(4), 560–5.

Neiderhiser, J.M., Reiss, D., Pedersen, N.L., Lichtenstein, P., Spotts, E.L., & Hansson, K (2004). Genetic and environmental influences on mothering of adolescents: A comparison of two samples. *Developmental Psychology, 40*, 335–351.

Nes, R.B., Roysamb, E., Harris, J.R., Czajkowski, N, & Tambs, K. (2010). Mates and marriage matter: Genetic and environmental influences on subjective wellbeing across marital status. *Twin Research and Human Genetics, 13*(4), 312–321.

Neumann, I. D. (2008). Brain oxytocin: A key regulator of emotional and social behaviours in both females and males. *Journal of Neuroendocrinololgy, 20*, 858–865.

Nicolaou, N., & Shane, S. (2009). Can genetic factors influence the likelihood of engaging in entrepreneurial activity? *Journal of Business Venturing, 24*, 1–22.

Nicolaou, N., Shane, S., Cherkas, L., Hunkin, J., & Spector, T. D. (2008a). Is the tendency to engage in entrepreneurship genetic? *Management Science, 54*, 167–179.

Nicolaou, N., Shane, S., Cherkas, L., & Spector, T. D. (2008b). The influence of sensation seeking in the heritability of entrepreneurship. *Strategic Entrepreneurship Journal, 2*, 7–21.

O'Connor, T. G., & Croft, C. M. (2001). A twin study of attachment in preschool children. *Child Development, 72*, 1501–1511.

O'Connor, T. G., Deater-Deckard, K., Fulker, D., Rutter, M., & Plomin, R. (1998). Genotype-environment correlations in late childhood and early adolescence: Antisocial behavioral problems and coercive parenting. *Developmental Psychology, 34*, 970–981.

Olson, J.M., Vernon, P.A., Harris, J.A., Jang, K.L. (2001). The heritability of attitudes: A study of twins. *Journal of Social and Personality Psychology, 80*(6), 845–860.

Ophir, A. G., Wolff, J. O., & Phelps, S. M. (2008). Variation in neural V1aR predicts sexual fidelity and space use among male prairie voles in semi-natural settings. *Proceedings of the National Academy of Sciences, 105*, 1249–1254.

Ottman, R. (1996). Gene-environment interaction: Definitions and study designs. *Preventive Medicine, 25*, 764–770.

Pedersen, N. L., Christensen, K., Dahl, A. K., Finkel, D., Franz, C. E, Gatz, M., et al. (2012). IGEMS: The Consortium on Interplay of Genes andEnvironment Across Multiple Studies. *Twin Research and Human Genetics, 16*(1), 481–489.

Perusse, D., Neale, M.C., Heath, A.C., & Eaves, L.J. (1994). Human parental behavior: Evidence genetic influence and potential implication for gene-culture transmission. *Behavior Genetics, 24*, 327–335.

Picardi, A., Fagnani, C., Nistico, L., & Stazi, M.A. (2011). A twin study of attachment style in young adults. *Journal of Personality 79*, 965–992.

Plomin, R., Reiss, D., Hetherington, E.M., & Howe, G.W. (1994). Nature and nurture: Genetic contributions to measures of the family environment. *Developmental Psychology, 30*, 32–43.

Prichard, Z. M., Mackinnon, A. J., Jorm, A. F., & Easteal, S. (2007). AVPR1A and OXTR polymorphisms are associated with sexual and reproductive behavioral phenotypes in humans. Mutation in brief no. 981 Online. *Human Mutation, 28*, 1150.

Rhyne, D. (1981). Bases of marital satisfaction among men and women. *Journal of Marriage and the Family, 43*, 941–955.

Robins, R.W., Caspi, A., & Moffitt, T.E. (2002). It's not just who you're with, it's who you are: Personality and relationship experiences across multiple relationships. *Journal of Personality, 70*, 925–964.

Roisman, G. I., & Fraley, R. C. (2008). Behavior-Genetic study of parenting quality, infant-attachment security, and their covariation in a nationally representative sample. *Developmental Psychology, 44*, 831–839.

Rothbart, M. K., Ahadi, S. A., & Evans, D. E. (2000). Temperament and personality: Origins and outcomes. *Journal of Personality and Social Psychology, 78*, 122–135.

Rusting, C.L. (1998). Personality, mood, and cognitive processing of emotional information: Three conceptual frameworks. *Psychological Bulletin, 124*, 165–198.

Sanson,A., & Hemphill, S.A., & Smart, D. (2004). Connections between temperament and social development: A review. *Social Development,13*, 142–170.

Scarr, S., & McCartney, K. (1983). How people make their own environments: A theory of genotype→ environment effects. *Child Development, 54*, 424–435.

Shanahan, M. J., & Hofer, S. M. (2005). Social context in gene-environment interactions: Retrospect and prospect. *Journals of Gerontology: Series B, 60B*, 65–76.

Simpson, J.A., Collins, W.A., Tran, S., & Haydon, K.C. (2007). Attachment and the experience and expression of emotions in romantic relationships: A developmental perspective. *Journal of Personality and Social Psychology, 92*, 355–367.

South, S.C., Krueger, R.F., Johnson, W., & Iacono, W.G. (2008). Adolescent personality moderates genetic and environmental influences on relationships with parents. *Journal of Personality and Social Psychology, 94*, 899–912.

Spanier, G. B. (1976). Measuring dyadic adjustment: New scales for assessing quality of marriage and similar dyads. *Journal of Marriage and the Family, 38*, 15–28.

Spinath, F.M., & O'Connor, T.G. (2003). A Behavioral genetic study of the overlap between personality and parenting. *Journal of Personality, 71*, 785–808.

Spotts, E.L., Lichtenstein, P., Pedersen, N., Neiderhiser, J.M., Hansson, K., Cederblad, M., & Reiss, D. (2005). Personality and marital satisfaction: A behavioural genetic analysis. *European Journal of Personality, 19*, 205–227.

Spotts, E. L., Neiderhiser, J. M., Towers, H., Hansson, K., Lichtenstein, P., Cederblad, M., Pedersen, N. L., Elthammar, & Reiss, D. (2004). Genetic and environmental influences on marital relationships. *Journal of Family Psychology, 18*(1), 107–119.

Spotts, E. L., Prescott, C. A., Kendler, K. S. (2006). Examining the origins of gender differences in marital quality: A behavior genetic analysis. *Journal of Family Psychology, 20*(4), 605–613.

Torgersen, A. M., Grova, B. K., & Sommerstad, R. (2007). A pilot study of attachment patterns in adult twins. *Attachment and Human Development, 9*, 127–138.

Treboux, D., Crowell, J.A., & Waters, E. (2004). When "New" meets "Old": Configurations of adult attachment representations and their implications for marital functioning. *Developmental Psychology, 40*, 295–314.

Trumbetta, S.L. (2004) Middle age, marriage, and health habits of America's Greatest Generation: Twins as tools for causal analysis. In DiLalla, L.F.(Ed.) Behavior genetics principles: Perspectives in development, personality, and psychopathology. Decade of Behavior. (p. 59–70) APA: Washington, DC.

Trumbetta, S., & Gottesman, I. (2000). Endophenotypes for marital status in the NAS-NRC twin registry. In J. L. Rogers & D. C. Rowe (Eds.), *Genetic influences on human fertility and sexuality* (pp. 253–269). Boston: Kluwer Academic Publishers.

Trumbetta, S.L. & Gottesman, I.I. (1997). Pair-bonding deconstructed by twin studies of marital status: What is normative. In: N. L. Segal, G. E. Weisfeld, & C. C. Weisfeld (Ed.), *Uniting Psychology and Biology: Intergrative Perspectives on Human Develoment: Essays in honor of Daniel G. Freedman.* Washington (D.C.): American Psychological Association. (pp. 485–491).

Trumbetta, S.L., Markowitz, E.M. & Gottesman, I.I. (2007). Marriage and genetic variation across the lifespan: Not a steady relationship? *Behavior Genetics, 37*, 362–375.

Ulbricht, J.A., Ganiban, J.M., Button, T.M.M., Feinberg, M., Reiss, D., & Neiderhiser, J.M. (in press). Marital Adjustment as a Moderator for Genetic and Environmental Influences on Parenting. *Journal of Family Psychology.*

Van IJzendoorn, M.H. (1992) Intergenerational transmission of parenting: a review of studies in non--clinical populations. *Developmental Review, 12*, 76–99.

Van IJzendoorn, M.H., Bakersmans-Kranenburg, M.J., & Mesman, J. (2008). Dopamine genes associated with parenting in the context of daily hassles. *Genes, Brain and Behavior, 7*, 403–410.

Wade, T.D., & Kendler, K.S. (2000). The genetic epidemiology of parental discipline. *Psychological Medicine, 30*, 1303–1313.

Walum, H., Lichtenstein, P., Neiderhiser, J.M., Reiss, D., Ganiban, J.M., & Spotts, E.L. (2012). Variation in the oxytocin receptor gene is associated with pair-bonding and social behavior. *Biological Psychiatry, 71*, 419–426.

Walum, H., Westberg, L., Henningsson, S., Neiderhiser, J. M., Reiss, D., Igl, W., et al. (2008). Genetic variation in the vasopressin receptor 1a gene (AVPR1A) associates with pair-bonding behaviour in humans. *PNAS, 105*(37), 14153–14156.

Zuckerman, M. (2005). *Psychobiology of personality* (2nd ed.). Cambridge, U.K.: Cambridge University Press.

Chapter 9
Interpersonal Relationships in Late Adulthood

**Carol E. Franz, Ruth Murray McKenzie, Ana Ramundo,
Eric Landrum and Afrand Shahroudi**

More and more adults now experience what the popular literature calls a "second adulthood" after age 40 (Sheehy 1996) or perhaps what could now be considered a third adulthood. With longer life expectancies, many adults in the 21st century live on average 20–25 years after the age of 60. In the United States (U.S.), the percentage of adults age 65 and over more than tripled since 1900, to comprise 13.1 % of the population in 2010[1] (U.S. Department of Health and Human Services Administration on Aging 2011): this represents 40 million adults. In addition, the older population is growing older; the cohort of adults age 85 and older is 45 times larger than in 1900. Much of the focus on adult relationships in behavior genetics, however, has been on early adult life relationships, mating, procreation, and the early years of marriage rather than on relationships in later adulthood.

Although later life has many continuities with earlier adulthood, it also poses unique challenges to relationships. Living situations vary greatly in older adults. Approximately 95 % of men and women have been married at least once in their lifetime by age 65. Yet over two-thirds of American men over 65 live with a spouse (69.9 %) as compared with only 41 % of women. These rates reflect higher rates of re-marriage among men following divorce and/or widowhood as well as

---

[1]For simplicity we refer here to population statistics in the United States of America. Demographics vary worldwide but overall life expectancies have been increasing.

C.E. Franz (✉) · A. Ramundo · E. Landrum · A. Shahroudi
Department of Psychiatry, University of California, San Diego, San Diego, USA
e-mail: cfranz@ucsd.edu

C.E. Franz
Center for Behavioral Genomics, Twin Research Laboratory, UCSD School of Medicine, 9500 Gilman Drive, La Jolla, CA, USA

R.M. McKenzie
Boston University, Boston, USA

© Springer Science+Business Media New York 2015
B.N. Horwitz and J.M. Neiderhiser (eds.), *Gene-Environment Interplay in Interpersonal Relationships across the Lifespan*,
Advances in Behavior Genetics 3, DOI 10.1007/978-1-4939-2923-8_9

women's longevity. Regardless of previous marital status, a higher percentage of older women live alone compared with older men (Figs. 9.1 and 9.2). Among young adults, living alone is more commonly due to singleness while among older adults it is more likely due to widowhood. Only a small percentage of the elderly population (4.1 %) lives in institutional settings such as nursing homes; an equivalent percentage of older adults live in households that included grandchildren. Unlike interpersonal relationships in other periods of life, old age relationships inevitably end with death. For adults who are married, longer life expectancies

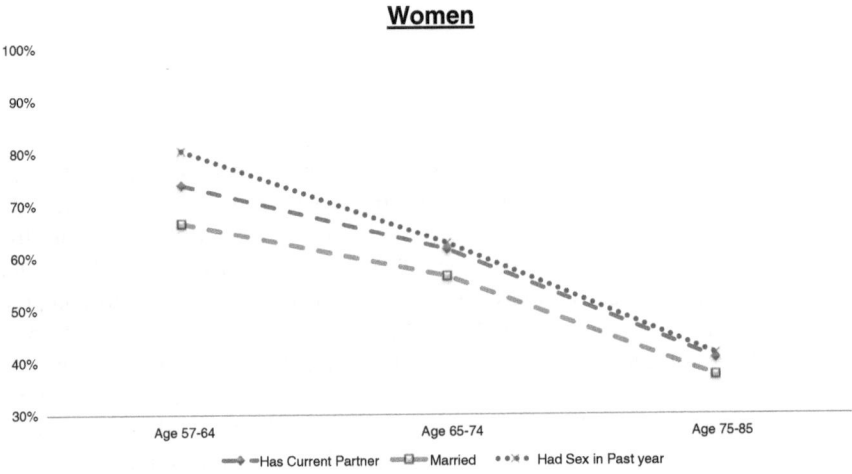

Fig. 9.1 Percentage of women married in intimate relationship and having sex in three age groups (Waite et al. 2009)

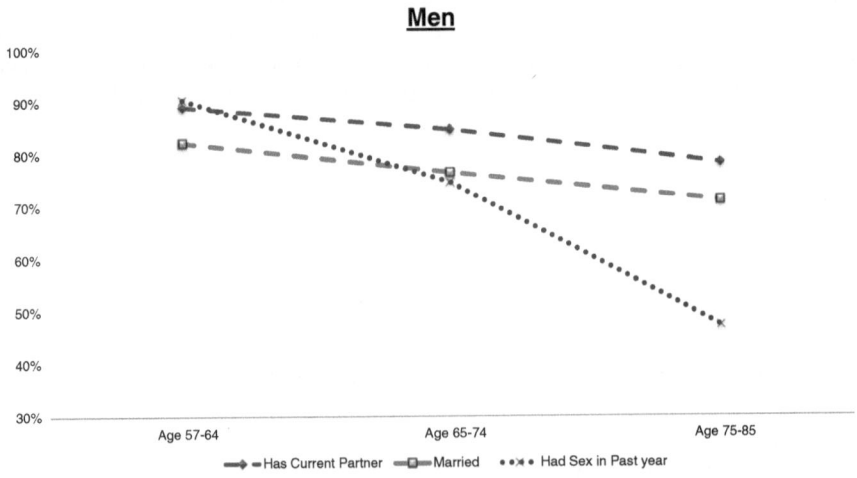

Fig. 9.2 Percentage of men married in intimate relationship and having sex in three age groups (Waite et al. 2009)

mean that marriages extend well past the years bearing and raising children or grandchildren. The longer a person lives discontinuities in important relationships are more likely to occur—but they may occur more frequently in some individuals compared with others and may be likely to occur for different reasons than interpersonal discontinuities earlier in life. Given the historical recency of the phenomenon of substantial longevity after age 65, the study of long-lasting and later-life interpersonal relationships provides interesting conceptual and analytic challenges for behavior genetics.

To add to the challenge of understanding interpersonal relationships in older adults, one of the most frequently studied relationships—marriage—has a complex history that requires the integration of evolutionary, cultural, psychosocial, and behavioral genetic perspectives. The concept of marriage as a legal/contractual relationship that identifies two people as spouses is relatively recent in evolutionary terms. Yet, across time and cultures some form of marriage-like social, legal and/or economic bonds exist that establish kinship. At a minimum, this kinship relationship functioned to provide an economically and psychologically secure base in which to procreate and/or rear offspring (Coontz 2005). In some countries, cultural and legal issues affecting marital status have changed radically during the past 100 years. A number of western countries, for instance, have experienced transformative alterations in laws and mores concerning divorce, intermarriage between racial groups and/or same sex couples, abortion, and women's rights (including property ownership, inheritance rights, and the right to vote). At the same time as social control over intimacy decreased in some cultures, it has increased in others (e.g., Iran). Access to birth control, in vitro fertilization, abortion, and "the sexual revolution" make it more possible to choose when to marry, when to procreate, and to be able to control family size. However, social and personal control over reproduction and intimacy has unknown ramifications for later life. Indeed, as sex, gender, and reproduction become more separated, one of the first steps in thinking about the behavior genetics of later life relationships is to recognize the heterogeneity and complexity of interpersonal phenotypes—even ones so deceptively simple and basic as marital status. As institutional/environmental infrastructures that regulate interpersonal relationships change, different patterns and forms of late-life relationships may emerge that can be examined in terms of behavioral genetics and adaptive processes.

The goal of this chapter is to review the current behavior genetic literature on interpersonal relationships in older adults and to propose areas for future study. A thorough review of the phenotypic literature on later-life relationships is beyond the scope of this chapter, however, we at times provide examples of this literature in order to provide a context or comparison for the behavior genetic literature. For this review, we define "older adults" as adults age 60 or older; younger adult relationships are addressed in other chapters. We begin by providing a theoretical framework for thinking about later life relationships, then examine the relevant behavior genetic research literature. Much of this research focuses on five main areas: marital phenotypes (e.g., marriage, divorce), other relationships (e.g., sibling, co-twin), social support, dyadic influences, and the extent to which relationships benefit health, wellbeing, or cognition in later life over and above genetic influences.

9.1 Theoretical Perspectives on Interpersonal Relationships in Older Adults

In this section, we present several alternative theoretical models that provide different insights into the etiology of and factors influencing later life relationships: these include Erikson's theory of psychosocial-development, Attachment theory, and Convoy theory. We then examine some evolutionary and behavior genetic theoretical contributions to understanding adult interpersonal relationships. This overview is not intended to present all possible theories of aging; we chose these particular models because they vary in the extent to which they integrate genetic, biological, psychological, cultural and life-course developmental perspectives on how relationships evolve across the life course.

9.1.1 Erikson's Theory of Lifespan Development

Eriksonian theory serves as one example of an attempt to integrate psychobiological, psychosocial, and cultural influences on normative development across the life course (Erikson 1963; Franz and White 1985). The developing person adapts to a series of age, life-stage, biological, and culturally determined challenges that include the resolution of biological imperatives such as attachment, sexuality, hormones, procreation, caring, and death. Development occurs in the context of ever-expanding interpersonal spheres and ever more complex intrapersonal development. Thus a person grows from dependence on its caregivers to interdependency with friends, lovers, family and community. Depending on life choices, ongoing adult relationships will likely include sexually intimate and committed relationship(s) as well as kinship/family, parental, friendship, and community/institutional connectedness. Each life stage involves negotiating developmental tasks and psychological maturation, ranging from developing a sense of trust in infancy to generativity (versus self-absorption) in middle adulthood to integrity (versus despair) in old age (Erikson 1963).

Sexual intimacy and procreation are seen as the developmental task/challenge of early adulthood. The interpersonal challenge for the midlife adult, Erikson writes, is to care: "I use 'care' in a sense which includes 'to care to do' something, to 'care for' somebody, to 'take care of' that which needs protection and attention, and to 'take care not to do' something destructive" (Evans 1967). For Erikson, middle and later adult life involves a new sense of "needing to be needed" in interrelationships. Erikson wrote little, however, about whether new or different qualities of relationships emerged in the last life stage, focusing instead on the emerging interiority and self-reflection that came as a result of relational losses and encroaching physical limitations of late life.

Erikson's theory has been criticized as being too Freudian, too sexist, and too rigid in its delineation of life stages. Its unique strength, however, is that it builds explicitly on epigenetic and psychobiological principles across the life course and assumes that there are ongoing important genetic influences on behavior through the influence of characteristics such as biology, temperament and personality (Franz and White 1985; Luyten and Blatt 2011; Vaillant et al. 1998; Vaillant 1977). His theory uniquely integrates the role of environment through the emphasis on the role of social and cultural forces interacting with individual characteristics. Most other developmental theories tend not to be as expansive or systematic about normative development across the life course and its implications for later life relationships as that of Erikson. Its expansiveness makes Eriksonian theory less useful for understanding individual differences. Despite their insufficiencies, lifespan theories can serve as rough roadmaps for what phenotypes might be important to examine when researchers do not yet have a way of thinking creatively beyond their experience.

9.1.2 Attachment Theory

Like Erikson, many developmental theorists identify infant attachment as the key foundation for *all* later development, whether or not they elaborate on its implications for later life interpersonal relationships. At the very heart of attachment theory lies the theoretical assumption that an infant has an evolutionarily based motivation to behave in ways to attract the help of a caregiver/attachment figure when it feels distressed and that attachment figures will be responsive to these cues (Ainsworth 1985; Bowlby 1969; Fraley et al. 2005). This sense of having a secure base and safe haven enables the child to begin the necessary developmental process of individuation and forms the foundation for ongoing relationships. Attachment, then, involves a complex interplay of genetic and environmental contributions from both sides of a dyadic relationship (Bowlby 1969; Gillath et al. 2008). Because attachment working models are created in the context of dyadic interpersonal interactions, other people play key roles in the development, maintenance, and change of attachment. Across the life course, shaped by ongoing interpersonal interactions and life experiences, attachment becomes consolidated into ways of thinking about and behaving towards other individuals who (possibly) mitigate stress and into predictable patterns of response to stress (Mikulincer and Shaver 2007; Nolte et al. 2011; Sbarra and Hazan 2008).

Current research and theory suggest that there is an evolutionary and neurobiological basis for attachment in adults. Secure attachment in adults helps to ensure the survival and development of secure attachment in the child. At the same time secure attachment in adults appears to reduces stress and increases feelings of security thereby contributing to adult health and wellbeing (Feldman et al. 2012; Fraley et al. 2005; Insel 2010; Schneiderman et al. 2012). Thus, the presence of trusted attachment figures increases the odds of survival for humans of all ages. Ainsworth and others differentiate between adult relationships involving affectional/sexual bonds (i.e. "pair-bonded" attachment relationships) and other close relationships. An adult affectional bond "is a relatively long-lived tie in which the partner is important as a unique individual, interchangeable with none other, from whom inexplicable, involuntary separation would cause distress, and whose loss would occasion grief" (Ainsworth 1985; p. 799). According to Bowlby, throughout the life course, "the capacity to make bonds with other individuals ... is regarded as a principal feature of effective personality functioning and mental health" (Bowlby 1988, p. 3). Alternatively, as Insel (2010) wrote: "We are, by nature, a highly affiliative species craving social contact. When social experience becomes a source of anxiety rather than a source of comfort, we have lost something fundamental." What Erikson, Bowlby and others were observing is what we now recognize as the electrical and biochemical "wiring" of the human brain to be connected with other people, and for that socialness to play a role in shaping the brain and later development (Hari and Kujala 2009; Insel 2010; Meyer-Lindenberg et al. 2011; Van Overwall et al. 2014).

In recent years, adult attachment has been examined primarily in the context of romantically linked adults ("sexual pair-bonded" adults) such as spouses, partners, boyfriends or girlfriends (Hazan and Shaver 1987). In adults, relationships differ in the extent to which they function as attachment relationships; adult attachment relationships involve more reciprocity, may be sexual, and may involve multiple people (Tancredy and Fraley 2006). In adults, secure attachment relationships (ones in which comfort and security can be found) with an intimate partner appear to be associated with better physical and psychological health (Hazan and Shaver 1987; Mikulincer and Shaver 2007; Mikulincer et al. 2009). According to Ainsworth, sexual pair-bonding is unique from other relationships because it involves three major behavioral systems: reproduction/mating; care-giving (for offspring and for each other) and the attachment system.

Attachment theory has not been extended systematically to later life relationships. Long term pair-bonded relationships, Ainsworth proposes, are likely to change over time; "as the relationship persists, the care-giving and attachment components are likely to become relatively more important and may sustain the relationship even though sexual interest has waned" (Ainsworth 1985, p. 805). Ainsworth proposed that social networks and social support systems, when they function well, do so because they are based on affectional bonds and attachment relationships (Ainsworth 1985). In particular she noted that adult relationships with siblings and parents (kinships) tend to outlast relationships with friends, or spouses. Family bonds can promote gene survival by favoring relatives, and assistance from relatives generally can be obtained more easily than from non-relatives. Shared history promotes similar interests, values, mutual understanding without excessive communication; long relationships (whether spouse, friends, relatives) appear to benefit from shared experience.

Although the theoretical applications of attachment theory to older life relationships remains under-developed and research focuses on romantic relationships, attachment theory provides some framework for understanding the development, quality, and functions of adult interpersonal relationships that can guide research on late life relationships. As does Eriksonian theory, but without the imposed exoskeleton of bio-psycho-social stage theorizing of normative development, it presupposes genetic and environmental influences on adult relationships. Because of attachment theory's more narrow focus on one aspect of development (attachment), however, the ontology of individual differences is more apparent. A key shortcoming of much traditional and behavior genetic research on adult attachment, however, is that its primary focus has been on attachment relationships in younger adults and on romantic relationships (Franz et al. 2011).

9.1.2.1 Social Convoy Theory

An alternative theoretical model examines life course relationships in terms of "social convoys" (Antonucci et al. 2010). According to Antonucci et al. 2010,

"the metaphor of the convoy was chosen to recognize social relations not as simple singular events but rather as linked interactions that accumulate and develop over time. According to the convoy model, individuals are surrounded by close social relationships at various stages of the life span and across the life course....a convoy is meant to protect and to socialize the individual by providing help and guidance with life's challenges" (p. 435).

Antonucci also explains that convoys exist across the life-span, and can be summarized by six key features: (1) convoys include multiple relationships whose main function is social support; (2) to be effective in providing support, convoys need to be flexible and adaptive; (3) convoys are shaped by individual and situational characteristics of the person—women and men, for instance, tend to have different size convoys; (4) convoys affect health and wellbeing through their influence on social support; (5) convoys are dynamic and change over time; and (6) people actively shape their own development throughout life by selecting and deselecting members of their convoys (Antonucci, et al. 2010). Research has found that older adults' convoys are smaller than younger adults', older adults tend to be more satisfied with their convoys, report them as less demanding/less stressful, and tend to use more avoidant strategies for dealing with stressful members of a convoy. Older adults appear to prefer to maintain relationships that are more positive and distance themselves from demanding or difficult relationships. These biases toward emotionally satisfying relationships, though, are not necessarily limited to old age but may also be based on time-left-to-live since dying younger adults also show socioemotional selectivity in their relationships. According to Antonucci, no single network structure is optimal at all stages of life (Antonucci et al. 2010). Each component of Antonucci's social convoy model could be examined with genetically informative research designs.

Unlike Eriksonian or attachment theories, convoy theory does not subscribe to underlying biological motivational principles to explain development and change in relationships across the life course. Rather convoy theory is primarily descriptive about the structures, function and quality of social relationships without theorizing about underlying causes for these behaviors. Regardless, all three theories propose that, at its very base, the life course is defined by social networks, that key relationships in these networks provide "social support" and that the nature of adult relationships changes across the life course. Each of these theories predicts that the adult interpersonal relationships are likely to change both quantitatively and qualitatively across the life course but are also contiguous with the past.

9.1.2.2 Insights from Evolutionary and Behavior Genetics Approaches

A major tenet of psychological evolutionary theory has been that people evolved to live in small social groups (Buss 2009; Confer et al. 2010). Within this view, most attention has been given to two aspects of relationships: the creation of relationships that support the long gestation and weaning of human offspring (pair-bonding), and the dynamics of sexual attraction (and conflict). Some authors suggest that pair-bonding is simply an extension or a by-product of a biologically-based attachment system (Buss et al. 1998; Diamond et al. 2010). Thus, qualities that contribute to successful attachment behavior are also selected for in pair-bonding whereby more dependable, emotionally committed (and longer surviving) caregivers are more likely to have surviving offspring who are also more likely to be securely attached. Thus, consistent with attachment theory or convoy theory, a small group of interrelated adults and children that provides emotional and physical sustenance for each other can act as an extended secure base or social support network. It is not difficult to envision, across evolution, that successful related social groups would contribute to longevity and to late life relationships characterized by complex bonds involving attachment, interdependence, mutual caring, social support, and sexual ties. Psychobiological research also suggests that sexual attraction and adult attachment appear to have somewhat distinct neurobiological trajectories (Diamond, et al. 2010; Porges 2003; Porges and Furman 2011).

An issue of relevance for understanding social and genetic causation in the intergenerational transmission of traits is the question of whether and why intimate partners and spouses resemble each other. As articulated by Eaves et al. mates may be similar for at least three reasons (Eaves et al. 2011). Phenotypic assortative mating reflects the likelihood that for some characteristics people choose mates who are like themselves. Assortative mating assumes that mate selection is nonrandom ("like tends to marry like"). An extension, or special case, of assortative mating is the phenomenon of social homogamy in which selection is assumed to be nonrandom. In social homogamy, selection is based on similar family or social background, making spouses appear more similar. Hypotheses regarding initial selection of a mate, however, may not inform later life intimate relationships. Most relevant to later life and to the process of unravelling mechanisms of social causation from genetic causation, mates may not start out as particularly similar but over time may become more similar to each other (spousal/mate interaction or dyadic effects). In social causation models, dyadic effects reflect increasing (or decreasing) interdependence or similarities within a relationship that may, for instance, affect characteristics such as health, behavior, cognitive status, attachment style (Kenny 1996).

Behavior geneticists have tended to focus more on early to middle-adult years and mating in their discussions of adult interpersonal relationships, with less emphasis on late life or other types of relationships. The study of mates is of intrinsic interest for questions such as the characteristics of long-term relationships (e.g., restricted versus unrestricted mating strategies, mate diversification, pair-bonding), similarities between spouses, or deciphering the effects of spouses on each other. Although one spouse is frequently depicted in non-genetically informed research as being an environmental influence on the other spouse, the extent of similarities between spouses may obscure, amplify, or interact with genetic influences. These notions of the consequences of mate selection have been examined in twin and non-twin studies in order to disentangle putative phenomena such as the physical and mental health benefits of mates. The unique contribution of twin studies to research on close adult relationships, however, is the ability to tease apart the effects of spouses from genetic influences.

Reiss et al. (Reiss et al. 2001; Reiss and Neiderhiser 2000), in particular, have attempted to integrate family systems models with models of reciprocal genetic and environmental processes on social behavior in order to account for outcomes such as maternal adjustment in younger adults. This model is described in greater detail in the chapter on younger adult relationships (see Chap. 7; this volume). In younger marriages, there is evidence that genetically influenced personal characteristics precede and predict marital processes and outcomes. In sum, heritable characteristics of one mate elicits and/or evokes responses from others (in this case, the spouse) which then play a role in heritable aspects of adjustment (Reiss et al. 2001). Due to the need for a large longitudinal sample that incorporates both twins and spouses, complex genetic models such as this have seldom been tested in older adults or long term relationships.

9.1.2.3 Summary

These theories and models all provide useful but somewhat different guidelines for examining late life relationships. Although Eriksonian and attachment theory are most explicitly developmental and open to the role of the interplay of genetic and environmental influences, Eriksonian theory appears rigid in its delineation of life stages, while attachment theory is weak beyond childhood or early adulthood in explaining ongoing intimate relationships and other non-attachment relationships. Convoy theory, while generating research on topics such as social convoys in later life and changes in convoys across the life course, is primarily descriptive, rather than explanatory. Evolutionary psychology tends to be silent about late life and long-term interrelationships. None of the theories provide much insight into dynamics of adult relational structures that do not involve children, marriage or mates. Finally, how and why early life genetic and environmentally influenced social interactional processes might play themselves out across the life course is not well articulated. Behavior genetics can benefit from these theories in a number of ways: to define relational phenotypes appropriate for different points in the lifecycle and different types of relationships, to evaluate whether different genes or environmental influences affect the same phenotype at different points in the lifecycle, to attend to issues of mate selection and their implications at different points in biological maturation and transitions, to examine interactions between genetic and environmental influences, and to recognize the role of social and culture forces. Longitudinal research of older twins in different types of relationships would allow for differentiation of genetic and environmental, reciprocal, and temporal effects of one person on the other's outcomes in long-term relationships.

9.2 Research on the Behavior Genetics of Later-Life Interpersonal Relationships

In the next section of this chapter, we examine research findings on the behavior genetics of interpersonal relationships and behavior in adults age 60 and older. Some studies we review will be of adults on the cusp of old age or will be longitudinal studies that include at least some older adults. As will be seen, most of these studies focus on marriage, divorce, social support, and spousal interaction; a few studies have examined other specific relationships such as those with siblings. Other chapters in this volume examine findings from twin studies regarding interpersonal relationships in younger adults as well as the current status of research examining genetic variants and social phenotypes (see Chap. 10) so those are not addressed here.

9.2.1 Marriage

Marriage is the most commonly studied area in the behavior genetics of adult interpersonal relationships. Although same-sex marriages and cohabitation are more common and more likely to be legally recognized, given the historical contexts of these studies, marriage is generally operationalized as a legal heterosexual relationship. The SATSA studies, however, include up to 18 % cohabiting couples (Spotts et al. 2005).

In younger adult twins (average age 40), heritabilities of ever being married ranged from 0.41 to 0.70 (Jerskey et al. 2010; Johnson et al. 2004; Kendler and Baker 2007; Middeldorp et al. 2005; Trumbetta and Gottesman 1997; Trumbetta et al. 2007). Heritabilities for having ever been divorced in younger adults (average age 40) ranged from approximately 0.3–0.59 (Jerskey et al. 2010; Jockin et al. 1996; McGue and Lykken 1992; Trumbetta and Gottesman 1997). Using a bivariate twin analysis, marriage and divorce were shown not to simply constitute opposite sides of the same coin in middle age men; rather, these two marital states were influenced by distinct genetic and environmental factors (Jerskey et al. 2010). It's important to note, however, that these studies often report best-fitting AE models in which genetic and common environmental influences (i.e., familial factors) are combined; this analytic approach tends to overstate heritabilities. Thus, it is difficult to ascertain whether the genetic component is significant on its own or if the results simply suggest familial influences. Curiously, some researchers view these heritabilities as accurately portraying heritability for marriage and divorce phenotypes because they believe that most participants have passed the point of experiencing marital status change by age 40.

To the authors' knowledge, only one study has examined heritability of marital phenotypes in adults over age 60—the NAS/NRC World War II sample (Trumbetta and Gottesman 1997; Trumbetta et al. 2007). A second study examined marital phenotypes in men at an average age of 55 (range 51–60; Franz 2013, unpublished data): the Vietnam Era Twin Study of Aging/VETSA sample (Kremen et al. 2012). The VETSA sample (born 1945—1955) was approximately seven years younger than the NAS/NRC cohort (born 1917–1927) at the time of these comparisons. Data collections occurred in 2003–2007 for the VETSA and in 1985 for the NAS/NRC). Both studies were of male twin pairs, participants in both studies were American war veterans and both studies had access to life history data from approximately age 40; the marital phenotype data reported here for the VETSA are unpublished. Despite similarities between the studies, the men were from different cohorts (about 30 years apart) with different legal and societal guidelines about marriage.

Comparisons of marital phenotypes in the NAS/NRC and VETSA cohorts find surprising consistency. In brief, heritability of ever having been married by the time one reaches approximately age 60 appears to be fairly consistent across the two cohorts (0.45, 0.47 NAS/NRC, VETSA respectively). The heritability of ever divorced is not significant in either sample but for different reasons: patterns of MZ/DZ correlations were quite different in the studies. For the NAS/NRC cohort at age 60, MZ correlations for ever divorced were 0.16 while DZ twins correlated 0.04; yielding a small and not significant heritability estimate. In VETSA twins at age 55, correlations were 0.30 and 0.43 for MZ and DZ twins respectively. Thus, with regard to divorce, VETSA DZ twins were as similar to each other as the MZ twins. It is unclear if these discrepancies are due to methodological issues, to age, or to cohort differences. With the greater freedom to divorce starting in the 1960s and 1970s in the United States, the VETSA twins (and their spouses) would have faced different legal and social contexts for divorce perhaps freeing them to more accurately display their behavioral tendencies. The NAS/NRC twins were also on average seven years older than the VETSA twins when heritability of divorce was examined, so they had had more opportunities for relationships to form and dissolve during that time. It is of interest that the heritability of marriage per se appeared to be relatively stable in adults regardless of cohort.

Cohort, gender, historical and life stage differences are seldom considered in these studies; marriage, divorce, and widow rates vary widely even within the U.S.. According to recent census figures, the percentage of people in the U.S. who are "never married" ranges from 85 % of 15–24 year olds to 34.5 % of adults age 25–34 to 11.6 % of adults age 35–59) and 4.5 % in adults 60 and over (U.S. Census Bureau 2000). The heritability of marriage per se may be a relatively uninformative phenotype in older adults. Divorce rates are lowest (9.9 %) and widowhood rates highest (27.1 %) in adults age 60 and over, compared with adults 25–59 (U.S. Census Bureau 2000). Heritability rates are likely to be influenced by the prevalence of a phenomenon in a group. This is especially apparent in Trumbetta and Gottesman's examination of the heritability of being married at each decade in the NAS/NRC sample. Heritabilities of "ever married" ranged from 0.26 at age 20, stabilizing at 0.38 at ages 30 and 40, then declining to be effectively zero at ages 60 and 70 (Trumbetta et al. 2007). The declines after age 50 were interpreted as due, in part, to spousal mortality and also to cultural shifts after the 1960s that contributed to more permissive attitudes toward divorce and cohabitation. More research is needed to determine whether these are cohort or age effects. The longer a person lives, the more likely they will exit and re-enter intimate relationships of different structures due to choice and fate. What is missing in the existing analyses is examination of patterns of relationships across the life course—patterns of behavior that may not become apparent until later adulthood.

Several marriage related phenotypes reflecting patterning of behavior were examined in the NAS/NRC and VETSA studies. Most of the theories we examined have some notion of pair-bonding. Lengthy pair-bonding is theoretically beneficial both for offspring and for the adults because of the mutual security and caregiving that can occur in those relationships. Heritability of a measure of pair-bonding up to age 60 for the NAS/NRC sample and up to age 55 for the VETSA sample was significant for both cohorts, and moderately strong [0.42 (0.33, 0.49)] for the NAS/NRC men [$h^2 = 0.27$ (0.16, 0.38) in the VETSA men] (Trumbetta and Gottesman 1997). Pair-bonding was operationalized as single/divorced/married only once; in late mid-life adults who had been married only one time were also in longer-term relationships than those with a history of divorce. Theoretically, stronger pair-bonding also could be operationalized in terms of number of marriages (if married), but having more marriages may also reflect processes of mate diversification or unrestricted mating strategies that can be potentially beneficial to the species. Heritability of the number of marriages was low but significant for both samples [0.28 (0.06, 0.37) and 0.17 (0.08, 0.26) NAS/NRC at age 60, and VETSA at age 55 respectively]. Age at first marriage—another indicator relevant to pair-bonding and reproduction—showed significant heritability in VETSA twins 0.45 (0.37, 0.52). Some ethological researchers view age of first marriage to be relevant to sexual pair-bonding since it would reflect genetic influences on procreation. Earlier marriage presumably allows more time to procreate (as well as more opportunity to divorce) whereas a first marriage at age 60 might reflect other motivations for marriage (Schleidt 1997).

Finally, findings on measures of marital quality in adults over 60 are rarely found in the behavior genetic literature on older adults. In the VETSA sample at age 55, however, none of the measures of marital conflict, marital agreement, and marital support were significantly heritable (heritabilities of 0.09, 0.07, 0.01 respectively) (Franz 2013, unpublished data). These results suggest that there are ways in which environmental influences may strongly impinge on some aspects of marital quality (e.g., the amount of discord or conflict in a relationship) whether or not it results in divorce. Most likely this is because the measures capture dyadic interactions (e.g., a marriage could be conflicted because of characteristics of the other person). Romantic attachment—feelings of trust toward a romantic relationship has not been examined in adults over 60, but is moderately heritable in VETSA men at age 55 (Franz et al. 2011). Insecure attachment in men 51–60 years old had strong genetic overlap with measures of anxiety and depression but also has specific genetic and environmental influences. There is also evidence that securely attached adults tend to be married to other securely attached adults and insecurely attached adults with other insecurely attached adults, thereby creating relational niches that will passively and actively affect both partners (Kirkpatrick and Davis 1994).

As discussed earlier, Ainsworth proposed that three biobehavioral systems comprise pair-bonded relationships: reproduction/mating; caregiving (for offspring, each other and others) and the attachment system (1985). Her work also suggested that characteristics by which mates are selected may vary by stage of life. Erikson proposed that by midlife, adult relationships would turn from preoccupation with intimacy (and sexuality) to ones characterized by caring and compassion. Compared with theory, the marital phenotypes that have been studied are relatively narrow in scope compared with the complexity of marital histories across the life course or with related phenotypes such as pair-bondedness, attachment, companionship, caregiving. We propose that greater attention needs to be paid to life-course dynamics in marriage related phenotypes. Although unexamined in these studies, biology (e.g., menopause, testosterone fluctuations) may also play a role in the behavior genetics of later life marital phenotypes. Little is known about whether these phenotypes share genetic or environmental influences, how life stage or marital characteristics modify the role of genes or environment on these phenotypes, and/or whether different genes might come into play at different times of life. Most studies do not attempt to deal with the possible cultural, social or legal influences on marriage or divorce. In addition, the focus on static, simplistic dichotomous phenotypes (e.g., ever married or divorced) ignores patterning of behavior across the life course that may not emerge until later life; for instance, persons in resilient relationships that survive despite life's ups and downs may be qualitatively and quantitatively different from those in relationships marked by discontinuities. Thus, behavior genetic studies that focus exclusively on early mating and early adult marital phenotypes may be unable to capture relevant inter- and intra- personal growth, change or development. Because of these factors, we expect that genetic and environmental influences on different types of relational phenotypes (including dissolution of relationships) will vary across the life course.

Clearly, studies incorporating larger longitudinal samples of men and women, from a variety of cultures, representing different cohorts are needed if we are to better tease apart genetic and environmental influences on phenotypes displayed in long and short term intimate relationships, pair-bonding, and contributions to the patterning of sexuality, intimacy, attachment and caring across the life course, specifically later life. These late-life relationships may also contribute meaningfully to the physical and mental wellbeing of children, grandchildren, and other kin.

9.2.2 Risk Factors for Divorce

Numerous non-twin studies have investigated risk factors for divorce in younger adults; these include factors such as marrying as a teenager, low SES, premarital cohabitation, premarital child birth, and growing up in a divorced household (among others). Some of these factors are more contested than others and, while they may predict divorce (at least in young couples), it cannot be assumed that they cause it and it cannot be presumed that these are risk factors in older couples. A much smaller number of studies have looked at risk factors for divorce in middle aged adults from a behavior genetics angle. One way in which genes might influence divorce risk is through their influence on other psychological variables such as personality.

There is little research on divorce in older adults. In a somewhat younger sample (average age 41 years old), Jockin et al. (1996) attempted to tease apart the complex relationship between genes and divorce by investigating the impact of personality on the heritability of divorce; 18 % had ever been divorced. They found that personality characteristics based on the Multidimensional Personality Questionnaire from one twin predicted the divorce risk of the co-twin almost as well as the personality of the twin (Jockin et al. 1996). The heritability of divorce was 0.59 for women and 0.55 for men; this value was a combination of the sum of additive genetic influences on personality (0.18 and 0.23 for women and men, respectively) and the additive genetic effects specific to divorce risk (0.41 for women and 0.31 for men). Nearly all of the non-shared environmental effects were specific to divorce risk. It should be noted that the divorce rates were relatively low in this sample.

Another study explored the links between genes, health status, and marriage in twin pairs discordant for marital status at late midlife (average age 57; born 1931–1952) (Osler et al. 2008). Among men discordant for marital status, the divorced/widowed twin had higher depression scores, lower cognitive functioning, and a higher prevalence of smoking than his married twin; among female twins discordant for marital status, a similar pattern was seen for depression and a trend for smoking (Osler et al. 2008). It appears, then, that environmental influences associated with depressive symptoms, smoking and (for men) cognitive status might account for differences in marital status. Additional research is necessary to further explicate the role of genes and environment on risk factors for divorce, marriage and patterns of marital relationships at different points in the life course.

9.2.3 Kinship relationships (e.g. siblings, twins)

Kinship relationships are among some of the most important interpersonal relationships that persist into late adult-life; siblings are said to be "the last remainder of one's family origin and, thus, represent important aspects of one's early and shared past" (Cicirelli 2010; Neyer 2002a). Although sibling relationships taper off in early adulthood, studies have shown increased proximity, contact, giving and receiving help among siblings in later life (White 2001).

Twin-twin relationships are thought to be a special kind of relationship; co-twins appear to report a greater closeness than that experienced by other siblings (Neyer 2002a; Tancredy and Fraley 2006). Observers of adult twins have noticed that a twin sometimes functions as an attachment figure for his/her co-twin and have examined attachment in twin relationships. In adults, attachment figures are those persons who provide feelings of security and dependability, comparable to significant caregivers in childhood. Using a novel measure of adult attachment, Tancredy and Fraley (2006) found that adult twins—especially older twins—placed each other higher in a hierarchy of attachment figures such as partners, friends, and parents than did non-twin siblings; partners and spouses rated the highest for both groups (Tancredy and Fraley 2006). In a study of loss of a co-twin in later adult life, Segal found that the surviving MZ twins' retrospective grief and loss scores were significantly higher than those of DZ twins (Segal et al. 2002). Although MZ and DZ twins did not differ when evaluated for current states of bereavement, the duration of feelings of loss was longer in MZ twins. It is unclear however, to what extent the closeness of twins is affected simply by genetic influences; for instance, in one study, twins reared together reported a close relationship throughout the lifespan, while twins reared apart showed less interest in maintaining a close relationship with a co-twin (Pielta et al. 2012).

Neyer examined differences between elderly MZ and DZ twin relationships from a developmental perspective (Neyer 2002a). Given that twins form important aspects of each others' environment and may be differentially "attracted" to each other due to genetics, Neyer hypothesized that relationships between elderly MZ twins would be closer than relationships between DZ twins. MZ twins reported greater frequency of contact, proximal and emotional closeness, and support than DZ twins across adulthood; attachment security, as well as relational satisfaction, were higher in MZ than DZ twins. Frequency of contact, emotional closeness, and support increased after midlife in all twins. These findings about adult twin attachment are consistent with Segal's findings that MZ twins experienced higher and longer levels of grief and loss than DZ twins when their co-twin died (Segal et al. 2002).

Neyer (2002b) also attempted to distinguish dyadic effects from assortative mating in co-twins compared with married couples. In elderly DZ dyads and young couples, but not elderly MZ dyads, feelings of attachment in one person correlated positively with feelings of attachment in the other person, and was related to frequency of contact. Among MZ dyads, feelings of attachment were independent of frequency of contact. Neyer interpreted the findings as suggesting that dyadic influences more strongly explained DZ twins' and couples' levels of closeness and attachment while individual differences (genetic and gene-environment influences) better explained the quality of the relationship between MZ twins.

Sibling relationships represent multiple aspects of interpersonal relationships not necessarily present in intimate relationships: kinship ties, friendship, interpersonal rivalries, social networks, social support, non-sexually pair-bonded attachments. As families become smaller and more mobile, it is unclear who or what replaces these kinship bonds that are theorized to be crucial for survival and wellbeing.

9.2.4 Social Support

The large amount of phenotypic research on social support reflects the general view that interpersonal relationships are important across the life course. Socially supportive relationships are defined as relationships that provide interpersonal, emotional, and/or potential material sustenance. Spouses and siblings can function as sources of support, yet most studies do not distinguish from whom support is received. Operationalizations of social support vary; some researchers examine the size of the social support networks or the amount of contact with socially supportive network members, while others focus more on self-reported perceptions of the quality of support provided. Historically, the majority of behavior genetic studies focus on social support in younger adults. Several studies, however, have wide age ranges that include adults over age 60.

Genetic and environmental influences on social support were examined longitudinally in Swedish Adoption Twin Study of Aging (SATSA) adults; age at the first time point ranged from 26 to 85 years old (mean = 58.8; SD = 12.7) (Bergeman et al. 2001). Follow-ups were conducted at three year intervals, up to two additional times. The heritability of social support varied by measure and time, ranging from 0.19 to 0.42. Non-shared environmental influences accounted for more than half the variance in any social support measure. Across time points, most of the genetic variance (0.74–0.95) was shared by the social support measures. Thus new genetic influences on social support did not come into play across the six years of the study, but environmental influences differed at each time point. Age trends were significant; feelings of family support and perceived adequacy of support increased with age. Significant gender differences also emerged; compared with women, men felt more supported by friends and less supported by family. Although the sample ranged in age from 26 to 85, the study did not address whether characteristics such as stage of life, personality, marital status, or life events (e.g. bereavement, moving, retirement etc.) modified these associations.

In an in-depth study of social support, Coventry et al. (2004) examined age and gender differences in perceived social support in a sample of twins ranging in age from 18 to 95 years old. Although their emphasis was not on genetic and environmental influences on support, their work fills a gap in twin research by examining different types of relationships. Perceived support from one's twin was highly stable across age groups while perceived support from other relationships varied by gender and age. Women scored marginally higher in perceived support than men. Perceived support from spouse, parent and friend declined with age for both men and women; while perceived support from children increased. Women were more likely to report having a confidant, however this decreased with age; having a confidant was more prevalent in older than younger men. For the most part, these results are quite consistent with convoy theory with regard to changes in the size and composition of convoys with age and point to the important and varying roles of multiple relationships in providing support.

Coventry et al. (2009) examined social support in MZ twins discordant for depression from the above sample (mean age 32, range 18–95); social support prior to the depressive episode was not significantly different within twin pairs. This suggests that social support is not an "environmental" factor contributing to vulnerability to depression, rather it shows that reports of social support and depression likely share some genes in common. However, measures of depression tend to include items that overlap with the construct of social support (e.g., "I felt I could not shake off the blues even with help from my family or friends," "I felt lonely," or "People were unfriendly"). Overlapping items would likely increase genetic associations (i.e., the "jingle/jangle" fallacy). Finally, better social support in older adult twins has been associated with lower mortality, better cognitive functioning and lower depressive symptomatology (McGue and Christensen 2007; Rasulo et al. 2005). A study of elderly Danish twins over the age of 75 found that social support—defined as having a spouse and feelings of close ties with friends and one's co-twin—were associated with increased likelihood of survival across six years (Rasulo et al. 2005). The frequency of contact with others benefited survival for women and MZ twins.

This handful of genetically informative studies of older adults demonstrates the role of genetic and environmental influences on social support across the life course. Genetic influences are fairly stable over time in multiple age groups. Across a short period of time (6 years) new genetic influences on social support did not emerge. To the extent that social support is genetic and stable, this points to the importance of individual differences as well as environmental features as predictors of social support across the life-course. In general, though, there are relatively few older adults in these samples, and the duration of longitudinal analyses is relatively brief. There is surprisingly little attention paid to gene—environment interactions, for instance, whether genetic influences on social support are moderated or mediated by major life transitions such as bereavement, health crises, or retirement that are common for older adults.

9.2.5 Benefits of Intimate Relationships

Epidemiological data show that single adults die at twice the rate of married adults and have the highest death rate among adults over age 55 (U.S. National Center for Health Statistics 2010). Some theories suggest that this is because marriage is a long-term source of security, mutual caring and meeting of needs; other views suggest marriage provides better access to resources and greater socioeconomic security; alternatively, men and women who marry may be physically and emotionally healthier than single people. Based on research findings, it is unclear whether there is something specific about marriage itself that is beneficial or if other factors account for the better outcomes of marriage. Non-twin studies find moderate levels of concordance among married/cohabiting couples for some health factors as well as for cognitive ability and education, suggesting—at a minimum—a role for assortative mating (Hoppmann et al. 2008; McCrae et al. 2008; Meyler et al. 2007). In this section, we examine the behavior genetic literature on the effects of spouses on physical and mental health, well-being, health habits, and cognition in older adults.

9.2.5.1 Marriage and Health

Spouses are concordant for a number of health risk factors, which may explain associations between marriage, morbidity and mortality in older adults. A meta-analysis of concordance for cardiovascular risk factors in non-twin studies found significant but modest spousal/partner concordance for blood pressure, smoking, glucose, low density lipoprotein (LDL) cholesterol, weight, BMI, and waist circumference (Di Castelnuovo et al. 2009; Jurj et al. 2006). Di Castelnuovo et al. found that duration of marriage had little effect on the strength of the associations (Di Castelnuovo et al. 2009). However, significant concordance for metabolic syndrome was found in Korean married couples age 60 and over, but not in couples younger than 40; the authors proposed that prolonged exposure to shared environmental influences such as sharing the same meals or activities contributed to greater similarity (Kim et al. 2006). Concordance for many health characteristics is mixed (Hippisley-Cox et al. 2002; Knuiman et al. 1996a, b; Knuiman and Vu 1996).

There is some phenotypic evidence that spouses influence each others' health behaviors. In one study of older adults, across four years, when one spouse changed his or her health behavior (i.e., smoking, drinking, exercising, cholesterol screening, getting a flu shot), the other did so as well (Falba and Sindelar 2008). Although these results might be construed as evidence for spousal/dyadic interaction, they also can be accounted for by assortative mating for related characteristics that might affect health behaviors (e.g., conscientiousness, education, cognitive ability). These results may have little to do with marriage per se—any set of people closely sharing the same space for a period of time may be more similar to each other for a variety of genetic and environmental reasons (Christakis 2004; Christakis and Fowler 2007). Most life course theories predict that spouses influence each other, however most theories do not account for mate selection dynamics.

Although concordance estimates provide some insight into marital/intimate partners' influence on each other, they are cannot differentiate between the role of genetic and environmental influences on the health characteristics and spousal influences. Twin studies that include couples—specifically longitudinal studies—permit a more direct test of the effects of selection, and mutual influence. In particular, incorporating behavior genetic methodology into marriage studies helps to address whether marriage promotes better health or better health behaviors than would otherwise be expected given the genetic predispositions of an individual. Given that many cardiovascular measures and health risks (e.g., BMI, waist-hip ratio, blood pressure, diabetes, pulmonary function, metabolic syndrome) are moderately-to-highly heritable ($h^2 = 0.25$–0.87) it is important to distinguish between spousal and genetic influences (or their interaction) in order to make causal inferences (Fava et al. 2004; Franz et al. 2007; Hottenga et al. 2005; Matteini et al. 2010; McClearn et al. 1994; Murabito et al. 2006; Scherrer et al. 2003, 2005; Seidlerova et al. 2008; Xian et al. 2010).

Trumbetta (2004) evaluated whether marriage buffered health risk behaviors in older married couples by examining health habits such as exercise, smoking, alcohol consumption, and diet in the 1972 cohort of the NAS/NRC when twin participants were 45–55 years old. Although, the sample is younger than the preferred age group for this chapter, we review these results because they reflect intra-couple influences just prior to this chapters' pre-defined age for later life relationships, thereby setting the stage for later life findings. With the exception of exercise, married men had better health habits than unmarried men; however, once genetic characteristics were accounted for, marriage did not buffer risky health-related behaviors. Rather, marriage appeared to amplify or reinforce the genetic predispositions of the twins, suggesting possible gene-by-environmental interactions or correlations. Some recent studies suggest that social contacts influence characteristics such as body mass index and weight gain over and above genetic influences; spouses are one form of social contact (Christakis and Fowler 2007; Herbert et al. 2006; McCaffery et al. 2011).

A recent study examined subjective health across the life course in 16,000 twins ages 24–95 (Finkel et al. 2014). Subjective health is typically assessed using a single item asking individuals to rate their overall health. The simplicity of the item can be misleading as research indicates that subjective health is a complex variable tapping personality and cognitive status, as well as physical health. Individuals who are married report better subjective health than those living alone. Behavior genetic approaches allowed us to examine gene by environment interplay in contributions to subjective health. Three measures of subjective health were collected by nine twin studies participating in the Consortium on Interplay of Genes and Environment across Multiple Studies (IGEMS) (Pedersen et al. 2013): overall self-rated health (SRH), health compared to others (COMP), and impact of health on activities (ACT). Marital status, coded as either married/cohabiting or living alone, was used as a marker of environmental resources that may impact subjective health perceptions. Age, sex, and marital status were examined as moderators of each type of subjective health. Results differed for the three subjective health items, indicating that they do not tap the same underlying construct. The strongest impact of marital status was found for ACT in men. Heritability of ACT was higher in single men (40 %) than married men (30 %) up to age 70, then the pattern reversed and heritability was higher in married men (26 %) than single men (10 %). Conversely, unique environmental influences increased for single men after age 70, but were relatively stable for married men. A similar pattern was found for SRH, but not COMP. Although the impact of marital status was smaller for women than for men, there was some suggestion of similar changes in genetic and environmental influences on ACT and SRH for women, but in the 50–60 age range. Results suggest gender differences in the role of marriage as a source of environmental resources and in the age of greatest impact.

To date, however, there is little behavior genetic evidence that older couples influence each other substantially over and above individual genetic vulnerabilities.

9.2.5.2 Marriage, Mental Health and Wellbeing

Twin studies find that both depression and wellbeing are moderately influenced by genes; heritabilities range from 16 to 37 % for depression to approximately 50 % for some types of wellbeing (Caprara et al. 2009; Franz et al. 2012; Franz et al. 2011; Kendler et al. 2006; Kendler et al. 1987; Nes et al. 2006). It is likely that most adults have a fairly stable "set-point" for wellbeing that can be accounted for primarily by genetic influences on personality (Tellegen et al. 1988). Spouses also tend to be moderately concordant for mental health (Meyler et al. 2007). Findings for the social transmission of wellbeing, though, are mixed in non-twin studies (Gerstorf et al. 2009; Larson and Almeida 1999; Strawbridge, et al. 2009; Walker et al. 2011).

Genetically-informative studies of wellbeing and mate influences have been conducted primarily in younger couples whose average age was less than 40. Results of two studies of younger adults provide some support for the influence of marriage on wellbeing. In brief, the magnitude of the heritability of subjective wellbeing was moderated by marital status in a study of twins ages 19–31 (Nes et al. 2010); genetic influences were smaller (environmental influences larger) for married/cohabiting twins than for unmarried participants. These results are consistent with Finkel et al. (2014) findings for subjective health in younger twins. Given that the Finkel et al. heritability results for married versus single adults reversed after age 70, care needs to be taken to not overgeneralize results from younger samples. In a separate study of young married couples, husbands' played an important role in their wives' mental health, even after accounting for genetic influences on wellbeing in the women (Spotts, et al. 2005). The same methods need to be applied to older couples.

Finally, using twins discordant for major depression, Kendler and Halberstadt (2012) compared reasons for major depression episodes based on the content of autobiographical interviews in 14 middle aged MZ twin pairs (age range 44–63) and found that, in the majority of pairs, the non-depressed twin had a long, stable and successful romantic relationship while the depressed twin had a history of romantic problems. Although the Kendler and Halberstadt sample is slightly younger than the designated age group for this chapter, we described it here because the study highlights a unique analytic approach of using qualitative data in the context of a twin design and also highlights the role of interpersonal relationships on mental health.

Longitudinal and cross-sectional non-twin studies of the life course find considerable evidence for a "U" shaped pattern for wellbeing—with wellbeing sinking from early to middle adulthood then rising during late midlife until the years of terminal decline; wellbeing for couples tends to be lowest when children are in the home (Blanchflower and Oswald 2008; Gerstorf et al. 2008a, b; Yang 2008). Longitudinal twin studies of older adults have yet to examine the extent to which genetic influences on wellbeing or mental health change over time (perhaps associated with hormonal changes, brain changes, or other processes of normal aging) or whether major social and interpersonal events such as changes in environmental "niches" that accompany aging (e.g., modifications of social convoys, children leaving home, bereavement, retirement) might modulate changes in wellbeing. Results from these studies suggest that wellbeing may be a fruitful area of research for understanding the interplay of genetic and environmental influences, particularly the role of evocative processes, in older adults' intimate relationships.

9.2.5.3 Marriage, Smoking and Alcohol Use

Smoking is significantly heritable across gender and age groups; studies find spouses to be moderately concordant for smoking and alcohol use (Carmelli et al. 1992; Kendler et al. 1999; Reynolds et al. 2006). Based on a complex twin-spouse/kinship design of a study of middle-aged Swedish couples, Reynolds et al. (Reynolds et al. 2006) reported mixed phenotypic assortment and social homogamy (i.e. evidence for shared environmental influences) underlying spousal similarity in alcohol consumption ($r \approx 0.23$). On the other hand, tobacco traits showed a more complex pattern: social homogamy was associated with tobacco use status but the amount of tobacco consumed per day was explained by phenotypic assortment. In the NAS/NRC late-middle-age twins, once genetic characteristics were accounted for, marriage did not buffer risky health-related behaviors (Trumbetta 2004). More subtle analyses of spousal interactions and alcohol consumption may help delineate the associations between heavy drinking, partner wellbeing, and marital stability across the life course (Leonard and Rothbard 1999). Without large longitudinal samples and detailed consumption histories, analyses of the independent and interactive influences of genes and environments on alcohol and tobacco use across the life course are likely to be extremely difficult.

9.2.5.4 Marriage and Other Habits (i.e., Activities, Exercise)

A body of literature has accumulated to support the idea that living a socially engaged lifestyle has a protective and positive effect on health and cognition as we age (Fried et al. 2004; Hoppmann et al. 2008; Lovden et al. 2005; McGue and Christensen 2007). In phenotypic analyses, older couples' social activity levels tend to be significantly correlated and stable over time; socially active partners have better cognitive performance than socially inactive partners (Hoppmann et al. 2008). The propensity to engage in some types of social activities may have genetic and environmental overlap with marital status or other relationship qualities.

Twin studies have shed some light on characteristics that may contribute to spousal influences on activity. A study of 147 male twin pairs found that higher engagement in cognitive activities at midlife (average age of 44.7 years) predicted a significant 26 % reduction in risk for developing dementia first within twin pairs 28–36 years later (Carlson et al. 2008). Many of the so-called cognitive activities were actually social in nature. This, however, may not be the entire picture. Other (genetic) influences on factors such as personality, brain structure, and/or neuro-cognitive-influencing disease processes, like depression, may play important roles in these associations (Lovden et al. 2005).

There is also evidence suggesting that correlations between social activity levels and healthy aging may be due to selection; this would explain why healthier individuals are more drawn to certain activities (including social activities) to begin with. A study of 1112 twin pairs ages 75 and older from the Longitudinal Study of Aging Danish Twins showed that genetic factors accounted for the majority of the association between social activity and several measures of aging such as physical functioning or strength, cognitive functioning, and depression symptomatology (McGue and Christensen 2007). Although level of social activity was associated with *initial* level of physical and cognitive functioning, it did not predict change in functioning over time. Comparisons of older monozygotic twins with discordant levels of social activities found that the more active twin was <u>not</u> less susceptible either to age-related declines in physical and cognitive functioning or to greater depression symptomatology. This further supports the idea that the relationship between social activity and later-life functioning is likely due to selection effects.

Physical activity/exercise also plays a significant role in maintaining physical and cognitive health over time and may be particularly important for everyday functioning in older adults. Modest spousal concordance on measures of physical activity/exercise has been reported mostly by researchers focusing on exercise adherence (Falba and Sindelar 2008; Macken et al. 2000; Speers et al. 1986). Genetic influences on adult physical activity/exercise (heritabilities ranging from 0.39 to 0.83) have been documented in the twin literature in young adults (Beunen and Thomis 1999; Heller et al. 1988; Lauderdale et al. 1997; Simonen et al. 2004, 2003). To our knowledge, however, older spouses' influence on each others' physical activity/exercise has not been studied in genetically-informative designs. Clearly more exploration of the role of behavioral and genetic influences on social activity in older adults is needed and a study design that includes spouses as well as twins would better address issues of causation.

9.2.5.5 Marriage, Cognition and Cognitive Change

Despite findings that couples tend to be concordant for cognitive ability and education (Hoppmann et al. 2008; McCrae et al. 2008; Meyler et al. 2007), few studies examine whether spouses influence each other's cognition over and above genetic influences. In non-twin studies, findings of spousal influence on cognition have been mixed (Gerstorf et al. 2009a, b; Gruber-Baldini et al. 1995; Strawbridge et al. 2009). Given the high stability of genetic influences on cognitive ability, the role of environmental influences—from a partner or other factors—are likely to be subtle (Heath et al. 1985; Lyons et al. 2009; Reynolds et al. 2002). Still, the study of spouses as mediators or moderators of cognitive change is a potentially important area for behavior genetic research in older adults.

9.2.5.6 Marriage and Socioeconomic Status (SES)

Strong associations have been found between marital status, SES, health outcomes, and health concordance in phenotypic studies. In the US Health and Retirement Study (HRS), late middle-aged couples in lower SES brackets were significantly more likely to have both members of the couple be in poor health (Wilson 2001). Partner concordance estimates for cardiovascular risk factors were still significant after adjusting for age, education, income, and behavioral risk factors (Wilson 2002). Such research, however, does not address the role of genetic influences on such outcomes. There is evidence that associations between adult social class and health are primarily due to genetic (not spousal) influences. In adult monozygotic (MZ) twins discordant for SES, higher SES MZ twins were no healthier than their lower SES co-twin (Osler et al. 2007). Given the prominent place given to SES in health research across the life course (Franz et al. 2014), a better understanding of the role of genetic and environmental influences on SES and related phenotypes is critical for informing public policy on these issues.

9.2.6 Interpersonal Relationships and Late-Life Transitions

To this point, we have focused on the presence of important interpersonal relationships such as spouses or kinships. Yet, two life events distinguish younger from older adults—caregiving for a disabled loved one, and loss. Older spouses are increasingly likely to become caregivers for each other as they age. Caregiving has well-documented negative consequences for the caregivers' morbidity and mortality with consequences potentially continuing for years beyond the end of care-giving (Robinson-Whelen et al. 2001; Schulz and Beach 1999; Zivin and Christakis 2007). Keeping in mind that depression and wellbeing are moderately influenced by genes, genetically informed research might be able to identify genetic and environmental risk and protective factors for caregivers and the cared-for spouse. For instance, it may be that caregivers with higher genetic risk for depression are more likely to experience prolonged negative affect during and after care-giving.

The loss of important interpersonal relationships to death is inevitable for older adults; caregiving and loss may also be closely linked. Phenotypic studies show that the risk for bereavement-related morbidity and mortality is greater for men than for women and for younger rather than older bereaved spouses (Burton et al. 2006; Stroebe et al. 2007). Results from non-twin studies show that following the death of a spouse, some former long term caregivers experience relief from stress and decreased negative affect; however, up to three years following the death of the spouse, scores on depression, loneliness, and positive affect remain similar to those who are still burdened with caring for an ailing spouse (Robinson-Whelen et al. 2001). Zivin and Christakis found that in couples over age 65, a spouse's hospitalization or death significantly increased the other spouse's subsequent risk for a mental health or substance abuse diagnosis and depression (Zivin and Christakis 2007). Depression prior to bereavement placed a person at higher risk for depression after the death; better outcomes following bereavement were associated with higher SES, self esteem, secure attachment, and optimism (Stroebe et al. 2007). Given that many of these characteristics are all heritable, we might predict that the more genetically resilient elderly adult will rebound from bereavement more quickly than other elderly adults. Related to the idea of genetic resilience, Kendler et al. (2008) examined whether bereavement related depression and major depression had similar etiology; young adult twin pairs discordant for either bereavement related depression or major depression (average age 36) were more similar than different, suggesting common etiology for the types of depression. Even though the Kendler et al. sample was young and the reasons for bereavement are likely different in younger adults than in older adults, these results suggest that genetic resilience and/or risk factors are important to identify across the life course.

9.3 Conclusion

This guy goes to a psychiatrist and says, 'Doc, my brother's crazy, he thinks he's a chicken.' And the doctor says, 'Well why don't you turn him in?' and the guy says, 'I would, but I need the eggs.' Well, I guess that's pretty much how I feel about relationships. They're totally irrational and crazy and absurd, but I guess we keep going through it because most of us need the eggs (Woody Allen, Annie Hall 1977).

Due to longer life expectancies, intimate relationships are now more likely to continue for decades beyond childbearing and childrearing years; friendships and other relationships navigate even more of life's twist and turns. Much of this expansion of the life course has occurred in the last two centuries. Longevity affects aspects of interpersonal relationships beyond marriage and intimacy. For adults experiencing retirement, more time may be available for social activities and interpersonal relationships, but financial considerations might require changes in lifestyle and relocations that disrupt long term relationships and support systems. Intergenerational relationships change such that a 60 year old adult might be caring for elderly parents as well as grandchildren, and 20 years later being cared for by children and grandchildren. The probability of experiencing a major chronic illness increases with age, putting strain and stress on existing relationships and—sometimes—creating new relationships (e.g., caregivers, residents in nursing homes). Loss of relationships due to deaths increases the probability of new intimate relationships and/or marriages as well as the possibility of living for years without a partner. Better health, longer life spans, and cohort differences in social/institutional control over relationships create new challenges for aging research. In behavior genetics the systematic study of later life interpersonal relationships is an even more recent area of research. This review suggests that genetically-informed studies of positive characteristics such as later life relationships have potential to contribute significantly to public health.

In the phenotypic literature, interpersonal relationships are found to be important resources across the life cycle. The fact that marital partners have similar health risk factors in later life has serious implications for health and wellbeing in later years (Wilson 2001) but it is easy to make incorrect inferences about sources of partner similarities based solely on phenotypic data. Genetically-informative designs allow us to distinguish the effects of genetic and other social influences from contributions of the partner and shared couple environment effects; such effects are confounded in non-twin studies. Much of our knowledge about the behavior genetics of later life relationships comes from only a handful of studies. Currently there are few large active twin studies of older adults that include spouses/partners and even fewer that incorporate other adults (e.g. friends, children, grandchildren etc.). Without studying other relationships the effect of intimate relationships cannot be differentiated from that of other close relationships. Genetically informed studies can help shed light on conditions under which genetic influences on important aspects of public health and aging (such as health, health behaviors, cognition, and mental health) are strongest, thereby allowing for improvement of our understanding of environmental exposures, and providing evidence for how influences on behavior are mediated and moderated by social relationships and networks.

One goal of our review was to illuminate the complexity of disentangling the influences on and effects of interpersonal relationships in older adults. In brief, behavior genetics still has a long way to go in the area of late life interpersonal relationships. Weiss (1974) described six types of key adult attachment relationships—that is, relationships which provide a sense of security and wellbeing.

These six types of interpersonal relationships range from spouses and lovers, to friends, family, teachers, therapists, and mentors. Depending on the specific relationship, these close adult relationships offer some combination of sex, companionship, shared experience, nurturing, caregiving. Some but not all relationships provide a sense of worth and competence. Yet, when we examine the bulk of research on late-life relationships, many of these important relationships remain understudied from a behavior genetic perspective. For example, there exists little to no twin research literature on relationships between romantic partners in lesbian/gay/bisexual/transgender couples, cross generational relationships (grandparents-grandchild; extended families), friendships, caregivers, single adults, cohabiting adults, widowhood or even spiritual relationships. It is difficult to find genetically informative studies that examine influences on and effects of major late life transitions such as widowhood, retirement, remarriage or relocations. What about hate, compassion, companionship, generosity, jealousy, passion, sacrifice, abuse, conflict, regret? Much of the behavior genetic research that exists on interpersonal relationships has been conducted on couples younger than age 60. The existing research largely focuses on white adults from western cultural heritages and a western view of marriage.

With regard to the most frequently examined phenotype of marital status, naïve phenotypes (e.g. married versus not married) still dominate. Many basic questions remain unaddressed as to the contribution of long-term marriages to aging, whether partner selection, or the effects of partner interaction vary across the life course. Is the choice of a mate at age 20 driven by the same qualities as choosing a mate at 75? How are they different? Are there developmental points, circumstances or cultures where some aspects of selection (e.g., homogamy) might overrule other aspects? Are there time periods during which dyadic/partner interaction effects might be more powerful than others? What genetic and environmental influences contribute to the maintenance of long term social bonds in different social contexts?

As he reviewed the behavior genetic literature, a young student lamented that long lasting marriages might just be epiphenomena/exaptations of increased longevity or perhaps vestigial evolutionary remnants (like a withered appendix) of earlier assortative mating and homogamy. Without greater insight into positive aspects of aging it can be difficult to identify what might be important, different, or useful about later life relationships. In light of attachment theory, it appears that the qualities that make someone a good choice as a "mate" also contribute to being a good "partner" in later life thereby contributing directly and indirectly to physical and psychological health in old age as well as to the wellbeing of one's kin. Across the life course, even relatively small statistical effects due to the influence of interpersonal relationships may add up to meaningful differences in years of healthy and happy aging. It is clear that there is substantial uncharted territory still to explore in the realm of genetic and environmental influences on later life relationships.

9.4 Future Directions

We propose five areas for future investigations into the behavior genetics of later life interpersonal relationships and associated outcomes.

1. In 1997, Segal wrote that the area of developmental social genetics was under-explored (Segal 1997). Although significant progress has been made in the past two decades, if we are to better understand the role of genetic and environmental influences on interpersonal relationships in later life we need to develop phenotypes that more accurately reflect life course and life history developmental social dynamics. What is missing in existing analyses is the examination of patterns of relationships across the life course—patterns that may not become apparent until later adulthood. For instance, applying behavior genetics to the study of adults in long-term monogamous marriages compared with adults with multiple divorces may provide insight into underlying dynamics of transformative (self-restoring) relationships (Fincham et al. 2007). Fincham et al. (2007) argue that the study of marriage has focused for too long on conflict and dissolution rather than on factors within happy marriages that affect long term marital outcomes and regulate marital homeostasis. They point out that *all* intimate relationships are beset by multiple stresses and strains but some couples are better able to "self-repair" relationships than others. Research finds several characteristics contribute to self-repairing marriages: the balance of positive and negative emotion, forgiveness, the sense of commitment/sacrifice and sanctification (i.e. religion/institutional affirmation) (Fincham et al. 2007). Although the behavior genetics of transformative processes have not been studied, some of the separate characteristics have been shown to be significantly heritable. Positive and negative affect, for instance, are recognized as being associated with different genes and different neurophysiological pathways (Watson et al. 1999). Theories of development suggest that successful, long-lasting, intimate relationships may be determined by social or individual selection that favor, for instance, mates who are securely attached, caring, and conscientious.

   Comparisons of attachment processes in different types of later life relationships (marriages, friendships, grandparenting) could provide more insight into the role of kinships and social convoys as well as into intergenerational transmission of traits. Unmarried adults tend to be depicted as a homogenous (almost pathological) group yet reasons to not marry are plentiful; what comprises and influences healthy relationships and their outcomes in unmarried adults is seldom examined. Finally, clearly more work is needed on later life interpersonal dyads of all types (e.g., friendships, siblings, partners) in order to examine whether and how different types of relationships influence old-age outcomes over and above genetic influences. Ideally, the examination of relationships other than spouses will shed light on the extent to which spousal relationships differ from other types of close relationships. It is clear that behavior genetic research has only begun to scratch the surface of the role of genetic and environmental influences or interactions in marital phenotypes, focusing instead on phenotypes predominantly occurring in early adulthood. Given that prolonged successful long-term intimate relationships potentially benefit future generations as well as the intimate couple, the lack of research on the ontology of such relationships is surprising.

2. Loss and bereavement are important issues in aging research because of their increased prevalence with age and because of the known associations between depressive symptoms, cognitive function, morbidity and mortality. A few behavior genetic studies examine responses to bereavement but much of this literature focuses on younger adults in whom reasons for bereavement are likely to be different from losses in older life. Behavior genetics can help to distinguish bereavement from depression, and may be able to help identify individuals more at risk for poor outcomes following a major loss. Whether or how loss or major transitions may mediate genetic influences on depression, wellbeing, health, cognition, or social support in later life is understudied. A better understanding of the mechanisms underlying these associations could contribute to the development of better interventions for depressed older adults.

3. There is serious need for multivariate twin analyses of different relational quality measures (e.g., social support, attachment) and outcomes such as depression in older and younger adults (see, for example, Franz et al. 2011). For instance, behavior genetics research can further the study of social support by identifying the role of genetic influences, its overlap with other constructs, direction of causality, reasons for stability or instability over time and whether these change over time or across life events. In addition, little is known about the extent to which social support from different relationships (e.g., an intimate partner, a close friendship, a co-twin, a therapist, or a sibling) is driven by similar or different underlying influences. New research designs, such as in-depth interviewing of discordant twins, may help to identify subtle differences between twins which then can be used in hypothesis generation and measure development (Kendler and Halberstadt 2012). Greater clarity about the genetic and environmental architecture of commonly used relational phenotypes might help reconcile inconsistent and/or redundant results.

In addition, because of the complexity of dyadic relationships over time and in different contexts, life course relationships maybe an especially rich source of information about the role of gene environment correlations and gene environment interactions. For instance, studies of peer deviance in younger people have elucidated ways in which individuals play active roles in selecting risky and protective aspects of an environment. Although in younger people, there is evidence that genetic factors assume increasing importance in selection and creation of social environments (e.g., choosing friends), little attention has been paid to whether there are points or events in the lifecycle where genetic influences on relational choices change in importance—possibly when health or other life events impinge on choices of relationships. With regard to gene-environment interactions, too little is known about the extent to which genetic influences vary in importance as a function of environmental conditions. For instance, do genetic influences on a health outcome such as Type 2 diabetes vary as a function of marital status? Does religiosity reduce the importance of genetic influences on divorce (given the greater social control/constraint of religious environments)? Too few behavior genetic studies have examined the influence of age, gender, and social history on relational status or outcomes associated with relationships.

4. Hormonal and biological influences on aging and relationships are important but understudied. Testosterone, for instance, tends to change with age. The direction of causality, however, between hormones and relationships is not clear and the role of testosterone in older relationships is understudied from a behavior genetic perspective. As was seen in Figs. 9.1 and 9.2, both men and women are sexually active in later life. Men are sexually active for longer, however, and sexual activity is more highly associated with wellbeing in men than in women (Waite et al. 2009). Men also continue to procreate later in life than women. In studies of younger men, married men or men in committed relationships have lower testosterone than men who are not married (Booth et al. 1993; Burnham et al. 2003; Gray et al. 2004). In one study of testosterone in 4462 men in their 30s and 40s, men with higher testosterone were less likely to marry and more likely to divorce (Booth et al. 1993). Panizzon has found associations between brain structure and testosterone in VETSA men; with high testosterone having more deleterious effects (Panizzon et al. 2010, 2012). Most of these studies have been conducted in younger men, however, they suggest that the study of hormones and behavior in the context of late-life interpersonal relationships may be an important area for future research linking behavior genetic with biological processes in later life.

5. Some researchers have started to examine ways in which new passionate and long-term romantic relationships are similar or different. It has been known that passionate love activates mesolimbic, dopamine-rich reward system areas of the brain in younger adults and in young mothers (Bartels and Zeki 2004; Strathearn et al. 2008). In one study, seventeen participants, ages 39–67 were shown pictures of their romantic partner/spouse to compare brain activation in adults in long-term relationships with that of adults in new intense romantic relationships using functional magnetic resonance imaging (Acevedo et al. 2012). Pictures of close friends and familiar acquaintances were used as controls. The same brain regions were activated in adults in long term love relationships as in the adults in new intense love relationships when they viewed the person's photo. These same regions have been implicated in previous rodent, pair-bond studies of maternal love behavior (attachment behaviors such as protecting and nurturing). Activations in dopamine-rich parts of the brain may be part of the reward system that reinforces and maintains long term attachment bonds and behaviors. We also now recognize that there electrical and biochemical "wiring" of the human brain to be connected with other people, and that social connectness plays a role in shaping the brain and later development (Hari and Kujala 2009; Insel 2010; Meyer-Lindenberg et al. 2011; Van Overwall et al. 2014). It is also evident that many brain structures are highly heritable (Kremen et al. 2013). Conducting such research in the context

of genetically informative designs could offer insight into the behavioral genetics of neural pathways involved in the development and maintenance of long-term attachments necessary to interpersonal relationships in older adult life.

References

Acevedo, B. P., Aron, A., Fisher, H. E., & Brown, L. L. (2012). Neural correlates of long-term intense romantic love. *Social Cognitive and Affective Neuroscience, 7*(2), 145–159.

Ainsworth, M. D. (1985). Attachments across the life span. *Bulletin of the New York Academy of Medicine, 61*, 792–812.

Antonucci, T. C., Fiori, K. L., Kirditt, K., & Jackey, L. M. H. (2010). Convoys of social relationships: Integrating life-span and life-course perspectives. In M. Lamb & A. Freund (Eds.), *The handbook of life-span development: Social and emotional development* (Vol. 2, pp. 434–473). Hoboken: Wiley.

Bartels, A., & Zeki, S. (2004). The neural correlates of maternal and romantic love. *Neuroimage, 21*(3), 1155–1166.

Bergeman, C. S., Neiderhiser, J. M., Pedersen, N. L., & Plomin, R. (2001). Genetic and environmental influences on social support in later life: A longitudinal analysis. *International Journal of Aging and Human Development, 53*(2), 107–135.

Beunen, G., & Thomis, M. (1999). Genetic determinants of sports participation and daily physical activity. *International Journal of Obesity and Related Metabolic Disorders, 23*(Suppl 3), S55–S63.

Blanchflower, D. G., & Oswald, A. J. (2008). Is well-being U-shaped over the life cycle? *Social Science and Medicine, 66*(8), 1733–1749.

Booth, A., Mazur, A. C., & Dabbs, J. M, Jr. (1993). Endogenous testosterone and competition: The effect of "fasting". *Steroids, 58*(8), 348–350.

Bowlby, J. (1969). *Attachment and loss* (Vol. 1). New York: Basic Books.

Bowlby, J. (1988). *A secure base: Clinical applications of attachment theory*. East Sussex: Brunner-Routledge.

Burnham, T. C., Chapman, J. F., Gray, P. B., McIntyre, M. H., Lipson, S. F., & Ellison, P. T. (2003). Men in committed, romantic relationships have lower testosterone. *Hormones and Behavior, 44*(2), 119–122.

Burton, A. M., Haley, W. E., & Small, B. J. (2006). Bereavement after caregiving or unexpected death: effects on elderly spouses. *Aging & Mental Health, 10*(3), 319–326.

Buss, D. M. (2009). The great struggles of life: Darwin and the emergence of evolutionary psychology. *American Psychologist, 64*(2), 140–148.

Buss, D. M., Haselton, M. G., Shackelford, T. K., Bleske, A. L., & Wakefield, J. C. (1998). Adaptations, exaptations, and spandrels. *American Psychologist, 53*(5), 533–548.

Caprara, G. V., Fagnani, C., Alessandri, G., Steca, P., Gigantesco, A., Cavalli Sforza, L. L., et al. (2009). Human optimal functioning: The genetics of positive orientation towards self, life, and the future. *Behavior Genetics, 39*(3), 277–284.

Carlson, M. C., Helms, M. J., Steffens, D. C., Burke, J. R., Potter, G. G., & Plassman, B. L. (2008). Midlife activity predicts risk of dementia in older male twin pairs. *Alzheimers Dement, 4*(5), 324–331.

Carmelli, D., Swan, G. E., Robinette, D., & Fabsitz, R. (1992). Genetic influence on smoking—A study of male twins. *New England Journal of Medicine, 327*(12), 829–833.

Christakis, N. A. (2004). Social networks and collateral health effects. *BMJ, 329*(7459), 184–185.

Christakis, N. A., & Fowler, J. H. (2007). The spread of obesity in a large social network over 32 years. *New England Journal of Medicine, 357*(4), 370–379.

Cicirelli, V. G. (2010). Attachment relationships in old age. *Journal of Social and Personal Relationships, 27*, 191–199.

Confer, J. C., Easton, J. A., Fleischman, D. S., Goetz, C. D., Lewis, D. M., Perilloux, C., et al. (2010). Evolutionary psychology. Controversies, questions, prospects, and limitations. *American Psychologist, 65*(2), 110–126.

Coontz, S. (2005). *Marriage, a history: From obedience to intimacy or how love conquered marriage.* New York: Viking.

Coventry, W. L., Gillespie, N. A., Heath, A. C., & Martin, N. G. (2004). Perceived social support in a large community sample—age and sex differences. *Social Psychiatry and Psychiatric Epidemiology, 39*, 625–636.

Coventry, W. L., Medland, S. E., Wray, N. R., Thorsteinsson, E. B., Heath, A. C., & Byrne, B. (2009). Phenotypic and discordant-monozygotic analyses of stress and perceived social support as antecedents to or sequelae of risk for depression. *Twin Research and Human Genetics, 12*, 469–488.

Di Castelnuovo, A., Quacquaruccio, G., Donati, M. B., de Gaetano, G., & Iacoviello, L. (2009). Spousal concordance for major coronary risk factors: A systematic review and meta-analysis. *American Journal of Epidemiology, 169*(1), 1–8.

Diamond, L. M., Fagundes, C. P., & Butterworth, M. R. (2010). Intimate relationships across the life span. In M. Lamb & A. Freund (Eds.), *The handbook of life-span development: social and emotional development* (Vol. 2). Hoboken, NJ: Wiley.

Eaves, L. J., Hatemi, P., Heath, A. C., & Martin, N. G. (Eds.). (2011). *Modeling the biological and cultural inheritance of social and political behavior in twins and families.* Chicago, IL: University of Chicago Press.

Erikson, E. (1963). *Childhood and society* (2nd ed.). New York: Norton.

Evans, R. (1967). *Dialogue with Erik Erikson.* New York: Harper & Row.

Falba, T. A., & Sindelar, J. L. (2008). Spousal concordance in health behavior change. *Health Services Research, 43*(1 Pt 1), 96–116.

Fava, C., Burri, P., Almgren, P., Groop, L., Hulthen, U. L., & Melander, O. (2004). Heritability of ambulatory and office blood pressure phenotypes in Swedish families. *Journal of Hypertension, 22*(9), 1717–1721.

Feldman, R., Zagoory-Sharon, O., Weisman, O., Schneiderman, I., Gordon, I., Maoz, R., et al. (2012). Sensitive parenting is associated with plasma oxytocin and polymorphisms in the OXTR and CD38 genes. *Biological Psychiatry, 72*, 175.

Fincham, F. D., Stanley, S. M., & Beach, S. R. (2007). Transformative processes in marriage: An analysis of emerging trends. *Journal of Marriage & Family, 69*(2), 275–292.

Finkel, D., Franz, C. E., & Horwitz, B., & Members of the IGEMS consortium. (2014). Gender and age differences in the impact of marital status on genetic and environmental influences on subjective health. Paper presented at the Annual Behavior Genetics Association meeting Charlottesville, VA June 18–21, 2014.

Fraley, R. C., Brumbaugh, C. C., & Marks, M. J. (2005). The evolution and function of adult attachment: A comparative and phylogenetic analysis. *Journal of Personality and Social Psychology, 89*(5), 731–746.

Franz, C. E. (2013). Genetic influences on intimate relationships in middle aged men. University of California San Diego, unpublished data.

Franz, C. E., Grant, M. D., Jacobson, K. C., Kremen, W. S., Eisen, S. A., Xian, H., et al. (2007). Genetics of body mass stability and risk for chronic disease: A 28-year longitudinal study. *Twin Research and Human Genetics, 10*, 537–545.

Franz, C. E., Panizzon, M. S., Eaves, L. J., Thompson, W., Lyons, M. J., Jacobson, K. C., et al. (2012). Genetic and environmental multidimensionality of well- and ill-being in middle aged twin men. *Behavior Genetics, 42*(4), 579–591.

Franz, C. E., Spoon, K., Thompson, W., Hauger, R. L., Hellhammer, D., Jacobson, K. C., et al. (2014). Adult cognitive ability and socioeconomic disadvantage as mediators of the effects of childhood disadvantage on salivary cortisol in aging adults. *Psychoneuroendocrinology, 38*, 2127.

Franz, C. E., & White, K. W. (1985). Individuation and attachment in personality development: Extending Erikson's theory. *Journal of Personality, 53*, 224–256.

Franz, C. E., York, T. P., Eaves, L. J., Prom-Wormley, E., Jacobson, K. C., Lyons, M. J., et al. (2011). Adult romantic attachment, negative emotionality, and depressive symptoms in middle aged men: A multivariate genetic analysis. *Behavior Genetics, 41*(4), 488–498.

Fried, L. P., Carlson, M. C., Freedman, M., Frick, K. D., Glass, T. A., Hill, J., et al. (2004). A social model for health promotion for an aging population: initial evidence on the Experience Corps model. *Journal of Urban Health, 81*(1), 64–78.

Gerstorf, D., Hoppmann, C. A., Anstey, K. J., & Luszcz, M. A. (2009a). Dynamic links of cognitive functioning among married couples: Longitudinal evidence from the Australian Longitudinal Study of Ageing. *Psychology and Aging, 24*(2), 296–309.

Gerstorf, D., Hoppmann, C. A., Kadlec, K. M., & McArdle, J. J. (2009b). Memory and depressive symptoms are dynamically linked among married couples: longitudinal evidence from the AHEAD study. *Developmental Psychology, 45*(6), 1595–1610.

Gerstorf, D., Ram, N., Estabrook, R., Schupp, J., Wagner, G. G., & Lindenberger, U. (2008a). Life satisfaction shows terminal decline in old age: Longitudinal evidence from the German Socio-Economic Panel Study (SOEP). *Developmental Psychology, 44*(4), 1148–1159.

Gerstorf, D., Ram, N., Rocke, C., Lindenberger, U., & Smith, J. (2008b). Decline in life satisfaction in old age: longitudinal evidence for links to distance-to-death. *Psychology and Aging, 23*(1), 154–168.

Gillath, O., Shaver, P. R., Baek, J. M., & Chun, D. S. (2008). Genetic correlates of adult attachment style. *Personality and Social Psychology Bulletin, 34*(10), 1396–1405.

Gray, P. B., Campbell, B. C., Marlowe, F. W., Lipson, S. F., & Ellison, P. T. (2004). Social variables predict between-subject but not day-to-day variation in the testosterone of US men. *Psychoneuroendocrinology, 29*(9), 1153–1162.

Gruber-Baldini, A. L., Schaie, K. W., & Willis, S. L. (1995). Similarity in married couples: A longitudinal study of mental abilities and rigidity-flexibility. *Journal of Personality and Social Psychology, 69*(1), 191–203.

Hari, R., & Kujala, M. V. (2009). Brain basis of human social interaction: from concepts to brain imaging. *Physiological Reviews, 89*(2), 453–479.

Hazan, C., & Shaver, P. (1987). Romantic love conceptualized as an attachment process. *Journal of Personality and Social Psychology, 52*, 511–524.

Heath, A. C., Berg, K., Eaves, L. J., Solaas, M. H., Corey, L. A., Sundet, J., et al. (1985). Educational policy and the heritability of educational attainment. *Nature, 314*, 734–736.

Heller, R. F., O'Connell, D. L., Roberts, D. C., Allen, J. R., Knapp, J. C., Steele, P. L., et al. (1988). Lifestyle factors in monozygotic and dizygotic twins. *Genetic Epidemiology, 5*(5), 311–321.

Herbert, A., Liu, C., Karamohamed, S., Liu, J., Manning, A., Fox, C. S., et al. (2006). BMI modifies associations of IL-6 genotypes with insulin resistance: The Framingham Study. *Obesity (Silver Spring), 14*(8), 1454–1461.

Hippisley-Cox, J., Coupland, C., Pringle, M., Crown, N., & Hammersley, V. (2002). Married couples' risk of same disease: Cross sectional study. *BMJ, 325*(7365), 636.

Hoppmann, C. A., Gerstorf, D., & Luszcz, M. (2008). Spousal social activity trajectories in the Australian longitudinal study of ageing in the context of cognitive, physical, and affective resources. *Journals of Gerontology. Series B, Psychological Sciences and Social Sciences, 63*(1), P41–P50.

Hottenga, J. J., Boomsma, D. I., Kupper, N., Posthuma, D., Snieder, H., Willemsen, G., et al. (2005). Heritability and stability of resting blood pressure. *Twin Research and Human Genetics, 8*(5), 499–508.

Insel, T. R. (2010). The challenge of translation in social neuroscience: A review of oxytocin, vasopressin, and affiliative behavior. *Neuron, 65*(6), 768–779.

Jerskey, B. A., Panizzon, M. S., Jacobson, K. C., Neale, M. C., Grant, M. D., Schultz, M., et al. (2010). Marriage and Divorce: A genetic perspective. *Pers Individ Dif, 49*(5), 473–478.

Jockin, V., McGue, M., & Lykken, D. T. (1996). Personality and divorce: A genetic analysis. *Journal of Personality and Social Psychology, 71*(2), 288–299.

Johnson, W., McGue, M., Krueger, R. F., & Bouchard, T. J, Jr. (2004). Marriage and personality: A genetic analysis. *Journal of Personality and Social Psychology, 86*(2), 285–294.

Jurj, A. L., Wen, W., Li, H. L., Zheng, W., Yang, G., Xiang, Y. B., et al. (2006). Spousal correlations for lifestyle factors and selected diseases in Chinese couples. *Annals of Epidemiology, 16*(4), 285–291.

Kendler, K. S., & Baker, J. H. (2007). Genetic influences on measures of the environment: A systematic review. *Psychological Medicine, 37*(5), 615–626.

Kendler, K. S., Gatz, M., Gardner, C. O., & Pedersen, N. L. (2006). Personality and major depression: A Swedish longitudinal, population-based twin study. *Archives of General Psychiatry, 63*, 1113–1120.

Kendler, K. S., & Halberstadt, L. J. (2012). The road not taken: life experiences in monozygotic twin pairs discordant for major depression. *Mol Psychiatry*.

Kendler, K. S., Heath, A. C., Martin, N. G., & Eaves, L. J. (1987). Symptoms of anxiety and depression in a volunteer twin population: The etiologic role of genetic and environmental factors. *Archives of General Psychiatry, 43*, 213–221.

Kendler, K. S., Myers, J., & Zisook, S. (2008). Does bereavement-related major depression differ from major depression associated with other stressful life events? *American Journal of Psychiatry, 165*(11), 1449–1455.

Kendler, K. S., Neale, M. C., Sullivan, P., Corey, L. A., Gardner, C. O., & Prescott, C. A. (1999). A population-based twin study in women of smoking initiation and nicotine dependence. *Psychological Medicine, 29*, 299–308.

Kenny, D. A. (1996). The design and analysis of social-interaction research. *Annual Review of Psychology, 47*, 59–86.

Kim, H. C., Kang, D. R., Choi, K. S., Nam, C. M., Thomas, G. N., & Suh, I. (2006). Spousal concordance of metabolic syndrome in 3141 Korean couples: A nationwide survey. *Annals of Epidemiology, 16*(4), 292–298.

Kirkpatrick, L. A., & Davis, K. E. (1994). Attachment style, gender, and relationship stability: A longitudinal analysis. *Journal of Personality and Social Psychology, 66*(3), 502–512.

Knuiman, M. W., Divitini, M. L., Bartholomew, H. C., & Welborn, T. A. (1996a). Spouse correlations in cardiovascular risk factors and the effect of marriage duration. *American Journal of Epidemiology, 143*(1), 48–53.

Knuiman, M. W., Divitini, M. L., Welborn, T. A., & Bartholomew, H. C. (1996b). Familial correlations, cohabitation effects, and heritability for cardiovascular risk factors. *Annals of Epidemiology, 6*(3), 188–194.

Knuiman, M. W., & Vu, H. T. (1996). Risk factors for stroke mortality in men and women: The Busselton study. *Journal of Cardiovascular Risk, 3*(5), 447–452.

Kremen, W. S., Franz, C. E., & Lyons, M. J. (2012). VETSA: The Vietnam era twin study of aging. *Twin Research and Human Genetics*, 1–4.

Kremen, W. S., Fennema-Notestine, C., Eyler, L. T., Panizzon, M. S., Chen, C.-H., Franz, C. E., et al. (2013). Genetics of brain structure: Contributions from the Vietnam era twin study of aging. *American Journal of Medical Genetics Part B Neuropsychiatric Genetics., 162*, 751–761.

Larson, R. W., & Almeida, D. M. (1999). Emotional transmission in the daily lives of families: A new paradigm for studying family process. *Journal of Marriage and the Family, 61*, 5–20.

Lauderdale, D. S., Fabsitz, R., Meyer, J. M., Sholinsky, P., Ramakrishnan, V., & Goldberg, J. (1997). Familial determinants of moderate and intense physical activity: A twin study. *Medicine and Science in Sports and Exercise, 29*, 1062–1068.

Leonard, K. E., & Rothbard, J. C. (1999). Alcohol and the marriage effect. *Journal of Studies on Alcohol. Supplement, 13*, 139–146.

Lovden, M., Ghisletta, P., & Lindenberger, U. (2005). Social participation attenuates decline in perceptual speed in old and very old age. *Psychology and Aging, 20*(3), 423–434.

Luyten, P., & Blatt, S. J. (2011). Integrating theory-driven and empirically-derived models of personality development and psychopathology: A proposal for DSM V. *Clinical Psychology Review, 31*(1), 52–68.

Lyons, M. J., York, T. P., Franz, C. E., Grant, M. D., Eaves, L. J., Jacobson, K. C., et al. (2009). Genes determine stability and environment determines change in cognitive ability during 35 years of adulthood. *Psychological Science, 11*, 1146–1152.

Macken, L. C., Yates, B., & Blancher, S. (2000). Concordance of risk factors in female spouses of male patients with coronary heart disease. *J Cardiopulm Rehabil, 20*(6), 361–368.

Matteini, A. M., Fallin, M. D., Kammerer, C. M., Schupf, N., Yashin, A. I., Christensen, K., et al. (2010). Heritability estimates of endophenotypes of long and health life: The long life family study. *Journal of Gerontology Series A: Biological Sciences and Medical Sciences, 65*(12), 1375–1379.

McCaffery, J. M., Franz, C. E., Jacobson, K., Leahey, T. M., Xian, H., Wing, R. R., et al. (2011). Effects of social contact and zygosity on 21-y weight change in male twins. *American Journal of Clinical Nutrition, 94*(2), 404–409.

McClearn, G. E., Svartengren, M., Pedersen, N. L., Heller, D. A., & Plomin, R. (1994). Genetic and environmental influences on pulmonary function in aging Swedish twins. *Journal of Gerontology, 49*, 264–268.

McCrae, R. R., Martin, T. A., Hrebickova, M., Urbanek, T., Boomsma, D. I., Willemsen, G., et al. (2008). Personality trait similarity between spouses in four cultures. *Journal of Personality, 76*(5), 1137–1164.

McGue, M., & Christensen, K. (2007). Social activity and healthy aging: A study of aging Danish twins. *Twin Research and Human Genetics, 10*(2), 255–265.

McGue, M., & Lykken, D. T. (1992). Genetic influence on risk of divorce. *Psychological Science, 3*(6), 368–373.

Meyer-Lindenberg, A., Domes, G., Kirsch, P., & Heinrichs, M. (2011). Oxytocin and vasopressin in the human brain: social neuropeptides for translational medicine. *Nature Reviews Neuroscience, 12*(9), 524–538.

Meyler, D., Stimpson, J. P., & Peek, M. K. (2007). Health concordance within couples: A systematic review. *Social Science and Medicine, 64*(11), 2297–2310.

Middeldorp, C. M., Cath, D. C., Vink, J. M., & Boomsma, D. I. (2005). Twin and genetic effects on life events. *Twin Research and Human Genetics, 8*(3), 224–231.

Mikulincer, M., & Shaver, P. R. (2007). *Attachment in adulthood: Structure, dynamics and change*. New York: Guilford Press.

Mikulincer, M., Shaver, P. R., Sapir-Lavid, Y., & Avihou-Kanza, N. (2009). What's inside the minds of securely and insecurely attached people? The secure-base script and its associations with attachment-style dimensions. *Journal of Personality and Social Psychology, 97*(4), 615–633.

Murabito, J. M., Guo, C. Y., Fox, C. S., & D'Agostino, R. B. (2006). Heritability of the ankle-brachial index: The Framingham Offspring study. *American Journal of Epidemiology, 164*(10), 963–968.

Nes, R. B., Roysamb, E., Harris, J. R., Czajkowski, N., & Tambs, K. (2010). Mates and marriage matter: genetic and environmental influences on subjective wellbeing across marital status. *Twin Research and Human Genetics, 13*(4), 312–321.

Nes, R. B., Roysamb, E., Tambs, K., Harris, J. R., & Reichborn-Kjennerud, T. (2006). Subjective well-being: genetic and environmental contributions to stability and change. *Psychological Medicine, 36*(7), 1033–1042.

Neyer, F. J. (2002a). Twin relationships in old age: A developmental perspective. *Journal of Social and Personal Relationships, 19*, 155–177.

Neyer, F. J. (2002b). The dyadic interdependence of attachment security and dependency: A Conceptual replication across older twin pairs and younger couples. *Journal of Social and Personal Relationships, 19*, 483–503.

Nolte, T., Guiney, J., Fonagy, P., Mayes, L. C., & Luyten, P. (2011). Interpersonal stress regulation and the development of anxiety disorders: An attachment-based developmental framework. *Frontiers in Behavioral Neuroscience, 5*, 55.

Osler, M., McGue, M., & Christensen, K. (2007). Socioeconomic position and twins' health: A life-course analysis of 1266 pairs of middle-aged Danish twins. *International Journal of Epidemiology, 36*(1), 77–83.

Osler, M., McGue, M., Lund, R., & Christensen, K. (2008). Marital status and twins' health and behavior: An analysis of middle-aged Danish twins. *Psychosomatic Medicine, 70*(4), 482–487.

Panizzon, M. S., Hauger, R., Dale, A. M., Eaves, L. J., Eyler, L. T., Fischl, B., et al. (2010). Testosterone modifies the effect of APOE genotype on hippocampal volume in middle-aged men. *Neurology, 75*(10), 874–880.

Panizzon, M. S., Hauger, R. L., Eaves, L. J., Chen, C. H., Dale, A. M., Eyler, L. T., et al. (2012). Genetic influences on hippocampal volume differ as a function of testosterone level in middle-aged men. *Neuroimage, 59*(2), 1123–1131.

Pedersen, N. L., Christensen, K., Dahl, A. K., Finkel, D., Franz, C. E., Gatz, M., et al. (2013). IGEMS: The consortium on interplay of genes and environment across multiple studies. *Twin Research and Human Genetics, 16*(1), 481–489.

Pielta, S., Bjorklund, A., & Bulow, P. (2012). Older twins' experiences of the relationship with their co-twin over the life-course. *Journal of Aging Studies, 26*, 119–128.

Porges, S. W. (2003). Social engagement and attachment: A phylogenetic perspective. *Annals of the New York Academy of Sciences, 1008*, 31–47.

Porges, S. W., & Furman, S. A. (2011). The early development of the autonomic nervous system provides a neural platform for social behavior: A polyvagal perspective. *Infant Child Dev, 20*(1), 106–118.

Rasulo, D., Christensen, K., & Tomassini, C. (2005). The influence of social relations on mortality in later life: A study on elderly Danish twins. *Gerontologist, 45*(5), 601–608.

Reiss, D., Cederblad, M., Pedersen, N. L., Lichtenstein, P., Elthammar, O., Neiderhiser, J. M., et al. (2001). Genetic probes of three theories of maternal adjustment: II. *Genetic and environmental influences. Fam Process, 40*(3), 261–272.

Reiss, D., & Neiderhiser, J. M. (2000). The interplay of genetic influences and social processes in developmental theory: Specific mechanisms are coming into view. *Development and Psychopathology, 12*(3), 357–374.

Reynolds, C. A., Barlow, T., & Pedersen, N. L. (2006). Alcohol, tobacco and caffeine use: Spouse similarity processes. *Behavior Genetics, 36*(2), 201–215.

Reynolds, C. A., Finkel, D., Gatz, M., & Pedersen, N. L. (2002). Sources of influence on rate of cognitive change over time in Swedish twins: An application of latent growth models. *Experimental Aging Research, 28*(4), 407–433.

Robinson-Whelen, S., Tada, Y., MacCallum, R. C., McGuire, L., & Kiecolt-Glaser, J. K. (2001). Long-term caregiving: What happens when it ends? *Journal of Abnormal Psychology, 110*(4), 573–584.

Sbarra, D. A., & Hazan, C. (2008). Coregulation, dysregulation, self-regulation: An integrative analysis and empirical agenda for understanding adult attachment, separation, loss, and recovery. *Pers Soc Psychol Rev, 12*(2), 141–167.

Scherrer, J. F., Xian, H., Bucholz, K. K., Eisen, S. A., Lyons, M. J., Goldberg, J., et al. (2003). A twin study of depression symptoms, hypertension, and heart disease in middle-aged men. *Psychosomatic Medicine, 65*, 548–557.

Scherrer, J. F., Xian, H., Shah, K. R., Volberg, R., Slutske, W., & Eisen, S. A. (2005). Effect of genes, environment, and lifetime co-occurring disorders on health-related quality of life in problem and pathological gamblers. *Archives of General Psychiatry, 62*, 677–683.

Schleidt, W. M. (1997). An ethological perspective on normal behavior especially as it relates to mating systems. In N. Segal, G. E. Weisfeld, & C. C. Weisfeld (Eds.), *Uniting psychology and biology: Integrative perspectives on human development* (pp. 493–506). Washington DC.: American Psychological Association.

Schneiderman, I., Zagoory-Sharon, O., Leckman, J. F., & Feldman, R. (2012). Oxytocin during the initial stages of romantic attachment: Relations to couples' interactive reciprocity. *Psychoneuroendocrinology*.

Schulz, R., & Beach, S. R. (1999). Caregiving as a risk factor for mortality: The caregiver health effects study. *JAMA, 282*(23), 2215–2219.

Segal, N. (1997). Twin research perspective on human development. In N. Segal, G. E. Weisfeld, & C. C. Weisfeld (Eds.), *Uniting psychology and biology* (pp. 145–173). Washington DC.: American Psychological Association.

Segal, N. L., Sussman, L. J., Marelich, W. D., Mearns, J., & Blozis, S. A. (2002). Monozygotic and dizygotic twins' retrospective and current bereavement-related behaviors: An evolutionary perspective. *Twin Research, 5*(3), 188–195.

Seidlerova, J., Bochud, M., Staessen, J. A., Cwynar, M., Dolejsova, M., Kuznetsova, T., et al. (2008). Heritability and intrafamilial aggregation of arterial characteristics. *Journal of Hypertension, 26*(4), 721–728.

Sheehy, G. (1996). *New passages: mapping your life across time*: Random House.

Simonen, R. L., Levalahti, E., Kaprio, J., Videman, T., & Battie, M. C. (2004). Multivariate genetic analysis of lifetime exercise and environmental factors. *Medicine and Science in Sports and Exercise, 36*, 1559–1566.

Simonen, R. L., Videman, T., Kaprio, J., Levalahti, E., & Battie, M. C. (2003). Factors associated with exercise lifestyle—a study of monozygotic twins. *International Journal of Sports Medicine, 24*(7), 499–505.

Speers, M. A., Kasl, S. V., Freeman, D. H, Jr, & Ostfeld, A. M. (1986). Blood pressure concordance between spouses. *American Journal of Epidemiology, 123*(5), 818–829.

Spotts, E. L., Pederson, N. L., Neiderhiser, J. M., Reiss, D., Lichtenstein, P., Hansson, K., et al. (2005). Genetic effects on women's positive mental health: Do marital relationships and social support matter? *Journal of Family Psychology, 19*(3), 339–349.

Strathearn, L., Li, J., Fonagy, P., & Montague, P. R. (2008). What's in a smile? Maternal brain responses to infant facial cues. *Pediatrics, 122*(1), 40–51.

Strawbridge, W. J., Wallhagen, M. I., Thai, J. N., & Shema, S. (2009). The influence of spouse lower cognitive function on partner health and well-being among community-dwelling older couples: Moderating roles of gender and marital problems. *Aging & Mental Health, 13*(4), 530–536.

Stroebe, M., Schut, H., & Stroebe, W. (2007). Health outcomes of bereavement. *Lancet, 370*(9603), 1960–1973.

Tancredy, C. M., & Fraley, R. C. (2006). The nature of adult twin relationships: an attachment-theoretical perspective. *Journal of Personality and Social Psychology, 90*(1), 78–93.

Tellegen, A., Lykken, D. T., Bouchard, T. J, Jr, Wilcox, K. J., Segal, N. L., & Rich, S. (1988). Personality similarity in twins reared apart and together. *Journal of Personality and Social Psychology, 54*(6), 1031–1039.

Trumbetta, S. L. (2004). Middle age, marriage, and health habits of America's greatest generation: Twins as tools for causal analysis. In L. F. DiLalla (Ed.), *Behavior genetics principles: Perspectives in development, personality, and psychopathology*. Washington DC.: American Psychological Association.

Trumbetta, S. L., & Gottesman, I. (1997). Pair-bonding deconstructed by twin studies of marital status: What is normative? In N. L. Segal, G. E. Weisfeld, & C. C. Weisfeld (Eds.), *Uniting psychology and biology: Integrative perspectives on human development*. Washington D.C.: American Psychological Association.

Trumbetta, S. L., Markowitz, E. M., & Gottesman, I. I. (2007). Marriage and genetic variation across the lifespan: Not a steady relationship? *Behavior Genetics, 37*(2), 362–375.

U.S. Department of Health and Human Services Administration on Aging. (2011). A profile of older Americans: 2011.

U.S. Census Bureau. (2000). Marital Status for the Population 15 Years and Over by Age for the United States, Regions, States, and Puerto Rico: 2000; Census 2000 Summary File 3.

U.S. National Center for Health Statistics. (2010). Vital statistics of the United States from Centers for Disease Control and Prevention: http://www.cdc.gov/nchs/nvss/mortality_tables.html.

Vaillant, GE (1977) *Adaptation to life*. Boston: Little, Brown.

Vaillant, G. E., Meyer, S. E., Mukamal, K., & Soldz, S. (1998). Are social supports in late midlife a cause or a result of successful physical aging? *Psychological Medicine, 28*, 1159–1168.

Van Overwall, F., Baetens, K., Marien, P., & Vandekerckhove, M. (2014). Social cognition and the cerebellum: A meta-analysis of over 350 fMRI studies. *Neuroimage, 86*, 554–572.

Waite, L. J., Laumann, E. O., Das, A., & Schumm, L. P. (2009). Sexuality: Measures of partnerships, practices, attitudes, and problems in the National Social Life, Health, and Aging Study. *Journals of Gerontology. Series B, Psychological Sciences and Social Sciences, 64*(Suppl 1), i56–i66.

Walker, R., Luszcz, M., Gerstorf, D., & Hoppmann, C. (2011). Subjective well-being dynamics in couples from the Australian longitudinal study of aging. *Gerontology, 57*(2), 153–160.

Watson, D., Wiese, D., Vaidye, J., & Tellegen, A. (1999). The two general activation systems of affect: Structural findings, evolutionary considerations, and psychobiological evidence. *Journal of Personality and Social Psychology, 76*, 820–838.

Weiss, R. (1974). The provisions of social relationships. In Z. Rubin (Ed.), *Doing unto others* (pp. 17–26). Englewood Cliffs, NJ: Prentice Hall.

White, L. (2001). Sibling relationships over the life course: A panel analysis. *Journal of Marriage and Family, 63*, 555–568.

Wilson, S. E. (2001). Socioeconomic status and the prevalence of health problems among married couples in late midlife. *American Journal of Public Health, 91*(1), 131–135.

Wilson, S. E. (2002). The health capital of families: An investigation of the inter-spousal correlation in health status. *Social Science and Medicine, 55*(7), 1157–1172.

Xian, H., Scherrer, J. F., Franz, C. E., McCaffery, J., Stein, P. K., Lyons, M. J., et al. (2010). Genetic vulnerability and phenotypic expression of depression and risk for ischemic heart disease in the Vietnam era twin study of aging. *Psychosomatic Medicine, 72*(4), 370–375.

Yang, Y. (2008). Long and happy living: Trends and patterns of happy life expectancy in the U.S., 1970-2000. *Social Science Research, 37*(4), 1235–1252.

Zivin, K., & Christakis, N. A. (2007). The emotional toll of spousal morbidity and mortality. *Am J Geriatr Psychiatry, 15*(9), 772–779.

Chapter 10
The Family System as a Unit of Clinical Care: The Role of Genetic Systems

David Reiss

Data are accumulating on the public health advantages of focusing on the family as a primary unit of both assessment and treatment in a range of psychiatric and medical problems. For example, therapeutic work with the family as a system is the preferred treatment for serious conduct problems and substance abuse in adolescents (Psychiatry 2005). Family-based treatment of diabetes in children and adolescents and of cardiovascular disease in adults is reliably related to improvement in the medical condition, in comparison to treatment programs that ignore the family and focus only on the family member with the illness. Indeed this approach to the family as a unit may prolong survival in several adult chronic illnesses (Armour et al. 2005; Hartmann et al. 2010; Martire 2005; Martire et al. 2004, 2010, 2011). Moreover, a focus on the family unit has yielded impressive findings on the prevention of psychological problems (e.g., Dadds et al. 1999; Dishion et al. 2002; Feinberg et al. 2009). Data also suggest that characteristics of the family system may play a decisive role in educational attainment and occupational and financial success of its members (e.g., Melby and Conger 1996; Teasdale and Owen 1984). In turn, there is more than ample published data showing the very favorable impact of educational attainment and occupational success on both mental and physical health across the entire span of life (e.g., Kern and Friedman 2008).

D. Reiss (✉)
School of Medicine, Yale Child Study Center, Yale University,
P.O. Box 207900, New Haven, CT 06519, USA
e-mail: david.reiss@yale.edu

© Springer Science+Business Media New York 2015
B.N. Horwitz and J.M. Neiderhiser (eds.), *Gene-Environment Interplay in Interpersonal Relationships across the Lifespan*,
Advances in Behavior Genetics 3, DOI 10.1007/978-1-4939-2923-8_10

10.1 Central Features of the Family as a System

The family has been an attractive target for therapeutic intervention because its workings seem to underlie the etiology of many clinically important behavioral syndromes or the clinical course of those syndromes. Two of these properties will be the focus of this chapter.

The first concerns the *stability of patterns in families across time*. Negative patterns have, of course, attracted most clinical attention and include severe hostility and aggression as well ineffective problem solving and withdrawal and neglect. In many cases families do not seem to be able to correct these patterns on their own thus enhancing the urgency of clinical intervention (Reiss and Emde 2003). Clinicians believe that working with families to enhance their own power of self-correction may ameliorate some current clinical problems and prevent future ones.

A second attractive property of the family system is *experience conserving or memorial function*. Clinicians have observed that severe trauma experienced by children, particularly abuse and neglect by their caretakers, reappears when these children become adults and form their own families. Evidence has centered on two outcomes of this conservation. The first is that many children, abused and neglected by their parents when they were children repeat this pattern when they become adults (e.g., Berlin et al. 2011; Thornberry and Henry 2013). Second, many of these same children develop lifelong patterns of poor self health care (e.g., (Springer 2009) and are much more likely to develop disabling or fatal illnesses when they become adults (Danese et al. 2009; Widom et al. 2012). Clinicians are intrigued by these phenomena of conservation. They imply that proper treatment of a maltreating parent–child relationship in one generation may prevent both a cycle of abuse and ill health in subsequent generations.

Family clinical researchers could be expected to examine first the psychosocial mechanisms that might account for stability of relationship patterns across a span of several years as well as conservation of adverse experience across an entire generation. This chapter begins an exploration of the ways research on genetics may help clarify mechanisms underlying stability and conservation in the family. I explore in particular the role of a biological systems approach to genetic influences and consider the role of genetic differences among individuals, the traditional domain of behavioral genetics, but also include two other closely related cellular mechanisms: gene expression and telomere function.

This same analysis could be applied to other important features of the family system which include (1) The reciprocal relationships among subsystems in the family: the marital, parent–child and sibling subsystems; (2) the reciprocal relationships between child and adult psychological development on the one hand and family relational patterns on the other; (3) the intergenerational transmission of family patterns such as marital discord or parent–child seductive behavior whether or not trauma is involved; and (4) the occasional but all the more remarkable capacity of families to shift course either within critical developmental periods such as the birth of the first child or across generations where children establish family patterns very different than those they knew as children. Genetics have added to our understanding of these features of family life as well but a full consideration of them is beyond the scope of this chapter.

10.2 Stability of Family Relationships Across Time

While there are notable examples of discontinuity in these patterns the persistence across extended periods of time of both negative and positive features of family interaction patterns has been noted repeatedly for both parent–child and marital subsystems (e.g., Dunn et al. 2005; Else-Quest et al. 2011; Erel and Burman 1995; Gerard et al. 2006; Mannering et al. 2011; Roskam and Meunier 2012; Stover et al. 2012). Because of this persistence researchers have, from many different vantage points, attempted to identify distinct styles or dimensions of parenting with only modest consensus. For example, for parent–child relationships, researchers have attempted to define common parenting practices across cultures (e.g., Deater-Deckard et al. 2011), distinguishing among dimensions by specific determinants of parenting styles such as personality (e.g., Prinzie et al. 2009) or common consequences (e.g., Skinner et al. 2005). Parental warmth—defined as the parent's pleasurable acceptance and support of their child—and control—usually defined as efforts to monitor and modify a child's behavior—regularly appear in such efforts. However, specific determinants or consequences for these and other dimensions remain elusive. Part of this stability is attributable *to persistent resonance of relationship qualities across family subsystems.* Thus, marital distress is reliably associated with strains in the parent child and sibling relationships within a family. The likelihood of ongoing reciprocal relationships among these subsystems is high (Dunn et al. 2005; Erel and Burman 1995; Gerard et al. 2006; Stover et al. 2012).

10.3 The Conservation of Early, Traumatic Parent–Child Relationships

A second striking feature of family system is, as noted, its apparent capacity to conserve early traumatic experiences of children in their family of origin. When children in such a family become adults they often repeat the pattern of abuse by becoming abusers themselves. They also are more likely to fall ill from psychiatric and medical illnesses that impair their capacity to develop and sustain the many roles and responsibilities of adult life, including the parental role.

There are many studies of the "transmission" of child abuse: parents who were abused as children become—in many instances—abusive themselves when they become parents (Dixon et al. 2005a, b, 2009; Pears and Capaldi 2001). However, the mechanisms accounting for this transmission are unclear. For example, some studies find that parental psychopathology may facilitate or mediate the transmission (Dixon et al. 2005b), others fail to find such a role (Berlin et al. 2011). Abuse and neglect in childhood is also strongly associated with problems in adult partner relationships. Compared with controls, adults abused in childhood are more likely to cohabit with partners rather than marry, to abruptly walk out on these relationships, are less likely to be sexually faithful and—if they marry—are more likely to get divorced (Colman and Widom 2004).

Features of current social life for parents abused as a child may moderate the risk of their becoming abusive parents. For example, low social support, social isolation and living with another abusive adult may either mediate or exacerbate the effects of mother's childhood history of abuse on her own parenting (Berlin et al. 2011; Dixon et al. 2005a, b). Correspondingly, a favorable marital match for a parent abused in childhood can also buffer the effects of this early developmental history of parents in regard to their own parenting (Cohn et al. 1992). The data thus far suggests the possibility of a reciprocal relationship between the current features of the family system and the long-term adverse effect of childhood trauma on current family relationships.

More recent research is now delineating notable effects of child abuse on physical health of these children, as they grow older. For example, Felitti and colleagues—in studies first reported almost two decades ago—reported a much greater incidence of child abuse and related early adversities among adults who showed poor self health care—including smoking and inactivity—and who developed a range of life threatening illness including cardiovascular disease and some cancers (Felitti et al. 1998). Other investigators questioned whether retrospective reports of early abuse might be influenced by the adults' ill health (Widom et al. 2004). Felitti and his colleagues addressed these concerns in two ways. First, they followed a group of individuals who reported on child adversity but were not yet ill and noted prospectively the development of lung cancer, significantly more frequent in those who were abused. Second, they investigated premature death in the relatives of adult victims of child abuse on the assumption those relative too were victims of early adverse experiences. There was a strong relationship between adverse experiences of the victim and premature death of a relative thus circumventing the biasing of retrospective reports by one's own current health. More recently, these results seem to have been confirmed by studies of maltreated children followed from childhood (Danese et al. 2009; Shalev et al. 2012; Widom et al. 2012).

Quite recently these two distinct lines of research are coming together around the concept of "second hit." Adults who report early maltreatment appear more sensitive to the stress of a caretaking role with the family as indexed by direct reports of stress (Pereira et al. 2012) or by stress related syndromes such as dissociation (Marysko et al. 2010). In a report on caretakers of demented spouses Kiecolt-Glaser et al. (2011) reported that the stress inherent in this role was already increasing risk factors for serious illness such as elevated inflammation. Although we have only preliminary data in the literature it seems plausible that adults who escape the effects of the first "hit"—child maltreatment—on their health may sustain health risk after a second hit: the enhanced stress of a caretaking role within the family of which they are adult members.

10.4 Genetic Systems and Family Systems

The initial enthusiasm for a family systems approach to clinical services drew heavily on the six systemic features of family life summarized above because an intervention that successfully changed the way the system operated might have many beneficial effects on all members and interrupt what were regarded as fundamental mechanisms that perpetuated dysfunctional patterns across time and across generations. Thus it is not surprising that initial empirical investigation of these characteristics of family systems focused almost exclusively on psychosocial mechanisms. For example, a series of well-conducted, very long-term studies firmly linked the relationship between parents and children in one generation to patterns of parent–child relationships a generation later when the children were themselves parents (Bailey et al. 2009; Kerr et al. 2009; Kovan et al. 2009; Neppl et al. 2009; Shaffer et al. 2009). These studies proposed mechanisms for such striking continuity centered on social learning or cognitive processes such as enduring social schemata and internal "working models" of social relationships. Published in the era of intense genetic investigation some allowed for the possibility of genetic mechanisms as well. However, these were not investigated.

In this chapter, I will examine the role of genetic processes in expanding our understanding of the family system. I will explore one of several contemporary approaches to these processes, an approach that is tune with the analysis of family systems. This approach utilizes biological systems approach to examine three aspects of genetic systems: genotypic differences among individuals that characterizes directly or indirectly the influences of both coding and non-coding regions of heritable DNA; the extent to which at any time and in particular tissues the gene is expressed (i.e. has an impact on some measurable biological or psychological characteristic); and finally the structure and function of the protective areas at each end of each chromosome called telomeres. All three operate interdependently on the chromosome of each cell and might be thought of as a "chromosomal system." I review this system very briefly here and expand our discussion of this system in the next section.

10.5 Genotypic Differences Among Individuals

As many chapters in the current volume illustrate, in the last two decades a growing number of researchers have combined careful assessments of family systems with estimates of genetic effects. Moreover, there has been a growing interest in social behavior in animal models that have yielded insights into the development of adult pair bonding and parenting; we now have illuminating efforts in comparative studies of humans and animals.

Many chapters in this book make effective use of some of the most classic designs in the study of family and other social relationships: the use of twins, siblings and children adopted at birth. These studies effectively focus on differences among individuals in the configuration and expression of their entire genome. Humans are mostly alike genetically but some genes have structural variants, affecting their function, that account for heritable differences among individuals. Twin designs, in particular, summate these genetic differences and have been particularly useful for establishing that some human traits, including the quality of parental and marital relationships they develop, are notably influenced by these cumulated genetic differences.

For example, as noted elsewhere, heritable characteristics of the child influence how their parents treat them (Narusyte et al. 2011; Neiderhiser et al. 2007; O'Connor et al. 1998; Plomin et al. 1994). Moreover, these methods have usefully identified genetic influences common to two different domains of behavior. Thus many of the same genetic influences on parenting, just mentioned, are the same that influence child conduct disorder (Pike et al. 1996; Reiss et al. 2000). Thus, genetic investigation has shed new light on mechanisms that may account for frequently observed associations between the quality of a subsystem (in this example, parenting) and individual characteristics (in this case, conduct problems in the offspring). Research designs that include the children of twins allow estimates of the role of genetics in the transmission of patterns of social behavior across generations (e.g., D'Onofrio et al. 2007). Taken together, these designs are the backbone of a subfield of genetics known as *quantitative genetics.*

More recently, the techniques of *molecular genetics* have begun to identify particular genes that may be involved in these processes. Since many of the genes so far identified have well-known roles in biological pathways, usually in the brain, these genetic studies—as I will show in the next section—provide early suggestions about brain mechanisms involved in family relationships.

Both quantitative and molecular genetics explore inherited individual differences among individuals and the role of these differences in the development and maintenance of key relationships within the family. However, in the last decade an entirely different function of genes is receiving major attention: the role of genes in recording and memorializing the exposure of individuals to environmental variation, including exposure to variation in the quality of family relationships. Two genetic processes have come to the fore.

10.6 Mechanisms of Gene Expression

The first of these is an increased understanding of the impact of family relationships on mechanisms by which genes are expressed, a development that follows from human and animal studies of mechanisms regulating whether a gene is silenced or active in particular tissues. Thus far, two distinct biological mechanisms have been identified as crucial in the expression of genes regulating the central nervous system. The first is an enzyme-catalyzed methylation (the adding of a methyl group CH_3) to sectors of DNA strands where the amino acid cytosine is bonded to a guanine (so-called CpG site). Ordinarily, methylation reduces the expression of a gene or completely silences it. However, in some cases, methylation may enhance expression. Thus, methylation studies are usually validated by a corresponding measure of the messenger RNA (mRNA) that is the immediate product of the DNA sequence. The second mechanism involves a modification of the proteins—called histones—to which the DNA strand is bound. The removal of an acetyl group from the histone core changes the conformation in ways that alter the expression of the genes wrapped around the affected histone (see a useful review, Sweatt 2009). While these mechanisms appear to be sensitive to experienced stress the exact mechanisms of this responsiveness remain under study.

10.6.1 Telomere Biology

Another mechanism for recording differences in environmental experience lies not in the genes themselves but at the tips of chromosomes called telomeres. Telomeres, visible in an ordinary light microscope, are extended strands of non-replicating DNA at both ends of each chromosome that serve to protect genes during the process of cell division. Since chromosomal replication is not fully complete, in each cell division, telomeres protect coding DNA by shortening; if they shorten too much chromosomal replication is impaired leading to cell death and a range of diseases. The length of telomeres is a dynamic process since the cell nucleus contains an enzyme, telomerase, that can repair and restore the length of telomeres. In 2004 Epel and colleagues reported that sustained stress shortened telomeres (Epel et al. 2004), a finding that has been replicated many times since.

10.6.2 The "Chromosome System"

Genotypic differences, patterns of gene expression and telomere erosion are all features of the structure and function of the human chromosome and their interrelationship is not only an important biological process but is critical to understanding the role of this system in moderating and mediating the impact of stressful circumstances on the family. For example, specific genes have been identified that influence telomere length (Mirabello et al. 2010) and the overall heritability of telomere length may be as high as 70 %. Estimates of the heritability of the expression of particular genes or gene locations are more complex because it may vary by anatomical location. Nonetheless, initial twin studies—documenting the importance of heritability of gene expression—reported it to be quite low (Bell and Spector 2012). More recent and more comprehensive studies suggest a much higher heritability of gene expression (Wright et al. 2014).

10.7 Genotypes, Gene Expression and Telomeres: Family Stability, and Its Memorial Functions, in a New Light

10.7.1 Persistence of Distinctive Interaction Patterns

Some versions of the family systems perspective see patterns of family relationships arising out of the unique history of the relationships itself. A given child may have a special meaning to a parent and the unique relationship with that child may grow from the distinctive circumstances surrounding the child's birth and early development. For example, Walsh reported the preponderance of grandparental death just before or after the birth of children destined to become schizophrenic; presumably the parental absorption in the death of their own parent tinged their relationship with their child providing relational grounds for eliciting the child's schizophrenic diathesis (Walsh 1978). Another psychosocial approach to understanding mechanisms accounting for both stability and style of family relational systems has been typical of research on marriage. In these studies there is less focus on the memorial functions of interaction in marriages and much more on the detailed, sequential patterns of affect and behavior as they unfold over time and predict marital satisfaction or dissolution; these patterns of affect are thought to "lock" unhappy marriages in self-defeating patterns of unhappiness (see review, Gottman and Notarius 2002).

While features of the family system that may perpetuate its function and dysfunction have attracted a good deal of attention, an extended program of research has suggested that other factors may influence the style of family relationships as they unfold but help to stabilize these features in more established systems. Indeed, we have known for some time that stable features of both the child's temperament and parental personality each play a role in differences among parents in parenting style (e.g., Barry et al. 2008; Bornstein et al. 2007, 2011; Kochanska 1995; Kochanska et al. 2004, 2007; Prinzie et al. 2009; Bornstein 1955).

A similar trend can be noted in marital research where personality of partners has been used to predict marital satisfaction, infidelity and breakup (Donnellan et al. 2004; Holtzworth-Munroe et al. 1997; Kinnunen et al. 2000; Shackelford et al. 2008). Thus, since both temperament and personality are, in part, heritable it is not surprising that there are substantial genetic effects on how parents and children perceive their relationships with one another (e.g., Rowe 1981; Plomin et al. 1994), how parental behavior is perceived by the child (Neiderhiser and Pike 1995) and how parents and children behave towards one another as videotaped and coded by a trained coder (O'Connor et al. 1995; Reiss et al. 2000). Thus, these genetic effects extend beyond the perceptions of parents or children; they influence observable styles of behavior.

The same findings are also true of marital satisfaction and break up; a substantial genetic influence has been reported (Bauer et al. 2007; D'Onofrio et al. 2007; Lykken 2002; McGue and Lykken 1992; Spotts et al. 2004, 2006). In several of these studies the influence of the wife's genotype extends to reports of marital satisfaction by the husband and vice versa.

The cumulative impact of these studies is to emphasize that attributes of individuals play an important part in giving shape and stability to their relationships with one another. Moreover, many of these traits are heritable. As noted elsewhere in this book, these traits or personality styles are the medium through which genetic influences have a notable effect in shaping and sustaining styles or patterns of interaction in the family system. However, as I will note, current concepts of both temperament and personality probably are not sufficient conceptual tools to understand fully the pathway from genes to family interaction patterns.

Until recently, most analyses of these genetic influences on both the parent–child and marital subsystems have used twin and adoption designs (e.g., Reiss et al. 2000; Dunn and Plomin 1986). However, as I will note, a recent series of molecular genetic studies promises to add to our understanding of these two family subsystems. As I will suggest, although these molecular genetic studies are in an early phase they promise to illumine much more completely the pathways from genes to family interaction patterns by identifying a range of biological systems—from neural circuits to endocrinological feedback systems to immune function—that link genes and family systems.

Despite a plethora of studies few have addressed the core question of the contribution of genetic influences to the *stability* of family patterns. In a broad variety of other domains genetic factors are the primary influence on stability of individual characteristics from depression in adult women (Kendler et al. 1993) to personality in adult men and women (Viken et al. 1994) to temperament in children (Plomin et al. 1993) and adolescents (Ganiban et al. 2008). These studies require longitudinal data collection in a genetically informed design and we have only a small set of longitudinal, genetically-informed studies of family relationships (Burt et al. 2005; Forget-Dubois et al. 2007; Larsson et al. 2008; Reiss et al. 2000). All of these studies used twins or twins and siblings who were *offspring*, not parents. All but one suggest that heritable characteristics of the child play a pre-eminent part in influencing the stability of parent–child relationships, particularly of negativity and conflict. The interesting exception is the study by Forget-Dubois and colleagues whose sample of twin children were far younger than in other studies; here genetic factors played a major role in *change* in hostile and negative parenting from 5 to 18 months. That is, genetic factors in the child that evoked irritable parenting at 5 months were uncorrelated with genetic factors that evoked the same parental behavior at 18 months. This finding is quite consistent with other data suggesting a dramatic change in what genes are expressed across the toddler periods [see, for example, a quantitative analysis (Plomin et al. 1993) and a summary of experience-dependent changes in gene expression, at the molecular level, in early childhood (Roth and Sweatt 2011)].

I could not locate any longitudinal study that directly examines the role of *parents* genes in the stability of parenting although there are a large number of cross-sectional studies, using twins, documenting a notable influence of parents' genotype on the parent child relationship as well (Kendler 1996; Neiderhiser et al. 2004, 2007; Perusse et al. 1994; Spinath and O'Connor 2003). For mothers, at least, there seems to be more consistent effect of her genotype on her warmth and support towards her children than on other attributes such as control and over-all negativity. From the child's perspective genetic influences of this kind would emerge as shared environmental influences because they tend to keep mother's consistent across children. We know from longitudinal data, where children are the twins (Reiss et al. 2000), that the pre-eminent influence of the persistence of warm parenting across adolescence is shared environment. Thus the variation among families of adolescents in the level of maternal warmth may be kept relatively sta-ble by her genes but variation in conflicts with her children persist because of her children's genes. Recently a number of publications suggest specific polymorphic genes that may be implicated in these effects (e.g., Avinun et al. 2012; Bakermans-Kranenburg and van Ijzendoorn 2008; Feldman et al. 2012). I will return to this molecular genetic work below.

For studies of genetic influences on both marital and parent–child subsystems heritable effects on personality partially but not completely mediate the genetic effects. Thus, the evocative effects of children's genotypes on negative relationship with their parents are mediated by their poor emotional self-regulation and aggres-sive traits (Ganiban et al. 2011; Narusyte et al. 2007; Spotts et al. 2005). Likewise, the impact of wives' genotype on her husband's marital satisfaction is also par-tially mediated by her aggressive traits (Spotts et al. 2005). The same is true for the genetic effects on ever marrying or never marrying (Johnson et al. 2004) and for the genetic effects on marital dissolution (Jocklin et al. 1996). However some investigators, searching for personality mediators on these relationships systems have come up empty handed (Spinath and O'Connor 2003).

These genetic findings contribute four ideas to our understanding of the stabil-ity of patterns in family systems across time. First, they underscore *the importance of persisting, heritable characteristics of individuals on the quality and duration of relationships.* We often call these characteristics "temperament" or "person-ality." In that sense recent genetic findings are, to some degree, an extension of what we already know about the role of personality in influencing marital rela-tionships and parent child relationships. However, personality is usually measured in adolescents and adults using self-report scales that tap the respondent's typi-cal behaviors (in some cases parents may report on their children using similar items). While these scales, especially those drawn from the Big Five approach to measurement, have exceptional predictive validity (e.g., Roberts et al. 2007) they are—for the most part—unsatisfactory leads to mechanistic explanations linking genetic and family process. By suggesting that personality is a mediator of genetic effects on family systems we may be "kicking the can down the road:" while we may be satisfied that measures of personality mediates some of the relationship between genes and family relationships we must then ask a question as to how genes influence personality and, in turn, precisely how personality influences pat-terns of family relationships. Almost certainly we will need simpler measures of the heritable traits such as aggressiveness and emotional lability to fashion a satis-fying mechanistic explanation linking genes and relationships. A very recent pilot study suggests one strategy. Lahey and colleagues, in a small sample of mothers of pre-schoolers, studied the association of a polymorphism regulating the density of estrogen receptors in the brain to observed negative parenting with their own children (Lahey et al. 2012). The areas of the brain involved—such as the left frontal gyrus and the right insula—had already been show to be engaged when mothers responded to photos of their own children. Thus, these brain areas seemed important in the development of mother child relationships and animal studies had already suggested that individual differences in maternal response to offspring might be influenced by the density of estrogen receptors (Champagne et al. 2006). In this pilot study, specific neural mechanisms are being "substituted" for personality in a preliminary delineation of a causal chain between genes and maternal parenting.

Second, these genetic data *amplify well-established findings on the role of "child effects:"* the role of children's characteristics in shaping relationships with their parents. However, these data us carry several steps further. They provide clearer evidence of the magnitude of these effects in the stability of parent–child relationships from early childhood onwards. Moreover, findings that child effects arise, in substantial measure, from genotypic differences among children helps resolve a core ambiguity in past "child effects" research: do child characteristics that drive current relationships with their parents simply reflect the influences of their parents in the past? While parental influences certainly shape patterns of gene expression, a topic I turn to shortly, they do not change a child's genotype. Sophisticated analytic models now allow us to assay the extent to which parent child relationships may exaggerate (e.g., Burt et al. 2005) or mute (Ulbricht et al. 2013) the evocative effect of child genotype but child genotypic main effects indisputably arise from a native property of the child.

The same reasoning applies to what can be termed "spouse effects:" native attributes that the spouse brings to a marital relationship. The smaller number of published studies on genetic influences on marital satisfaction and dissolution does add emphasis to the already well-establish influence of personality on marriage. In the same sense as with the parent child relationship, these genetic data make it incontrovertibly clear that these are characteristics brought by the spouse to the relationship, not somehow influenced by its earliest phases before researchers arrived on the scene. Moreover, at least two samples of adult twins show that, for a broad range of personality characteristics, the spouses of MZ twins are no more alike than spouses of DZ twins (Lykken 2002; Lykken and Tellegen 1993; Towers 2003). Thus, heritable characteristics of adults are unlikely to influence their marriage by their witting or unwitting selection of a spouse with similar personality. Thus, the genetic main effects on marital satisfaction and dissolution must operate after a spouse is selected and across a full range of personal characteristics, heritable and otherwise, of their spouse.

Taken together, these genetic data suggest that both differences among families in the quality and durability of their principal subsystems are notably influenced by genotypic differences with a surprising impact of the child on parent-offspring relationships. This leads to as third idea: *that genetic analyses may provide biological clues to different relational processes within family subsystems.* Although findings are preliminary they suggest a three-step process in this inquiry. First, some relational processes may have little or no genetic influence; by implications the dynamics and development of these processes are initiated by psychological and social processes although brains and gene function may be affected secondarily (see below). Current data suggest that secure attachment between infant and caregiver may be one such relational system. Four twin studies of genotypic differences among young children suggest this might be the case (Bokhorst et al. 2003; Fearon et al. 2006; O'Connor and Croft 2001; Roisman and Fraley 2008). However, by adolescence genetic factors appear to play a more decisive role in attachment processes (Fearon et al. 2013). Two studies have found associations between specific polymorphism and disorganization of attachment in children (Luijk et al. 2011; Spangler et al. 2009) or distress in the strange situation test of attachment (Raby et al. 2012). These data have been among those suggesting that disorganization of attachment and attachment distress may be processes quite separate from variations in attachment security.

Both positive and negative relational processes in sibling relationships also stand out as striking examples of relational systems that show little genetic influence. Indeed, among all relational systems studied, the stability and change across time—in families of adolescents—are influenced by experiences shared by siblings (Reiss et al. 2000), not their genotypes. Some of this shared experience may reflect levels of positivity in the wider social system including parents' relationships with each other or with them (Bussell et al. 1999; Feinberg et al. 2005). But the shared environment must also be the environment that siblings themselves create for one another. These "sibling-to-sibling" environments are highly charged with reciprocity so that positive behavior in one is responded to in kind and the same for negative behavior. This does not seem an artifact of twinships as these effects of the shared environment are equally apparent in comparisons of ordinary siblings in non-divorced families with full, half and blended sibs in step families (Reiss et al. 2000).

A second step in analysis then is to focus on those relational processes that show substantial genetic influence. Is there overlap or distinctiveness among the genetic factors that influences different relational systems? The few behavioral genetic studies of the role of parents' genes in their parenting styles or their response to offspring demands (Out et al. 2010) have not explored this issue although many are well positioned to do so. The genetic distinctiveness of child factors that elicit parental reaction has been explored (Loehlin et al. 2005), a subject I return in the next section.

In the last four years a large number of molecular genetic studies liking specific polymorphisms to observed parenting behaviors, usually with young children, have been reported. These polymorphisms are designated as "candidate genes" usually for one of two reasons. First, in some instances they are selected because of the association, in animal studies, of homologous genes, with parenting. Second, some are selected because these genes have been linked with other forms of social behavior in humans. As with many candidate gene studies, results have been inconsistent. For example, one study reports an association of the dopamine receptor gene DRD2 in the child with impaired parenting (Mileva-Seitz et al. 2012) and another does not (Mills-Koonce et al. 2007) and some studies show opposite effects of the same allelic variation (Bakermans-Kranenburg and van Ijzendoorn 2008; Mileva-Seitz et al. 2011). Moreover, many of the human data are drawn from complex, multivariable studies with an unknown number of analytic "tries" in efforts to find effects of allelic variation before a "hit." Indeed, many of the cautions underscored recently for candidate gene × environment studies (Duncan and Keller 2011) apply to this newly emerging field.

Nonetheless, it would be a mistake to dismiss these studies because if properly executed and reported their potential is substantial. Among the published sets of data three neuroregulatory systems have been the focus of attention: the dopamine system often explored in relation to attentional and memory function (Mileva-Seitz et al. 2012); the serotonin systems often involved in stress perception and stress modulation (Bakermans-Kranenburg and van Ijzendoorn 2008; McCormack et al. 2009; Sturge-Apple et al. 2012) and the oxytocin/vasopressin systems implicated in social motivation and bonding (Avinun et al. 2012; Bakermans-Kranenburg and van Ijzendoorn 2008; Bisceglia et al. 2012; Feldman et al. 2012).

The promise of these studies is to delineate distinctive biological mechanisms that underlie complex patterns of behavior entailed in measures of "maternal sensitivity", the most common dependent measure in these studies. As Feldman points out, studies of genetic variation promise to elucidate which neural systems may be involved in complex behaviors where direct access to these neural systems in human studies may be limited even with advanced neuroimaging techniques (Feldman et al. 2012). However, published studies have come considerably short of this for three remediable reasons. First, with few exceptions (Bakermans-Kranenburg and van Ijzendoorn 2008; Sturge-Apple et al. 2012) studies examine only one class of neuroregulatory systems at a time without attempts to show differential effects on parenting. Second, omnibus measures of parenting—such as maternal sensitivity—are drawn from a wide literature of family system research; they are "top–down" measures—derived from the complexity of family systems rather than the relatively simpler measures derived from studies of brain and behavior—and are therefore ill-suited to help distinguish among different components of competent parenting. Third, candidate genes studies would be better served by including more specific intermediate phenotypes that link allelic variation to parenting. For example, Mileva-Seitz et al. (2012) have argued that because allelic variation of two dopamine receptor genes influence attention and memory process they may interfere with maternal engagement and child focused verbalizations. However, in their study linking dopamine genes to these maternal behaviors the hypothesized intermediate phenotypes were not measured. I have already cited an example in pilot study of mothers of young children by Lahey et al. (2012). They linked allelic variation in a gene regulating estrogen receptors in the brain to maternal brain areas typically activated in response to pictures of the mothers' children. Activation of these areas, in turn, was linked to directly observe maternal parenting behavior The authors acknowledge the significant limitations of this study notably its small sample size and the failure to record brain activation before, or even at the same time, as observing parental interaction. Nonetheless, the study is an intriguing proof of concept.

10.8 The Memorial Functions of the Family Systems

As I have briefly reviewed in the section on family systems clinicians, clinical researchers and I have focused attention on how traumatic experiences of both children and parents may affect the family subsystems. Their concern, of course, has been with how these experiences, some of them in the distant past, are manifest in current patterns of affect and behavior in the current life of the family. Thus, they were concerned with understanding two mechanisms: the *immediate registration* of these experiences on the patterns of affect or behavior in the family as well as the *retention* of the effects on family patterns of those experiences long after the stressful circumstances may have occurred. This work did not attempt to bypass more conventional understanding of how stressful life experience is remembered or suppressed in the memories of individuals. Rather it sought to understand in what ways family interaction patterns played a distinctive and important role of their own in both the registration and retention process. Clinical family researchers contributed ideas and observations on how the operations of the family system *qua* system might both register and retain the effects of early adversity.

To pick from many examples, family clinical researchers have carefully studied family rituals—those that are borrowed from a general culture such as distinctive family celebrations of holidays such as Christmas—or those that are fashioned purposely or inadvertently to memorialize a specific family event such as an annual family trip to the gravesite of a member who died in childhood. Rituals are attractive for the study of memorial function in families because they often serve to connect the family's past with the present (Fiese et al. 2002; Smit 2011). However, adverse experiences from early in the child's life such as parental alcoholism may be carried forward into a second generation, if the family ritual includes the parental drinking (e.g., an alcoholic father regularly gets drunk at the family celebration of Christmas). In such a circumstance the family of the child, now an adult, also becomes plagued with adult drinking (Bennett et al. 1987). The ritual in the child's family of origin *registers* the parental alcoholism; the continuing practice of a ritual that is tainted with drinking helps to retain the experience across a generation.

Studies of memorial processes in families were not meant to replace more conventional understanding of memory as a neuropsychological process but as a parallel process that might both sustain more conventionally-conceived individual as well as endow them with greater meaning and heightened bonding to the family of the past and present. It is in this same sense that current work on genotypic differences, gene expression and telomere biology, reviewed briefly in Sect. 10.2, serves as an extension of this effort to understand both the registration and retention of severely stressful circumstances on the life of the family.

Analysis of genotypic differences has been most useful in understanding the registration process. The genetic work grew out of early twin studies that showed that sensitivity to stressful circumstances could be inherited. For example, Kendler and Karkowski-Shuman (1997) showed that MZ twins whose co-twins had experienced depression were more likely than DZ twins to be become depressed themselves in response to stressful circumstances. Some children are sensitive to the effects of neglect, abuse and other childhood adversity whereas others are relatively insensitive. For example, Widom—in work I have already cited—followed up documented cases of child abuse and neglect thirty years later (Widom et al. 2012). She found that 73 % of the adults reported themselves in good health whereas 27 % did not. Kendler's work suggests that genotypic differences may account for the relative invulnerability of these adults to the effect of abuse on their adult health. Indeed, Widom in her large sample of abused children, in comparison to controls found that variations in the sex-linked MAOA gene conferred resilience on white children in a 20 year follow-up of their violent and antisocial behavior (Widom and Brzustowicz 2006). This finding confirmed a number of prior findings (Caspi et al. 2002; Kim-Cohen et al. 2006) and has been confirmed subsequently (e.g., Weder et al. 2009). Variations in a number of other genes have also been reported to moderate the impact of maltreatment on a broad range of constructs of child adaptation (e.g., Cicchetti and Rogosch 2012). Of particular interest is a recent report showing the role of variations in four genes on the impact of poor parental quality in adolescence, disengagement rather than abuse, on their parental function when they became adults (Beaver and Belsky 2012).

The role of epigenetic processes in registering and retaining early environmental experiences was first proposed over 50 years ago (Griffith and Mahler 1969) on logical rather than empirical grounds. Their theoretical work was remarkably prescient in specifying that the methyl molecule, attached to a promoter region of a gene could silence it and that demethylation of the gene would restore its function. However, Michael Meaney and his group did the leading empirical work in this field in rats. They reported that pups who were well not well cared for by their mothers (defined in rats by two simple behaviors: little maternal back arching during nursing to make her teats more accessible to her litter and infrequent maternal licking and grooming of her pups) showed reduced expression of a gene regulating the density of glucocorticoid receptors in the rats' hippocampus (Weaver et al. 2004). As a result, the hippocampus was much less sensitive to circulating levels of glucocorticoids resulting in less efficient down regulation of the glucocorticoid system during periods of experimental stress. There was a critical period of effect of maternal behavior: the pups' first seven days of life. However, the excessive stress-sensitivity of these animals was life long (Liu et al. 1997). Recent research has suggested that, in rodents, variation in maternal care affects many areas of the genome including coding (exnoic) and non-coding (intronic) genomic regions (McGowan et al. 2011).

Equally striking was the work of Frances Champagne on the link between maternal care of rat pups and the maternal behavior provided by those pups once they grew to maturity and had their own offspring. She showed that, in the same critical seven day period, inadequate maternal care led to reduced expression of estrogen receptors in a specific brain region (the medial preoptic area) of the pup and consequent impairment of maternal behavior when those pups became adults (Champagne et al. 2006). Thus, two closely-related lines of investigation suggested that the effects of inadequate maternal care early in childhood was reflected in adult life in at least two ways: by ongoing excessive stress sensitivity with increased risk of behavioral pathology and by impaired maternal behavior of the pup when she matured.

For almost a decade the applicability of this work to human families was uncertain for four reasons. First, there seemed *no analogous critical period so clearly defined* for the effects of inadequate human, maternal care. Second, the animal studies showed *reduced genetic expression in highly specific areas of the brain* ordinarily inaccessible in human investigation. Third, the role of genetic expression as the mediator could be demonstrated in animals *where direct observation of early childhood and maturity can be accomplished in about two months* and the life span is one to two years. Fourth, because patterns of genetic expression are also under genetic control, *experiments that are possible only with animals are necessary to establish the causal role of experience.* These remain serious obstacles in human studies but, despite them, progress has been made in three ways.

First, to gain access to specific brain areas related to glucocorticoid receptors McGowan and colleagues used post mortem brain samples in patients who committed suicide. They compared those with an early history of child abuse with those with no abuse and with controls and found a reduction in expression of the glucocorticoid receptor gene analogous to that in the rat with inadequate maternal care (McGowan et al. 2009). A recently published extension of the McGowan work has shown that a broad range of genes show altered profiles of expression in association with early childhood adversity in humans and deficient maternal care in rats. Of special interest in both species were altered expression of genes controlling synaptic function and neural connectivity suggesting that through the medium of genes expression early experience may lead to distinct change in the structure and function of neural circuits linked to the hippocampus (Suderman et al. 2012). Despite these very promising homologies across species, the role of early experience in human gene expression profiles remains uncertain because, as noted, lack of experimental controls.

Thus, a second line of investigation has assumed increasing importance: gene expression research in primates. As in other areas of developmental research, the use of non-human primates has assumed great importance because both behavioral syndromes at one point in development and developmental trajectories of behavioral syndromes seem quite similar across human and non-human primates. In the domain of gene expression primate research offers two additional advantages. First, infants can be experimentally assigned to favorable or adverse rearing circumstances. Second, systematic comparisons can be made between brain and peripheral tissue. Very recent findings by Suomi and his colleagues have addressed both issues. First, rearing conditions are strongly associated with widespread changes in gene expression in both in circulating leukocytes (Cole et al. 2012) and in brain (Provencal et al. 2012). Second, Suomi and his colleagues addressed differences between prefrontal cortex tissues and circulating T-cells. The overlap between the two is small but interesting. For example in both tissues there is a hypermethylation of the promoter region of the gene regulating the glucocorticoid receptors in the hippocampus of monkeys that were reared in adverse circumstances leading to its relative silencing in both tissues (Proven cal, op. cit.).

These data lend weight to human studies associating both childhood and adult adversity with changes in the expression of genes that regulate both corticoid and immune function (see, for example, Cole et al. 2007; Miller et al. 2009) and more recent experimental studies in humans. In one study breast cancer patients, shortly after surgery, were assigned to either cognitive behavioral treatment for stress management or a control group. The treated group showed a down regulation of genes relating to pro-inflammatory processes, genes similar to those in several non-experimental studies of humans subjects under stress (Antoni et al. 2012). In a second study, the levels of suppression of the glucocorticoid receptor gene in peripheral tissue (buccal cells) are inferred from methylation of receptor sites. The cortisol response to a standard stressor was greater in subjects with highly methyl- ated promoter sites, especially in women (Edelman et al. 2012). This finding is consistent with both the rat experimental studies as well as the human post mor- tem studies. However, the origins of these differences in gene expression were not determined in this study.

In sum, the pattern of both animal and human studies are such that early adverse experiences—particularly adverse maternal or the absence of maternal care—is causally linked alterations in gene expression and that there is a role for both brain and peripheral tissue in detecting these alterations. What is unknown in human studies is whether these changes in expression mediate the effects of early adverse parental care on the large number of behavioral and medical sequelae of early adversity. Thus far human studies have failed to clarify whether there are any demarcated critical periods of sensitivity to environmental adversity as they are in the rat models that gave birth to this domain of investigation. The potential signifi- cance of this work for understanding family systems is considerable. I turn next to telomeres.

The intellectual history of telomere research is quite different from that of gene expression. The field began with an influential report of human data on the impact of stress on telomere length (Epel et al. 2004) and the follow-up stud- ies have sought to understand the mechanism of this effect, the range of circum- stances that seem to produce it and the consequences of telomere shortening for mental and physical health. Thus, far animals studies have been little used to track the pathway from stress to telomere shortening to disease (for a rare exception see Kotrschal et al. 2007), there is little evidence of a critical period for the appar- ent effects of environmental adversity. Moreover, virtually all studies have used peripheral tissue, usually white cells, for assays. Since these assays have predicted disease onset in many organs it has been assumed that circulating white cells are a reliable window on cellular aging throughout the body.

This work began only eight years ago. In 2004, Elissa Epel and her colleagues first reported the association of perceived stress in women, reduced telomerase activity and shortened telomeres (Epel et al. 2004). Notably her subject were all mothers and those showing the largest impact on telomere length and telomerase were those enduring the stress of caring for a chronically ill child across a long period of time who also perceived themselves as experience substantial stress. Since reduced telomerase and shortened telomeres are associated with cellular aging and a number of illnesses of mid life and beyond, Epel and her colleagues reasoned that telomere biology might be a critical intermediate step between stresses. This paper stimulated three lines of research,

First, there has been an active search for the range of psychological circumstances associated with telomere shortening; these have included sleep problems in healthy men (Jackowska et al. 2012), low intelligence (Kingma et al. 2012), reduced physical fitness in cardiac patients (Krauss et al. 2011), pessimism (O'Donovan et al. 2009) and employment and work schedules in women (Parks et al. 2011) as well as work-related stress in both genders. However, there is a consistent thread pointing to family related stress: being reared in an orphanage (Drury et al. 2012), fetal exposure to maternal stress with effects lasting to adult life (Entringer et al. 2011), persistent effects of exposure to violence during childhood (Shalev et al. 2012), single marital status (Mainous et al. 2011), and stress engendered by caretaking of both ill children (Epel, op. cit.) and disabled spouses (Damjanovic et al. 2007). The results thus far suggest there is no critical or sensitive period for the effects of stress on telomere function. However, children exposed to severely adverse rearing conditions both prenatally and postnatally show enduring effects on telomere length; thus it is possible there is a sensitive period for enduring effects but extended prospective studies are required to confirm this speculation. Recently reported is an important "double hit" phenomenon: caretakers who, as children, were exposed to adverse rearing environments show greater telomere shortening when as older adults they become caretakers of a disable spouse (Kiecolt-Glaser et al. 2011).

Second, research has begun to examine the mechanisms linking these circumstances to telomere biology. Given the range of psychological circumstances associated with telomere shortening it seems likely that several mechanisms may be involved. In vivo (Epel et al. 2010; Kroenke et al. 2011; Tomiyama et al. 2012) and in vitro (Choi et al. 2008) studies suggest that elevated levels of responsiveness of the HPA axis and abnormal diurnal patterns of cortisol secretion may be one mechanism.

Third there has been a renewed interest in the health risk engendered by shortened telomeres and research has ranged from mild cognitive decline in aging subjects (Choi et al. 2008; Valdes et al. 2010; Yaffe et al. 2011), to cardiovascular disease (e.g., O'Donnell et al. 2008) to cancer (Ma et al. 2011). An overall pattern of findings across many studies does link shortened telomeres to the risk of disease but the findings are not always straightforward. For example, in a few studies shortened telomeres appeared *protective* in the progress of coronary artery calcification (Kroenke et al. 2012), the occurrence of non-Hodgkin's lymphoma (Lan et al. 2009; Shen et al. 2011), and of lung cancer (Shen et al. 2011).

In summary, research thus far suggests promising links between severe stress—especially that related to adverse family relationships—accelerated telomere shortening and the risk of serious illness in adult life. However, no single study has convincingly followed all three steps of this pathway. Indeed, further progress depends on overcoming three obstacles.

First, this hypothesized chain of events unfolds over broad reaches of time, particularly in tracking the effect of prenatal and postnatal childhood adversity.

At this juncture rodent or primate models are critical for testing the plausibility of this pathway.

Second, genetic factors have rarely been considering either in establishing the proposed causal links in this pathway. Telomere length varies significantly among individuals of the same age and health status and a notable proportion of this variation is due to genotypic differences among them (Broer et al. 2013). Since genetic factors are important in the etiology of both entering stressful circumstances and in reactions to stress (Caspi et al. 2010; Edelman et al. 2012; Kendler and Baker 2007; Kendler and Karkowski-Shuman 1997; Kessler et al. 1992; Out et al. 2010) it is plausible that some of the observed associations between stress and telomere length are due to genetic influences common to both. The same is true of genes that may be common to telomere biology and cancer; a particularly careful study showed that much of the association between telomere length and colon cancer (Pellatt et al. 2012) could be attributed to genes common to both.

A third obstacle is increasing evidence of other influences on telomere length that may be closely related to stress; two persistent factors are obesity both in children and adults (Buxton et al. 2011; Pellatt et al. 2012) and cigarette smoking (Babizhayev and Yegorov 2011; Pellatt et al. 2012). As research in this domain has matured these "life style" factors are often controlled statistically or in matching of subjects; however they are a reminder of the importance of rigorous design in drawing causal inferences from association between environmental factors and biological sequelae. Moreover, simply controlling for "life style" factors may be inappropriate if these factors *mediate* the effect of stress on telomere biology. Very recent efforts to use stress reduction interventions to test the role of stress in telomere biology are promising techniques for strengthening causal inferences, particularly if they show that reduction in psychological distress mediates the effect of the intervention [see, for example, a recent, preliminary study (Lavretsky et al. 2012)]. Moreover, a number of studies linking stress to telomere biology show dose-response relationships but such relationships, by themselves, do not rule out the effects of confounding genetic or life style factors.

Despite their limitations recent studies of both gene expression dynamics and telomere biology have opened a new world of research that points the way to understanding genetic mechanisms, conceived as cellular systems, that help explain long-lasting environmental effects, particularly the impact of adverse circumstances within the life of the family unit, on some outcomes of early adversity: particular physical health and illness. Some authors have exuberantly claimed these studies of gene expression and telomere biology have delineated a new, chromosome-based "memory" system for social experience (Hoffmann and Spengler 2012; Schury and Kolassa 2012). However, the vast majority of studies fail to establish these mechanisms as a memory system for four reasons. First, these systems do not allow for the memorialization of *specific experiences*: the work on telomere biology is now conspicuous for the broad array of environmental circumstances that lead to telomere shortening. Second, these system do not allow for *discriminating responses to experience*: current assay technique for gene expression profiles are regularly showing scores of if not hundred of differences between individuals exposed and not exposed to stressful life circumstances. Third, these systems do not yet explain or account for *active learning and skills acquisition* of individuals in response to environmental threat. Rather, they record only the passively acquired biological wounds of adversity than have not been avoided through skills that might have been acquired through the functions of other more fully developed memory systems.

Fourth, and following from the third, this work does not explain a fundamental feature noted in clinical family system research: *the repetition of patterns of parent–child abuse across generations.* On this issue most relevant work to date along these lines is that of Frances Champagne (Champagne, op. cit.). Her work suggests that gene expression mechanisms may be part of a causal chain that originates in a very specific experience (inadequate maternal care), effects the expression of a single gene (regulating the estrogen receptors in a very specific part of the brain) with a very differentiated response (inadequate maternal care when the affected pup reaches maturity). We may legitimately ask why this work, at least in the gene expression field, has not become more paradigmatic and whether in this domain—as in other areas of molecular genetic research the balance is tipping in favor of broad scans of the entire genome rather than theory-guided work on specific pathways linking specific experiences to differentiated outcomes.

Setting aside the exuberant claims for the discovery of new memorial systems we can begin to appreciate the vast possibilities in family systems research opened to us by current research on gene expression and telomere biology. Four are already clearly apparent.

First, these mechanisms promise to account for how adverse experiences in families, perhaps those that are suffered uniquely by young children and caretakers of ill members may give rise to biological impairments in almost every major organ system. In this sense we may properly regard both gene expression and telomere biology as delineating important though imprecise *translation mechanisms.*

Second, it broadens our perspective on the public health implication of healthy families that protect their children from adverse experiences and the restoration to the modern family of supportive social networks that buffer the stress of care for disabled and chronically ill adults, networks that have—in the past—often been readily available in traditional cultures (Limb et al. 2014; Whitbeck et al. 2002, 2004).

Third, of necessity it energizes the still nascent field of family interventions to address the risk of severe medical illness and its consequences, a field I briefly reviewed at the start of this chapter. Moreover, new biological tools might aid this field of intervention—now with a more preventive focus—as we understand better some of the intervening steps between adverse experience in the family and the development of illness. Equally important might be the additions to conventional goals of family intervention such as encouraging familial behavior than enhances exercise or changes dietary patterns.

Fourth, as our understanding grows of these translation systems we can ask a critical and precise question of successful interventions. Do they succeed because they reverse the underlying risk processes or because they successfully compensate for them. That is why the first interventions studies in this domain loom large though their results are quite preliminary (Antoni et al. 2012; Lavretsky et al. 2012).

10.9 Conclusion

I have drawn, at several points in this chapter, a number of implications already as I have reviewed ways in which data on the "chromosome system" is altering our understanding of the family system. Here it remains to step back and take a broader look.

First, as already noted, genetic analysis has added a great deal to our notions of the etiology of parenting and to individual differences among children in their response to abuse and neglect in their families. Moreover, again as noted, these studies suggest that some distinctions among different qualities of parenting may be possible because they are influenced by distinctly and different neurobiological mechanisms. However, this promise is, as yet, unfulfilled. Furthermore, genetic studies of the parent–child relationship draw continuing attention to the impact of children on their parents from infancy through adolescence. The impact is so substantial that parenting may be partially re-conceptualized not only as a style stemming only from the parents who genotype and personal history but as a strategy for coping with sustained challenges from their child. Two broader lessons can be drawn from the work on genotypic differences I have cited.

First, genes that have been reported to have an influence on parenting of adults have also been shown to influence the social behavior of much younger children. For example, Walum and his colleagues studied variation in the oxytocin receptor gene in girls age 8 and 9 and then again when these girls were 19 and 20. Variations in this gene, as I have already reported, influenced the quality of parenting in several studies. Walum found that it also influenced social competence at age 8–9 and the quality of bonding of the late adolescents to their romantic partners (Walum et al. 2012). Genetic findings like these lend emphasis to the perspective that parenting quality is a station in human development that has many prior ones. Moreover, it raises a possibility that some genetic influences on parenting are not concurrent with our measurement of the parenting phenotype in adulthood but may exert some or all of its influence through its impact on social development earlier in the life course.

Second, as we construct more accurate and more sophisticated models of the role of the "chromosome systems" in influencing family systems we will find animal models not only more useful as heuristics but more critical in resolving interpretive dilemmas in human data. We can return to the oxytocin receptor gene again. The initial research on the role of this gene in relationship quality was done in voles; that work was an heuristic (Donaldson and Young 2008; Francis et al. 2002). Once comparable findings were obtained in humans, further work in voles and other animal species deployed transgenic strains and brain dissection to work out plausible mechanisms for the effects of variation in this gene.

I have already noted the major implications for the family system of current studies on two other aspects of the "chromosome system:" telomere structure and genetic expression. Here a broader point can be made. This work reinforces psychosocial investigations to emphasize the role of the family as a conserver of experience across broad spans of time. Much of the literature, both psychosocial and biological, has focused on how the family conserves adverse experience generated by its own processes, particularly the abuse of children. But studies of other traumatic circumstances paint a picture of the family as the conserver of positive experiences as well. For example detailed studies of holocaust survivors, their children and grandchildren call our attention to the earliest phases of the recovery from the holocaust by those survivors lucky enough to have survived and reached a safe heaven. Here, their capacity to keep their traumatic experience contained and to grasp eagerly and with wonder at the opportunity for a new beginning seems to be conserved across at least one generation. These studies painted a picture of the conservation of resilience but with the holocaust survivor generation now almost all lost to ordinary mortality we have missed a chance to probe this process with modern-day neurobiology. (The literature on holocaust survivors is vast but see Barel et al. 2010; Dekel et al. 2012; Sagi-Schwartz et al. 2008; Shmotkin et al. 2011.)

Finally, I want to return to a small point earlier in this chapter and enlarge it here in a concluding paragraph. Earlier I noted that some aspects of family relationships seemed little influenced by genetic factors. Sibling relationships stand out but also so does secure attachment, at least in early childhood [genetic factors become more prominent in by adolescence (Fearon et al. 2013)]. Moreover, many of the genetic effects—particularly in molecular genetic studies—are small. Likewise, I reviewed the many features of family life of adults that might mitigate the effects of their early history of abuse and neglect. Finally, I called attention to the particularly promising research—rich but purely psychosocial—on family rituals. It is instructive that this line of research, to judge by recent literature searches, is now moribund. After a flurry of promising papers in the 80s and 90s and just barely into this millennium the research seems to have petered out. However, the displacement of research of this kind need not and should not be displaced by current enthusiasms over the "chromosome system" and related efforts in neurobiology. The family system is too interesting and important for that.

References

Antoni, M. H., Lutgendorf, S. K., Blomberg, B., Carver, C. S., Lechner, S., Diaz, A., et al. (2012). Cognitive-behavioral stress management reverses anxiety-related leukocyte transcriptional dynamics. *Biological Psychiatry, 71*(4), 366–372. doi:10.1016/j.biopsych.2011.10.007, S0006-3223(11)00965-6 [pii].

Armour, T. A., Norris, S. L., Jack, L., Jr., Zhang, X., & Fisher, L. (2005). The effectiveness of family interventions in people with diabetes mellitus: A systematic review. *Diabetic Medicine, 22*(10), 1295–1305. doi:10.1111/j.1464-5491.2005.01618.x, DME1618 [pii].

Avinun, R., Ebstein, R. P., & Knafo, A. (2012). Human maternal behaviour is associated with arginine vasopressin receptor 1A gene. *Biology Letters.* doi:10.1098/rsbl.2012.0492, rsbl.2012.0492 [pii].

Babizhayev, M. A., & Yegorov, Y. E. (2011). Smoking and health: Association between telomere length and factors impacting on human disease, quality of life and life span in a large population-based cohort under the effect of smoking duration. *Fundamental and Clinical Pharmacology, 25*(4), 425–442. doi:10.1111/j.1472-8206.2010.00866.x, FCP866 [pii].

Bailey, J. A., Hill, K. G., Oesterle, S., & Hawkins, D. J. (2009). Parenting practices and problem behavior across three generations: Monitoring, harsh discipline, and drug use in the intergenerational transmission of externalizing behavior. *Developmental Psychology, 45*(5), 1214–1226.

Bakermans-Kranenburg, M. J., & van Ijzendoorn, M. H. (2008). Oxytocin receptor (OXTR) and serotonin transporter (5-HTT) genes associated with observed parenting. *Social Cognitive and Affective Neuroscience, 3*(2), 128–134.

Barel, E., Van Ijzendoorn, M. H., Sagi-Schwartz, A., & Bakermans-Kranenburg, M. J. (2010). Surviving the Holocaust: A meta-analysis of the long-term sequelae of a genocide. *Psychological Bulletin, 136*(5), 677–698.

Barry, R. A., Kochanska, G., & Philibert, R. A. (2008). G × E interaction in the organization of attachment: Mothers' responsiveness as a moderator of children's genotypes. *Journal of Child Psychology and Psychiatry, 49*(12), 1313–1320.

Bauer, L. O., Covault, J., Harel, O., Das, S., Gelernter, J., Anton, R., & Kranzler, H. R. (2007). Variation in GABRA2 predicts drinking behavior in project MATCH subjects. *Alcoholism, Clinical and Experimental Research, 31*(11), 1780–1787.

Beaver, K. M., & Belsky, J. (2012). Gene-environment interaction and the intergenerational transmission of parenting: Testing the differential-susceptibility hypothesis. *Psychiatric Quarterly, 83*(1), 29–40. doi:10.1007/s11126-011-9180-4

Bell, J. T., & Spector, T. D. (2012). DNA methylation studies using twins: What are they telling us? *Genome Biology, 13*(10), 172. doi:10.1186/gb-2012-13-10-172

Bennett, L. A., Wolin, S. J., Reiss, D., & Teitelbaum, M. A. (1987). Couples at risk for transmission of alcoholism: Protective influences. *Family Process, 26*(1), 111–129.

Berlin, L. J., Appleyard, K., & Dodge, K. A. (2011). Intergenerational continuity in child maltreatment: Mediating mechanisms and implications for prevention. *Child Development, 82*(1), 162–176. doi:10.1111/j.1467-8624.2010.01547.x

Bisceglia, R., Jenkins, J. M., Wigg, K. G., O'Connor, T. G., Moran, G., & Barr, C. L. (2012). Arginine vasopressin 1a receptor gene and maternal behavior: evidence of association and moderation. *Genes, Brain and Behavior, 11*(3), 262–268. doi: 10.1111/j.1601-183X.2012.00769.x

Bokhorst, C. L., Bakermans-Kranenburg, M. J., Fearon, R., van Ijzendoorn, M. H., Fonagy, P., & Schuengel, C. (2003). The importance of shared environment in mother-infant attachment security: A behavioral genetic study. *Child Development, 74*(6), 1769–1782.

Bornstein, M. H., Hahn, C.-S., Haynes, O. M., Belsky, J., Azuma, H., Kwak, K., et al. (2007). Maternal personality and parenting cognitions in cross-cultural perspective. *International Journal of Behavioral Development, 31*(3), 193–209.

Bornstein, M. H., Hahn, C. S., & Haynes, O. M. (2011). Maternal personality, parenting cognitions and parenting practices. *Developmental Psychology, 47*(3), 658–675. doi:10.1037/a0023181

Broer, L., Codd, V., Nyholt, D. R., Deelen, J., Mangino, M., Willemsen, G., et al. (2013). Meta-analysis of telomere length in 19 713 subjects reveals high heritability, stronger maternal inheritance and a paternal age effect. *European Journal of Human Genetics.* doi:10.1038/e jhg.2012.303

Burt, S., McGue, M., Krueger, R. F., & Iacono, W. G. (2005). How are parent–child conflict and childhood externalizing symptoms related over time? Results from a genetically informative cross-lagged study. *Development and Psychopathology, 17*(1), 145–165.

Bussell, D. A., Neiderhiser, J. M., Pike, A., Plomin, R., Simmens, S., Howe, G. W., et al. (1999). Adolescents' relationships to siblings and mothers: A multivariate genetic analysis. *Developmental Psychology, 35*(5), 1248–1259.

Buxton, J. L., Walters, R. G., Visvikis-Siest, S., Meyre, D., Froguel, P., & Blakemore, A. I. (2011). Childhood obesity is associated with shorter leukocyte telomere length. *Journal of*

Clinical Endocrinology and Metabolism, 96(5), 1500–1505. doi:10.1210/jc.2010-2924, jc.2010-2924 [pii].

Caspi, A., McClay, J., Moffitt, T., Mill, J., Martin, J., Craig, I. W., et al. (2002). Role of genotype in the cycle of violence in maltreated children. *Science, 297*(5582), 851–854.

Caspi, A., Hariri, A. R., Holmes, A., Uher, R., & Moffitt, T. E. (2010). Genetic sensitivity to the environment: The case of the serotonin transporter gene and its implications for studying complex diseases and traits. *American Journal of Psychiatry, 167*(5), 509–527. doi:10.1176/appi.ajp.2010.09101452, appi.ajp.2010.09101452 [pii].

Champagne, F. A., Weaver, I. C. G., Diorio, J., Dymov, S., Szyf, M., & Meaney, M. J. (2006). Maternal care associated with methylation of the estrogen receptor-α1b promoter and estrogen receptor-α expression in the medial preoptic area of female offspring. *Endocrinology, 147*(6), 2909–2915.

Choi, J., Fauce, S. R., & Effros, R. B. (2008). Reduced telomerase activity in human T lymphocytes exposed to cortisol. *Brain, Behaviour and Immunity, 22*(4), 600–605. doi:10.1016/j.bbi.2007.12.004, S0889-1591(07)00337-6 [pii].

Cicchetti, D., & Rogosch, F. A. (2012). Gene × environment interaction and resilience: Effects of child maltreatment and serotonin, corticotropin releasing hormone, dopamine, and oxytocin genes. *Development* and *Psychopathology, 24*(2), 411–427. doi:10.1017/S0954579412000077

Cohn, D. A., Cowan, P. A., Cowan, C. P., & Pearson, J. (1992). Mothers' and fathers' working models of childhood attachment relationships, parenting styles and child behavior. *Development and Psychopathology, 4*, 417–431.

Cole, S. W., Hawkley, L. C., Arevalo, J. M., Sung, C., Rose, R. M., & Cacioppo, J. T. (2007). Social regulation of gene expression in human leukocytes. *Genome Biology, 8*(9).

Cole, S. W., Conti, G., Arevalo, J. M., Ruggiero, A. M., Heckman, J. J., & Suomi, S. J. (2012). Transcriptional modulation of the developing immune system by early life social adversity. *Proceedings of the National Academy of Sciences of the United States of America.* doi:10.10 73/pnas.1218253109, 1218253109 [pii].

Colman, R. A., & Widom, C. S. (2004). Childhood abuse and neglect and adult intimate relationships: A prospective study. *Child Abuse and Neglect, 28*(11), 1133–1151. doi:10.1016/j.chiabu.2004.02.005, S0145-2134(04)00237-6 [pii].

D'Onofrio, B. M., Turkheimer, E., Emery, R. E., Harden, K. P., Slutske, W. S., Heath, A. C., et al. (2007). A genetically informed study of the intergenerational transmission of marital instability. *Journal of Marriage and Family, 69*(3), 793–809.

Dadds, M. R., Holland, D. E., Laurens, K. R., Mullins, M., Barrett, P. M., & Spence, S. H. (1999). Early intervention and prevention of anxiety disorders in children: Results at 2-year follow-up. *Journal of Consulting and Clinical Psychology, 67*(1), 145–150.

Damjanovic, A. K., Yang, Y., Glaser, R., Kiecolt-Glaser, J. K., Nguyen, H., Laskowski, B., et al. (2007). Accelerated telomere erosion is associated with a declining immune function of caregivers of Alzheimer's disease patients. *Journal of Immunology, 179*(6), 4249–4254. 179/6/4249 [pii].

Danese, A., Moffitt, T. E., Harrington, H., Milne, B. J., Polanczyk, G., Pariante, C. M., et al. (2009). Adverse childhood experiences and adult risk factors for age-related disease: Depression, inflammation, and clustering of metabolic risk markers. *Archives of Pediatrics and Adolescent Medicine, 163*(12), 1135–1143. doi:10.1001/archpediatrics.2009.214, 163/12/1135 [pii].

Deater-Deckard, K., Lansford, J., Malone, P., Alampay, L., Sorbring, E., Bacchini, D., et al. (2011). The association between parental warmth and control in thirteen cultural groups. *Journal of Family Psychology : JFP: Journal of the Division of Family Psychology of the American Psychological Association (Division 43), 25*(5), 790–794. doi:10.1037/a0025120

Dekel, S., Solomon, Z., & Rozenstreich, E. (2012). Secondary salutogenic effects in veterans whose parents were Holocaust survivors? *Journal of Psychiatric Research* (e-pub). doi:http://dx.doi.org/10.1016/j.jpsychires.2012.10.013

Dishion, T. J., Kavanagh, K., Schneiger, A., Nelson, S., & Kaufman, N. K. (2002). Preventing early adolescent substance use: A family-centered strategy for the public middle school. *Prevention Science, 3*(3), 191–201.

Dixon, L., Browne, K., & Hamilton-Giachritsis, C. (2005a). Risk factors of parents abused as children: A mediational analysis of the intergenerational continuity of child maltreatment (Part I). *Journal of Child Psychology and Psychiatry, 46*(1), 47–57.

Dixon, L., Hamilton-Giachritsis, C., & Browne, K. (2005b). Attributions and behaviours of parents abused as children: A mediational analysis of the intergenerational continuity of child maltreatment (Part II). *Journal of Child Psychology and Psychiatry, 46*(1), 58–68.

Dixon, L., Browne, K., & Hamilton-Giachritsis, C. (2009). Patterns of risk and protective factors in the intergenerational cycle of maltreatment. *Journal of Family Violence, 24*(2), 111–122.

Donaldson, Z. R., & Young, L. J. (2008). Oxytocin, vasopressin, and the neurogenetics of sociality. *Science, 322*(5903), 900–904.

Donnellan, M., Conger, R. D., & Bryant, C. M. (2004). The Big Five and enduring marriages. *Journal of Research in Personality 38*(5), 481–504.

Drury, S., Theall, K., Gleason, M., Smyke, A., De Vivo, I., Wong, J. Y., et al. (2012). Telomere length and early severe social deprivation: Linking early adversity and cellular aging. *Molecular Psychiatry, 17*(7), 719–727. doi:10.1038/mp.2011.53

Duncan, L. E., & Keller, M. C. (2011). A critical review of the first 10 years of candidate gene-by-environment interaction research in psychiatry. *American Journal of Psychiatry, 168*(10), 1041–1049. doi:10.1176/appi.ajp.2011.11020191

Dunn, J., & Plomin, R. (1986). Determinants of maternal behaviour towards 3-year-old siblings. *British Journal of Developmental Psychology, 4*(2), 127–137.

Dunn, J., O'Connor, T. G., & Cheng, H. (2005). Children's responses to conflict between their different parents: Mothers, stepfathers, nonresident fathers, and nonresident stepmothers. *Journal of Clinical Child and Adolescent Psychology, 34*(2), 223–234.

Edelman, S., Shalev, I., Uzefovsky, F., Israel, S., Knafo, A., Kremer, I., et al. (2012). Epigenetic and genetic factors predict women's salivary cortisol following a threat to the social self. *PLoS ONE, 7*(11), e48597. doi:10.1371/journal.pone.0048597, PONE-D-12-13109 [pii].

Else-Quest, N. M., Clark, R., & Owen, M. T. (2011). Stability in mother-child interactions from infancy through adolescence. *Parenting: Science and Practice, 11*(4), 280–287.

Entringer, S., Epel, E. S., Kumsta, R., Lin, J., Hellhammer, D. H., Blackburn, E. H., et al. (2011). Stress exposure in intrauterine life is associated with shorter telomere length in young adulthood. *Proceedings of the National Academy of Sciences of the United States of America, 108*(33), E513–E518. doi:10.1073/pnas.1107759108, 1107759108 [pii].

Epel, E., Blackburn, E., Lin, J., Dhabhar, F., Adler, N., Morrow, J., et al. (2004). Accelerated telomere shortening in response to life stress. *Proceedings of the National Academy of Sciences of the United States of America, 101*(49), 17312–17315. doi:10.1073/pnas.0407162101

Epel, E. S., Lin, J., Dhabhar, F. S., Wolkowitz, O. M., Puterman, E., Karan, L., & Blackburn, E. H. (2010). Dynamics of telomerase activity in response to acute psychological stress. *Brain, Behavior, and Immunity, 24*(4), 531–539.

Erel, O., & Burman, B. (1995). Interrelatedness of marital relations and parent–child relations: A meta-analytic review. *Psychological Bulletin, 118*(1), 108–132.

Fearon, P., Shmueli-Goetz, Y., Viding, E., Fonagy, P., & Plomin, R. (2013). Genetic and environmental influences on adolescent attachment. *Journal of Child Psychology and Psychiatry*. doi:10.1111/jcpp.12171

Fearon, R. M. P., Van Ijzendoorn, M. H., Fonagy, P., Bakermans-Kranenburg, M. J., Schuengel, C., & Bokhorst, C. L. (2006). In search of shared and nonshared environmental factors in security of attachment: A behavior-genetic study of the association between sensitivity and attachment security. *Developmental Psychology, 42*(6), 1026–1040.

Feinberg, M. E., Kan, M. L., & Goslin, M. C. (2009). Enhancing coparenting, parenting, and child self-regulation: Effects of family foundations 1 year after birth. *Prevention Science, 10*(3), 276–285.

Feinberg, M. E., Reiss, D., Neiderhiser, J. M., & Hetherington, E. (2005). Differential association of family subsystem negativity on siblings' maladjustment: Using behavior genetic methods to test process theory. *Journal of Family Psychology, 19*(4), 601–610.

Feldman, R., Zagoory-Sharon, O., Weisman, O., Schneiderman, I., Gordon, I., Maoz, R., et al. (2012). Sensitive parenting is associated with plasma oxytocin and polymorphisms in the OXTR and CD38 genes. *Biological Psychiatry, 72*(3), 175–181. doi:10.1016/j.biopsych.2011.12.025, S0006-3223(12)00003-0 [pii].

Felitti, V. J., Anda, R. F., Nordenberg, D., Williamson, D. F., Spitz, A. M., Edwards, V., et al. (1998). Relationship of childhood abuse and household dysfunction to many of the leading causes of death in adults: The adverse childhood experiences (ACE) study. *American Journal of Preventive Medicine, 14*(4), 245–258.

Fiese, B. H., Tomcho, T. J., Douglas, M., Josephs, K., Poltrock, S., & Baker, T. (2002). A review of 50 years of research on naturally occurring family routines and rituals: Cause for celebration? *Journal of Family Psychology, 16*(4), 381–390.

Forget-Dubois, N., Boivin, M., Dionne, G., Pierce, T., Tremblay, R. E., & Perusse, D. (2007). A longitudinal twin study of the genetic and environmental etiology of maternal hostile-reactive behavior during infancy and toddlerhood. *Infant Behavior and Development, 30*(3), 453–465.

Francis, D. D., Young, L. J., Meaney, M. J., & Insel, T. R. (2002). Naturally occurring differences in maternal care are associated with the expression of oxytocin and vasopressin (V1a) receptors: Gender differences. *Journal of Neuroendocrinology, 14*(5), 349–353.

Ganiban, J. M., Saudino, K. J., Ulbricht, J., Neiderhiser, J. M., & Reiss, D. (2008). Stability and change in temperament during adolescence. *Journal of Personality and Social Psychology, 95*(1), 222–236.

Ganiban, J. M., Ulbricht, J., Saudino, K. J., Reiss, D., & Neiderhiser, J. M. (2011). Understanding child-based effects on parenting: Temperament as a moderator of genetic and environmental contributions to parenting. *Developmental Psychology, 47*(3), 676–692. doi:10.1037/a0021812, 2010-25349-001 [pii].

Gerard, J. M., Krishnakumar, A., & Buehler, C. (2006). Marital conflict, parent–child relations, and youth maladjustment: A longitudinal investigation of spillover effects. *Journal of Family Issues, 27*(7), 951–975.

Gottman, J. M., & Notarius, C. I. (2002). Marital research in the 20th century and a research agenda for the 21st century. *Family Process, 41*(2), 159–197.

Griffith, J. S., & Mahler, H. R. (1969). DNA ticketing theory of memory. *Nature, 223*(5206), 580–582.

Hartmann, M., Bäzner, E., Wild, B., Eisler, I., & Herzog, W. (2010). Effects of interventions involving the family in the treatment of adult patients with chronic physical diseases: A meta-analysis. *Psychotherapy and Psychosomatics, 79*(3), 136–148.

Hoffmann, A., & Spengler, D. (2012). DNA memories of early social life. Neuroscience. doi:10.1016/j.neuroscience.2012.04.003, S0306-4522(12)00302-8 [pii].

Holtzworth-Munroe, A., Stuart, G. L., & Hutchinson, G. (1997). Violent versus nonviolent husbands: Differences in attachment patterns, dependency, and jealousy. *Journal of Family Psychology, 11*(3), 314–331.

Jackowska, M., Hamer, M., Carvalho, L. A., Erusalimsky, J. D., Butcher, L., & Steptoe, A. (2012). Short sleep duration is associated with shorter telomere length in healthy men: Findings from the Whitehall II Cohort study. *PLoS ONE, 7*(10), e47292. doi:10.1371/journal.pone.0047292, PONE-D-12-19697 [pii].

Jocklin, V., McGue, M., & Lykken, D. T. (1996). Personality and divorce: A genetic analysis. *Journal of Personality and Social Psychology, 71*(2), 288–299.

Johnson, W., McGue, M., Krueger, R. F., & Bouchard, T. J., Jr. (2004). Marriage and personality: A genetic analysis. *Journal of Personality and Social Psychology, 86*(2), 285–294.

Kendler, K. S. (1996). Parenting: A genetic-epidemiologic perspective. *American Journal of Psychiatry, 153*(1), 11–20.

Kendler, K. S., & Baker, J. H. (2007). Genetic influences on measures of the environment: A systematic review. *Psychological Medicine, 37*(5), 615–626.

Kendler, K. S., & Karkowski-Shuman, L. (1997). Stressful life events and genetic liability to major depression: Genetic control of exposure to the environment? *Psychological Medicine, 27*(3), 539–547.

Kendler, K. S., Neale, M. C., Kessler, R. C., Heath, A. C., et al. (1993). A longitudinal twin study of 1-year prevalence of major depression in women. *Archives of General Psychiatry, 50*(11), 843–852.

Kern, M. L., & Friedman, H. S. (2008). Early educational milestones as predictors of lifelong academic achievement, midlife adjustment, and longevity. *Journal of Applied Developmental Psychology, 30*(4), 419–430. doi:10.1016/j.appdev.2008.12.025

Kerr, D. C., Capaldi, D. M., Pears, K. C., & Owen, L. D. (2009). A prospective three generational study of fathers' constructive parenting: Influences from family of origin, adolescent adjustment, and offspring temperament. *Developmental Psychology, 45*(5), 1257–1275.

Kessler, R. C., Kendler, K. S., Heath, A., Neale, M. C., & Eaves, L. (1992). Social support, depressed mood, and adjustment to stress: A genetic epidemiologic investigation. *Journal of Personality and Social Psychology, 62*(2), 257–272.

Kiecolt-Glaser, J. K., Gouin, J. P., Weng, N. P., Malarkey, W. B., Beversdorf, D. Q., & Glaser, R. (2011). Childhood adversity heightens the impact of later-life caregiving stress on telomere length and inflammation. *Psychosomatic Medicine, 73*(1), 16–22. doi:10.1097/PSY.0b013e3 1820573b6, PSY.0b013e31820573b6 [pii].

Kim-Cohen, J., Caspi, A., Taylor, A., Williams, B., Newcombe, R., Craig, I., & Moffitt, T. (2006). MAOA, maltreatment, and gene-environment interaction predicting children's mental health: New evidence and a meta-analysis. *Molecular Psychiatry, 11*(10), 903–913.

Kingma, E. M., de Jonge, P., van der Harst, P., Ormel, J., & Rosmalen, J. G. (2012). The association between intelligence and telomere length: A longitudinal population based study. *PLoS ONE, 7*(11), e49356. doi:10.1371/journal.pone.0049356, PONE-D-12-22189 [pii].

Kinnunen, U., Rytkonen, O., Miettinen, N., & Pulkkinen, L. (2000). Personality and marriage: Personality characteristics as predictors of marital quality and stability. *Psykologia, 35*(4), 332–345.

Kochanska, G. (1995). Children's temperament, mothers' discipline, and security of attachment: Multiple pathways to emerging internalization. *Child Development, 66*, 597–615.

Kochanska, G., Aksan, N., Penney, S. J., & Boldt, L. J. (2007). Parental personality as an inner resource that moderates the impact of ecological adversity on parenting. *Journal of Personality and Social Psychology, 92*(1), 136–150.

Kochanska, G., Friesenborg, A. E., Lange, L. A., & Martel, M. M. (2004). Parents' personality and infants' temperament as contributors to their emerging relationship. *Journal of Personality and Social Psychology, 86*(5), 744–759. doi:10.1037/0022-3514.86.5.744, 2004-13298-007 [pii].

Kotrschal, A., Ilmonen, P., & Penn, D. J. (2007). Stress impacts telomere dynamics. *Biology Letters, 3*(2), 128–130. doi:10.1098/rsbl.2006.0594, C33K527360234125 [pii].

Kovan, N. M., Chung, A. L., & Sroufe, A. L. (2009). The intergenerational continuity of observed early parenting: A prospective, longitudinal study. *Developmental Psychology, 45*(5), 1205–1213.

Krauss, J., Farzaneh-Far, R., Puterman, E., Na, B., Lin, J., Epel, E., et al. (2011). Physical fitness and telomere length in patients with coronary heart disease: Findings from the heart and soul study. *PLoS ONE, 6*(11), e26983. doi:10.1371/journal.pone.0026983, PONE-D-11-09446 [pii].

Kroenke, C. H., Epel, E., Adler, N., Bush, N. R., Obradovic, J., Lin, J., et al. (2011). Autonomic and adrenocortical reactivity and buccal cell telomere length in kindergarten children. *Psychosomatic Medicine, 73*(7), 533–540. doi:10.1097/PSY.0b013e318229acfc, PSY.0b013e318229acfc [pii].

Kroenke, C. H., Pletcher, M. J., Lin, J., Blackburn, E., Adler, N., Matthews, K., et al. (2012). Telomerase, telomere length, and coronary artery calcium in black and white men in the CARDIA study. *Atherosclerosis, 220*(2), 506–512. doi:10.1016/j.atherosclerosis.2011.10.041, S0021-9150(11)01065-3 [pii].

Lahey, B. B., Michalska, K. J., Liu, C., Chen, Q., Hipwell, A. E., Chronis-Tuscano, A., et al. (2012). Preliminary genetic imaging study of the association between estrogen receptor-alpha gene polymorphisms and harsh human maternal parenting. *Neuroscience Letters, 525*(1), 17–22. doi:10.1016/j.neulet.2012.07.016, S0304-3940(12)00939-1 [pii].

Lan, Q., Cawthon, R., Shen, M., Weinstein, S. J., Virtamo, J., Lim, U., et al. (2009). A prospective study of telomere length measured by monochrome multiplex quantitative PCR and risk of non-Hodgkin lymphoma. *Clinical Cancer Research, 15*(23), 7429–7433. doi:10.1158/1078-0432.CCR-09-0845, 1078-0432.CCR-09-0845 [pii].

Larsson, H., Viding, E., Rijsdijk, F. V., & Plomin, R. (2008). Relationships between parental negativity and childhood antisocial behavior over time: A bidirectional effects model in a longitudinal genetically informative design. *Journal of Abnormal Child Psychology, 36*(5), 633–645.

Lavretsky, H., Epel, E. S., Siddarth, P., Nazarian, N., Cyr, N. S., Khalsa, D. S., et al. (2012). A pilot study of yogic meditation for family dementia caregivers with depressive symptoms: Effects on mental health, cognition, and telomerase activity. *International Journal of Geriatric Psychiatry*. doi:10.1002/gps.3790

Limb, G. E., Shafer, K., & Sandoval, K. (2014). The impact of kin support on urban American Indian families. *Child and Family Social Work. 19*(4), 432–442.

Liu, D., Diorio, J., Tannenbaum, B., Caldji, C., Francis, D., Freedman, A., et al. (1997). Maternal care, hippocampal glucocorticoid receptors, and hypothalamic-pituitary-adrenal responses to stress. *Science, 277*(5332), 1659–1662.

Loehlin, J. C., Neiderhiser, J. M., & Reiss, D. (2005). Genetic and environmental components of adolescent adjustment and parental behavior: A multivariate analysis. *Child Development, 76*(5), 1104–1115.

Luijk, M. P., Roisman, G. I., Haltigan, J. D., Tiemeier, H., Booth-Laforce, C., van Ijzendoorn, M. H., et al. (2011). Dopaminergic, serotonergic, and oxytonergic candidate genes associated with infant attachment security and disorganization? In search of main and interaction effects. *Journal of Child Psychology and Psychiatry, 52*(12), 1295–1307. doi:10.1111/j.1469-7610.2011.02440.x

Lykken, D. T. (2002). How relationships begin and end: A genetic perspective. In A. L. Vangelisti & H. T. Reis, et al. (Eds.), *Stability and change in relationships* (*Advances in personal relationships*) (pp. 83–102). New York, NY: Cambridge University Press.

Lykken, D. T., & Tellegen, A. (1993). Is human mating adventitious or the result of lawful choice? A twin study of mate selection. *Journal of Personality and Social Psychology, 65*(1), 56–68.

Ma, H., Zhou, Z., Wei, S., Liu, Z., Pooley, K. A., Dunning, A. M., et al. (2011). Shortened telomere length is associated with increased risk of cancer: A meta-analysis. *PLoS ONE, 6*(6), e20466. doi:10.1371/journal.pone.0020466, PONE-D-11-04747 [pii].

Mainous, A. G., III, Everett, C. J., Diaz, V. A., Baker, R., Mangino, M., Codd, V., et al. (2011). Leukocyte telomere length and marital status among middle-aged adults. *Age Ageing, 40*(1), 73–78. doi:10.1093/ageing/afq118, afq118 [pii].

Mannering, A. M., Harold, G. T., Leve, L. D., Shelton, K. H., Shaw, D. S., Conger, R. D., et al. (2011). Longitudinal associations between marital instability and child sleep problems across infancy and toddlerhood in adoptive families. *Child Development, 82*(4), 1252–1266.

Martire, L. M. (2005). The "relative" efficacy of involving family in psychosocial interventions for chronic illness: Are there added benefits to patients and family members? *Families, Systems, and Health, 23*(3), 312–328.

Martire, L. M., Lustig, A. P., Schulz, R., Miller, G. E., & Helgeson, V. S. (2004). Is it beneficial to involve a family member? A meta-analysis of psychosocial interventions for chronic illness. *Health Psychology, 23*(6), 599–611.

Martire, L. M., Schulz, R., Helgeson, V. S., Small, B. J., & Saghafi, E. M. (2010). Review and meta-analysis of couple-oriented interventions for chronic illness. *Annals of Behavioral Medicine, 40*(3), 325–342.

Martire, L. M., Stephens, M. A. P., & Schulz, R. (2011). Independence centrality as a modera-
     tor of the effects of spousal support on patient well-being and physical functioning. *Health
     Psychology, 30*(5), 651–655.
Marysko, M., Reck, C., Mattheis, V., Finke, P., Resch, F., & Moehler, E. (2010). History of child-
     hood abuse is accompanied by increased dissociation in young mothers five months postna-
     tally. *Psychopathology, 43*(2), 104–109.
McCormack, K., Newman, T., Higley, J., Maestripieri, D., & Sanchez, M. (2009). Serotonin
     transporter gene variation, infant abuse, and responsiveness to stress in rhesus macaque
     mothers and infants. *Hormones and Behavior, 55*(4), 538–547.
McGowan, P. O., Sasaki, A., D'Alessio, A. C., Dymov, S., Labonte, B., Szyf, M., et al. (2009).
     Epigenetic regulation of the glucocorticoid receptor in human brain associates with child-
     hood abuse. *Nature Neuroscience, 12*(3), 342–348. doi:10.1038/nn.2270, nn.2270 [pii].
McGowan, P. O., Suderman, M., Sasaki, A., Huang, T. C., Hallett, M., Meaney, M. J., et al.
     (2011). Broad epigenetic signature of maternal care in the brain of adult rats. *PLoS ONE,
     6*(2), e14739. doi:10.1371/journal.pone.0014739
McGue, M., & Lykken, D. T. (1992). Genetic influence on risk of divorce. *Psychological
     Science, 3*(6), 368–373.
Melby, J. N., & Conger, R. D. (1996). Parental behaviors and adolescent academic performance:
     A longitudinal analysis. *Journal of Research on Adolescence, 6*(1), 113–137.
Mileva-Seitz, V., Fleming, A. S., Meaney, M. J., Mastroianni, A., Sinnwell, J. P., Steiner, M.,
     et al. (2012). Dopamine receptors D1 and D2 are related to observed maternal behavior.
     Genes, Brain and Behavior, 11(6), 684–694. doi:10.1111/j.1601-183X.2012.00804.x
Mileva-Seitz, V., Kennedy, J., Atkinson, L., Steiner, M., Levitan, R., Matthews, S. G., et al.
     (2011). Serotonin transporter allelic variation in mothers predicts maternal sensitivity,
     behavior and attitudes toward 6-month-old infants. *Genes, Brain and Behavior, 10*(3), 325–
     333. doi:10.1111/j.1601-183X.2010.00671.x
Miller, G. E., Chen, E., Fok, A. K., Walker, H., Lim, A., Nicholls, E. F., et al. (2009). Low early-
     life social class leaves a biological residue manifested by decreased glucocorticoid and
     increased proinflammatory signaling. *Proceedings of the National Academy of Sciences of
     the United States of America, 106*(34), 14716–14721.
Mills-Koonce, W. R., Propper, C. B., Gariepy, J.-L., Blair, C., Garrett-Peters, P., & Cox, M. J.
     (2007). Bidirectional genetic and environmental influences on mother and child behavior:
     The family system as the unit of analyses. *Development and Psychopathology, 19*(04),
     1073–1087.
Mirabello, L., Yu, K., Kraft, P., De Vivo, I., Hunter, D. J., Prescott, J., et al. (2010). The associa-
     tion of telomere length and genetic variation in telomere biology genes. *Human Mutation,
     31*(9), 1050–1058. doi:10.1002/humu.21314
Narusyte, J., Andershed, A. K., Neiderhiser, J. M., & Lichtenstein, P. (2007). Aggression as a medi-
     ator of genetic contributions from negative parent–child relationships and
     adolescent antisocial behavior. *European Child and Adolescent Psychiatry, 16*(2), 128–137.
Narusyte, J., Neiderhiser, J. M., Andershed, A. K., D'Onofrio, B. M., Reiss, D., Spotts, E., et al.
     (2011). Parental criticism and externalizing behavior problems in adolescents: The role of
     environment and genotype-environment correlation. *Journal of Abnormal Psychology.*
     doi:10.1037/a0021815, 2011-01600-001 [pii].
Neiderhiser, J. M., & Pike, A. (1995). Mediation of adolescents' perception on the relation-
     ship between parenting and adolescent adjustment: Genetic and environmental influences
     (abstract). *Behavior Genetics, 25*(3), 281.
Neiderhiser, J. M., Reiss, D., Lichtenstein, P., Spotts, E. L., & Ganiban, J. (2007). Father-
     adolescent relationships and the role of genotype-environment correlation. *Journal of Family
     Psychology, 21*(4), 560–571. doi:10.1037/0893-3200.21.4.560, 2007-18728-002 [pii].
Neiderhiser, J. M., Reiss, D., Pedersen, N. L., Lichtenstein, P., Spotts, E. L., Hansson, K., et al.
     (2004). Genetic and environmental influences on mothering of adolescents: A comparison of
     two samples. *Developmental Psychology, 40*(3), 335–351. doi:10.1037/0012-1649.40.3.335,
     2004-13591-002 [pii].

Neppl, T. K., Conger, R. D., Scaramella, L. V., & Ontai, L. L. (2009). Intergenerational continuity in parenting behavior: Mediating pathways and child effects. *Developmental Psychology, 45*(5), 1241–1256.

O'Connor, T. G., & Croft, C. M. (2001). A twin study of attachment in preschool children. *Child Development, 72*(5), 1501–1511.

O'Connor, T. G., Deater-Deckard, K., Fulker, D., Rutter, M., & Plomin, R. (1998). Genotype-environment correlations in late childhood and early adolescence: Antisocial behavioral problems and coercive parenting. *Developmental Psychology, 34*(5), 970–981.

O'Connor, T. G., Hetherington, E. M., Reiss, D., & Plomin, R. (1995). A twin-sibling study of observed parent-adolescent interactions. *Child Development, 66*(3), 812–829.

O'Donnell, C. J., Demissie, S., Kimura, M., Levy, D., Gardner, J. P., White, C., et al. (2008). Leukocyte telomere length and carotid artery intimal medial thickness: The Framingham Heart Study. *Arteriosclerosis, Thrombosis, and Vascular Biology, 28*(6), 1165–1171. doi:10.1161/ATVBAHA.107.154849, ATVBAHA.107.154849 [pii].

O'Donovan, A., Lin, J., Tillie, J., Dhabhar, F. S., Wolkowitz, O. M., Blackburn, E. H., et al. (2009). Pessimism correlates with leukocyte telomere shortness and elevated interleukin-6 in post-menopausal women. *Brain, Behavior, and Immunity, 23*(4), 446–449. doi:10.1016/j.bbi.2008.11.006, S0889-1591(08)00427-3 [pii].

Out, D., Pieper, S., Bakermans-Kranenburg, M. J., & van Ijzendoorn, M. H. (2010). Physiological reactivity to infant crying: A behavioral genetic study. *Genes, Brain and Behavior, 9*(8), 868–876. doi:10.1111/j.1601-183X.2010.00624.x, GBB624 [pii].

Parks, C. G., DeRoo, L. A., Miller, D. B., McCanlies, E. C., Cawthon, R. M., & Sandler, D. P. (2011). Employment and work schedule are related to telomere length in women. *Occupational and Environmental Medicine, 68*(8), 582–589. doi:10.1136/oem.2010.063214, oem.2010.063214 [pii].

Pears, K. C., & Capaldi, D. M. (2001). Intergenerational transmission of abuse: A two-generational prospective study of an at-risk sample. *Child Abuse and Neglect, 25*(11), 1439–1461.

Pellatt, A. J., Wolff, R. K., Lundgreen, A., Cawthon, R., & Slattery, M. L. (2012). Genetic and lifestyle influence on telomere length and subsequent risk of colon cancer in a case control study. *International Journal of Molecular Epidemiology and Genetics, 3*(3), 184–194.

Pereira, J., Vickers, K., Atkinson, L., Gonzalez, A., Wekerle, C., & Levitan, R. (2012). Parenting stress mediates between maternal maltreatment history and maternal sensitivity in a community sample. *Child Abuse and Neglect, 36*(5), 433–437.

Perusse, D., Neale, M. C., Heath, A. C., & Eaves, L. J. (1994). Human parental behavior: Evidence for genetic influence and potential implication for gene-culture transmission. *Behavior Genetics, 24*(4), 327–335.

Pike, A., McGuire, S., Hetherington, E. M., Reiss, D., & Plomin, R. (1996). Family environment and adolescent depression and antisocial behavior: A multivariate genetic analysis. *Developmental Psychology, 32*(4), 590–603.

Plomin, R., Emde, R. N., Braungart, J. M., Campos, J., Corley, R., Fulker, D. W., et al. (1993). Genetic change and continuity from fourteen to twenty months: The MacArthur longitudinal twin study. *Child Development, 64*, 1354–1376.

Plomin, R., Reiss, D., Hetherington, E., & Howe, G. W. (1994). Nature and nurture: Genetic contributions to measures of the family environment. *Developmental Psychology, 30*(1), 32–43.

Prinzie, P., Stams, G. J., Dekovic, M., Reijntjes, A. H., & Belsky, J. (2009). The relations between parents' big five personality factors and parenting: A meta-analytic review. *Journal of Personality and Social Psychology, 97*(2), 351–362. doi:10.1037/a0015823, 2009-10712-010 [pii].

Provencal, N., Suderman, M. J., Guillemin, C., Massart, R., Ruggiero, A., Wang, D., et al. (2012). The signature of maternal rearing in the methylome in rhesus macaque prefrontal cortex and T cells. *Journal of Neuroscience, 32*(44), 15626–15642. doi:10.1523/JNEUROSCI.1470-12.2012, 32/44/15626 [pii].

Psychiatry, A Ao Ca A. (2005). Practice parameter for the assessment and treatment of children and adolescents with substance use disorders. *Journal of the Academy of child and Adolescent Psychiatry, 44*(6), 609–621.

Bornstein, B. Fragment of an analysis of an obsessional child. The first six months of analysis. (1955). *International Journal of Psychoanalysis, 36*(79), 313–332. (The Psychoanalytic Study of the Child (An Annual) 8, 1953)

Raby, K., Cicchetti, D., Carlson, E. A., Cutuli, J., Englund, M. M., & Egeland, B. (2012). Genetic and caregiving-based contributions to infant attachment: Unique associations with distress reactivity and attachment security. *Psychological Science, 23*(9), 1016–1023.

Reiss, D., & Emde, R. N. (2003). Relationship disorders are psychiatric disorders: Five reasons they were not included in the DSM-IV. In K. A. Phillips & M. B. First, et al. (Eds.), *Advancing DSM: Dilemmas in psychiatric diagnosis* (pp. 191–223). Washington, DC: American Psychiatric Association.

Reiss, D., Neiderhiser, J., Hetherington, E. M., & Plomin, R. (2000). *The relationship code: Deciphering genetic and social patterns in adolescent development.* Cambridge, MA: Harvard University Press.

Roberts, B. W., Kuncel, N. R., Shiner, R., Caspi, A., & Goldberg, L. R. (2007). The power of personality: The comparative validity of personality traits, socioeconomic status, and cognitive ability for predicting important life outcomes. *Perspectives on Psychological Science, 2*(4), 313–345.

Roisman, G. I., & Fraley, R. (2008). A behavior-genetic study of parenting quality, infant attachment security, and their covariation in a nationally representative sample. *Developmental Psychology, 44*(3), 831–839.

Roskam, I., & Meunier, J. C. (2012). The determinants of parental childrearing behavior trajectories: The effects of parental and child time-varying and time-invariant predictors. *International Journal of Behavioral Development, 36*(3), 186–196.

Roth, T. L., & Sweatt, J. D. (2011). Annual Research Review: Epigenetic mechanisms and environmental shaping of the brain during sensitive periods of development. *Journal of Child Psychology and Psychiatry, 52*(4), 398–408. doi:10.1111/j.1469-7610.2010.02282.x, JCPP2282 [pii].

Rowe, D. C. (1981). Environmental and genetic influences on dimensions of perceived parenting: A twin study. *Developmental Psychology, 17*(2), 203–208.

Sagi-Schwartz, A., van Ijzendoorn, M. H., & Bakermans-Kranenburg, M. J. (2008). Does intergenerational transmission of trauma skip a generation? No meta-analytic evidence for tertiary traumatization with third generation of Holocaust survivors. *Attachment and Human Development, 10*(2), 105–121.

Schury, K., & Kolassa, I. T. (2012). Biological memory of childhood maltreatment: Current knowledge and recommendations for future research. *Annals of the New York Academy of Sciences, 1262*, 93–100. doi:10.1111/j.1749-6632.2012.06617.x

Shackelford, T. K., Besser, A., & Goetz, A. T. (2008). 2008. *Personality, martial satisfaction, and probability of marital infidelity: Individual differences research, 6*(1), 13–25.

Shaffer, A., Burt, K. B., Obradovic, J., Herbers, J. E., & Masten, A. S. (2009). Intergenerational continuity in parenting quality: The mediating role of social competence. *Developmental Psychology, 45*(5), 1227–1240.

Shalev, I., Moffitt, T. E., Sugden, K., Williams, B., Houts, R. M., Danese, A., et al. (2012). Exposure to violence during childhood is associated with telomere erosion from 5 to 10 years of age: A longitudinal study. *Molecular Psychiatry.* doi:10.1038/mp.2012.32, mp201232 [pii].

Shen, M., Cawthon, R., Rothman, N., Weinstein, S. J., Virtamo, J., Hosgood, H. D., III, et al. (2011). A prospective study of telomere length measured by monochrome multiplex quantitative PCR and risk of lung cancer. *Lung Cancer, 73*(2), 133–137. doi:10.1016/j.lungcan.2010.11.009, S0169-5002(10)00544-1 [pii].

Shmotkin, D., Shrira, A., Goldberg, S. C., & Palgi, Y. (2011). Resilience and vulnerability among aging Holocaust survivors and their families: An intergenerational overview. *Journal of Intergenerational Relationships, 9*(1), 7–21.

Skinner, E., Johnson, S., & Snyder, T. (2005). Six dimensions of parenting: A motivational model. *Parenting, 5.* doi:10.1207/s15327922par0502_3

Smit, R. (2011). Maintaining family memories through symbolic action: Young adults' perceptions of family rituals in their families of origin. *Journal of Comparative Family Studies, 42*(3), 355–368.

Spangler, G., Johann, M., Ronai, Z., & Zimmermann, P. (2009). Genetic and environmental influence on attachment disorganization. *Journal of Child Psychology and Psychiatry, 50*(8), 952–961.

Spinath, F. M., & O'Connor, T. G. (2003). A behavioral genetic study of the overlap between personality and parenting. *Journal of Personality, 71*(5), 785–808.

Spotts, E. L., Lichtenstein, P., Pedersen, N., Neiderhiser, J. M., Hansson, K., Cederblad, M., & Reiss, D. (2005). Personality and marital satisfaction: A behavioural genetic analysis. *European Journal of Personality, 19*(3), 205–227.

Spotts, E. L., Neiderhiser, J. M., Towers, H., Hansson, K., Lichtenstein, P., Cederblad, M., et al. (2004). Genetic and environmental influences on marital relationships. *Journal of Family Psychology, 18*(1), 107–119.

Spotts, E. L., Prescott, C., & Kendler, K. (2006). Examining the origins of gender differences in marital quality: A behavior genetic analysis. *Journal of Family Psychology, 20*(4), 605–613.

Springer, K. W. (2009). Childhood physical abuse and midlife physical health: Testing a multi-pathway life course model. *Social, Science and Medicine, 69*(1), 138–146. doi:10.1016/j.socscimed.2009.04.011

Stover, C. S., Connell, C. M., Leve, L. D., Neiderhiser, J. M., Shaw, D. S., Scaramella, L. V., et al. (2012). Fathering and mothering in the family system: Linking marital hostility and aggression in adopted toddlers. *Journal of Child Psychology and Psychiatry, 53*(4), 401–409. doi:10.1111/j.1469-7610.2011.02510.x

Sturge-Apple, M. L., Cicchetti, D., Davies, P. T., & Suor, J. H. (2012). Differential susceptibility in spillover between interparental conflict and maternal parenting practices: Evidence for OXTR and 5-HTT genes. *Journal of Family Psychology, 26*(3), 431–442. doi:10.1037/a0028302, 2012-12089-001 [pii].

Suderman, M., McGowan, P. O., Sasaki, A., Huang, T. C., Hallett, M. T., Meaney, M. J., et al. (2012). Conserved epigenetic sensitivity to early life experience in the rat and human hippocampus. *Proceedings of the National Academy of Sciences of the United States of America, 109*(Suppl 2), 17266–17272. doi:10.1073/pnas.1121260109, 1121260109 [pii].

Sweatt, J. (2009). Experience-dependent epigenetic modifications in the central nervous system. *Biological Psychiatry, 65*(3), 191–197. doi:10.1016/j.biopsych.2008.09.002

Teasdale, T., & Owen, D. R. (1984). Heredity and familial environment in intelligence and educational level: A sibling study. *Nature 309*(5969), 620–622.

Thornberry, T. P., & Henry, K. L. (2013). Intergenerational continuity in maltreatment. *Journal of Abnormal Child Psychology, 41*(4), 555–569. doi:10.1007/s10802-012-9697-5

Tomiyama, A. J., O'Donovan, A., Lin, J., Puterman, E., Lazaro, A., Chan, J., et al. (2012). Does cellular aging relate to patterns of allostasis? An examination of basal and stress reactive HPA axis activity and telomere length. *Physiology   & Behavior, 106*(1), 40–45. doi:10.1016/j.physbeh.2011.11.016, S0031-9384(11)00536-1 [pii].

Towers, H. (2003). Contributions of current family factors and life events to women's adjustment: Nonshared environmental pathways. *Dissertation Abstracts International: Section B: The Sciences and Engineering, 63*(12-B), 6123.

Ulbricht, J. A., Ganiban, J., Button, T. M. M., Feinberg, M., Reiss, D., & Neiderhiser, J. (2013). Marital adjustment as a moderator for genetic and environmental influences on parenting. *Journal of Family Psychology, 27*(1), 42–52.

Valdes, A. M., Deary, I. J., Gardner, J., Kimura, M., Lu, X., Spector, T. D., et al. (2010). Leukocyte telomere length is associated with cognitive performance in healthy women. *Neurobiology of Aging, 31*(6), 986–992. doi:10.1016/j.neurobiolaging.2008.07.012, S0197-4580(08)00256-X [pii].

Viken, R. J., Rose, R. J., Kaprio, J., & Koskenvuo, M. (1994). A developmental genetic analysis of adult personality: Extraversion and neuroticism from 18 to 59 years of age. *Journal of Personality and Social Psychology, 66*(4), 722–730.

Walsh, F. W. (1978). Concurrent grandparent death and birth of schizophrenic offspring: An intriguing finding. *Family Process, 17*(4), 457–463.

Walum, H., Lichtenstein, P., Neiderhiser, J. M., Reiss, D., Ganiban, J. M., Spotts, E. L., et al. (2012). Variation in the oxytocin receptor gene is associated with pair-bonding and social behavior. *Biological Psychiatry, 71*(5), 419–426. doi:10.1016/j.biopsych.2011.09.002, S0006-3223(11)00860-2 [pii].

Weaver, I. C., Cervoni, N., Champagne, F. A., D'Alessio, A. C., Sharma, S., Seckl, J. R., et al. (2004). Epigenetic programming by maternal behavior. *Nature Neuroscience, 7*(8), 847–854. doi:10.1038/nn1276

Weder, N., Yang, B. Z., Douglas-Palumberi, H., Massey, J., Krystal, J. H., Gelernter, J., et al. (2009). MAOA genotype, maltreatment, and aggressive behavior: The changing impact of genotype at varying levels of trauma. *Biological Psychiatry, 65*(5), 417–424.

Whitbeck, L. B., Chen, X., Hoyt, D. R., & Adams, G. W. (2004). Discrimination, historical loss and enculturation: Culturally specific risk and resiliency factors for alcohol abuse among American Indians. *Journal of Studies on Alcohol, 65*(4), 409–418.

Whitbeck, L. B., McMorris, B. J., Hoyt, D. R., Stubben, J. D., & LaFromboise, T. (2002). Perceived discrimination, traditional practices, and depressive symptoms among American Indians in the upper Midwest. *Journal of Health and Social Behavior, 43*(4), 400–418.

Widom, C. S., & Brzustowicz, L. M. (2006). MAOA and the "cycle of violence:" Childhood abuse and neglect, MAOA genotype, and risk for violent and antisocial behavior. *Biological Psychiatry, 60*(7), 684–689.

Widom, C. S., Czaja, S. J., Bentley, T., & Johnson, M. S. (2012). A prospective investigation of physical health outcomes in abused and neglected children: New findings from a 30-year follow-up. *American Journal of Public Health, 102*(6), 1135–1144. doi:10.2105/AJPH.2011.300636

Widom, C. S., Raphael, K. G., & DuMont, K. A. (2004). The case for prospective longitudinal studies in child maltreatment research: Commentary on Dube, Williamson, Thompson, Felitti, and Anda. *Child Abuse and Neglect, 28*(7), 715–722.

Wright, F. A., Sullivan, P. F., Brooks, A. I., Zou, F., Sun, W., Xia, K., et al. (2014). Heritability and genomics of gene expression in peripheral blood. *Nature Genetics, 46*(5), 430–437. doi:10.1038/ng.2951

Yaffe, K., Lindquist, K., Kluse, M., Cawthon, R., Harris, T., Hsueh, W. C., et al. (2011). Telomere length and cognitive function in community-dwelling elders: Findings from the Health ABC Study. *Neurobiology of Aging, 32*(11), 2055–2060. doi:10.1016/j.neurobiolaging.2009.12.006, S0197-4580(09)00398-4 [pii].

Index

A

Abortion, 205
Academic achievement, 30
Adolescence, 131–145
  family as system, 141–143
  gene environment processes in, 131
  marital relationship, 139–141
  parent-adolescent relationship, 133–136
  personal factors influence on quality, 132–133
  sibling relationship, 137–139
Adoption studies
  of early parenting, 25, 31, 34–35
  of infant attachment, 20–21
Adult Attachment Scale, 178
Adult relationships
  attachment relationships, 177–179
  determinants of, 172–173
  marital quality, 179–183
  marital status, 176–177
  marriage, 175–176
  quality, genetic and environmental contributions to, 173–175
  relationship with parents, 184–188
  work relationships, 193–194
Adulthood. *See also* Adult relationships
  gene-environment transactions in, 97–121
  interpersonal relationships in, 6–7
Affect
  negative, 152, 162, 225, 226, 229
  positive, 18, 162, 226
Aging, 63–64
Aggression, 98, 101, 103–105, 132, 152
  physical, 88, 91, 106, 108, 109, 112, 114, 157
  relational, 108–109, 114

  social, 106, 109
  verbal, 88, 91, 112
Agoraphobia, 13
Alcohol consumption
  marital quality and, 182, 195, 221, 223
  by peers, 108
  by siblings, 137
Alpha amylase reactivity, 40
Altruism, 131, 180, 193
Angry-oppositional behavior, 100
Antisocial behavior, 99, 141, 143, 157, 158
  child effects, 30, 190, 191
  corporal punishment and, 67
  friends behavior, 106
  MAOA activity and, 40, 255
  via parental negative feelings, 29, 67
  physical maltreatment and, 29–30
  sibling socialization effect, 91
Anxious attachment, 178
Arginine vasopressin receptor 1A (*AVPR1A*), 26
Assortative mating, 4, 7, 140, 210, 217, 219, 220, 228
Attachment
  anxious, 178
  avoidant, 178
  disorganization, 13, 17, 19, 21, 23, 36, 39–41, 245
  distress, 245
  and environmental causation, 15–17
  infant. *See* Infant attachment
  and personality disorders, 178
  relationships, during adulthood, 177–179
  representation, 20, 39
  romantic. *See* Romantic attachment
  security. *See* Attachment security

© Springer Science+Business Media New York 2015
B.N. Horwitz and J.M. Neiderhiser (eds.), *Gene-Environment Interplay in Interpersonal Relationships across the Lifespan*, Advances in Behavior Genetics 3, DOI 10.1007/978-1-4939-2923-8

Attachment Q-Sort (AQS), 18, 20
Attachment security, 18, 19, 20, 22, 23, 28,
    36, 252
    child-based research, 188
    and reduction in stress, 207
Attachment theory, 15, 207–212
Autism, 180
Avoidant attachment, 178
AVPR1A polymorphism, 180–181

B

Behavioral inhibition, 23, 100
Behavioral withdrawal, 100
Bereavement, 7, 217–219, 223, 225–226, 229
Big Five personality traits, 132, 136, 172, 178,
    250
Birth control, 205
Birth weight discordance, 30
Bivariate genetic analysis, 65–67, 68–69
Bonding
    offspring bonding, 210
    pair bonding, 176, 177, 181, 207–208, 211,
        214

C

Caregiver sensitivity and attachment, 15–16,
    28
Causal Contingent Common (CCC) model,
    177
Child-based designs, 188–193
Child effects, 14, 16, 20, 23, 24, 25, 38, 42,
    251
    antisocial behavior, 30
    parental negativity, 32
    role in parenting, 27, 29, 67
    in theory, 135, 191
Childhood
    gene-environment transactions in, 97–121
    interpersonal relationships in, 4–6
    middle, 5, 111, 144
    parenting in, 56–77
Children, 3. *See also* Childhood
    marriage, learning about, 176–177
    parenting. *See* Young children, parenting
    popular peer affiliation in, 155–156
Chromosome system, 247
Cognition/cognitive change, marriage and,
    224–225
Cohabitation, 183, 212, 214, 216
Colorado Adoption Project (CAP), 63

Communication
    affective, 36
    parent-child, instructive vs. informal, 33
COMT, 36–38, 182, 41–42
Conduct or behavior problems, 33, 35
    parental negativity and, 26, 29–32
Corporal punishment, 30, 32
Cortisol
    reactivity, 34
    stress response, 37
Co-workers, 171–196
Cyber-bullying, 100

D

DAT1, 38–39, 140, 179
Delinquent behavior, 105, 132
    alcohol consumption, 108, 137, 182, 195,
        221, 223
    childhood, 93
    nonviolent, 99
    violent, 99
Depression, 8, 13, 25, 98, 100, 103, 114, 115,
    117, 118, 132
    bereavement and, 226, 229
    major, 180, 222, 226
    maternal, 16, 30, 35, 58, 59
    social support and, 218–219
Developmental and sex differences, 100–101
Deviant peer affiliation, 98–99, 100, 101
    gene × environment interactions, 107–109
    genotype-environment correlation,
        104–107
Diathesis-stress model of psychopathology,
    33, 34, 98, 103, 116, 175
Differential Susceptibility Hypothesis, 120,
    175
Divorce, 6, 85, 132, 140, 142, 173, 231
    behavior genetics studies, 205, 212–215
    environmental influences, 176–177
    multiple divorces, 229
    risk factors of, 216
    threat of, 181, 183
Dizygotic (DZ) twins, 3, 17, 18, 19, 28, 57,
    84, 85, 101, 104, 106, 108, 119, 176
Dopamine receptor D2 gene (DRD2), 38, 40,
    140, 245
Dopamine receptor D4 gene (DRD4), 21–23,
    36–38, 140, 179, 182
Duke Dementia Study, 177
Dyadic Adjustment Scale (DAS), 179

E

Early parenting
    adoptions studies of, 25, 31, 34–35
    genotype-environment correlation in,
        27–42
    linkage studies of, 25–27, 35–36
    multi-causal model of, 27
    as pure environmental factor, 23–24
    step-family studies of, 31–32
    twin studies of, 24–25, 27–31, 33–34
Electrodermal reactivity, 40
Emerging adulthood, 5, 6, 153, 156, 158, 159,
        163, 165, 166
Emotional arousal, 21, 23, 38, 40
Emotional resilience, 37, 42
Environmental influences
    on adolescent family relationships,
        131–145
    on family systems, 241–262
    on late adulthood, 203–231
    on parenting in childhood, 57–77
    on parenting young children, 13–46
    on popularity, 151–167
    on problematic peer relationships, 97–121
    shared. *See* Shared environmental
        influences
    on sibling relationships, 83–93
Environmental Risk (E-Risk) Longitudinal
        Twin Study, 112
Erikson's theory of lifespan development,
        206–207
Experiences in Close Relationships Inventory
        (ECRI), 178
Extended Children of Twins (ECOT) model,
        192
Extended transmission disequilibrium tests
        (ETDT), 22
Externalizing problems, 4, 8, 13, 29, 31, 99,
        115, 117, 132
    frustrating events, 35
    genetic and environmental contributions
        to, 66, 67
    harsh parenting, 33
    linkage studies, 25, 36
    molecular genetic studies, 36

F

Familial adversity, 34
Family Environment Scale (FES), 142
Family processes, 31, 141, 143, 250
Family relationships research, in era of
        genomics, 143–144

Family systems, 141–143, 241–262
    central features of, 242–243
    gene expression, mechanisms of, 246–247
    genetic processes, role of, 245
    genotypic differences among individuals,
        245–246
    memorial functions of, 254–260
    parent-child relationships, conservation of,
        243–244
    relationships across time, stability of, 243
    stability of, 248–253
    theory of, 1, 2
FKBP5, 36, 41
Fraternal twins. *See* Dizygotic (DZ) twins
Friendships, 92, 93, 98, 99, 114, 134, 139,
        172, 226, 229
    abusive, 118
    nomination, 105, 106
    rupture of, 114

G

Gene-environment correlation (rGE), 102.
        See also Genotype-environment
        correlation (*r*GE)
    linking deviant peer affiliation
        and development, 104–107
    peer victimization, 109–115
Gene × environment interactions (G×E), 2,
        3, 7–8, 16
    contextual suppression, 103–104
    contextual trigger, 103
    deviant peer affiliation, 107–109
    future directions of, 44–46
    peer relationships, 103–104
    peer victimization, 115–118
Gene-environment interplay
    from behavioral genetic studies, 104–109
    mechanisms of, 101–104
    in parenting young children, 13–46
    in peer relationships, 98
    peer victimization and development,
        109–118
Gene-environment transactions in, 97–98
    deviant peer affiliation, 98–99
    gene-environment interaction, 107–109,
        115–118. *See also* Gene × environment
        interactions (G×E)
    gene-environment interplay. *See* Gene-
        environment interplay
    peer victimization, 99–100
    problematic peer relations, 100–101
Gene expression, mechanisms of, 246–247

chromosome system, 247
telomere biology, 247
Genetic influences, on popular peer affiliation,
    155–156
traits/behaviors, types of, 157–159
Genetics
    molecular, 246
    quantitative, 246
Genome wide association studies (GWAS),
    144
Genotype-environment correlation (rGE), 2,
    3–7
    active, 4–6, 82, 102, 134, 168–169
    adulthood, interpersonal relationships in,
        6–7
    childhood, interpersonal relationships in,
        4–6
    deviant peer affiliation, 104–107
    in early parenting, 27–42
    evocative, 4–6, 8, 10, 27, 31, 82, 102, 109,
        111–112, 134, 135, 152
    passive, 4–6, 10, 27, 82, 102–103, 106,
        109, 134, 135
    peer relationships, 102–103
    peer victimization, 109–115
Genotypic differences among individuals,
    245–246
Glucocorticoid receptor (GR), 40
Gossiping, 109. See also Rumor spreading

H
Harshness, 27, 29, 35, 39
Health, marriage and, 220–222
Home Observation for Measurement of the
    Environment (HOME), 25
Hormonal influences, on aging
    and relationships, 230–231
Hostile reactive behavior, 24
5-HTT, 21, 25, 26, 39–40, 142, 179
5-HTTLPR, 22, 25, 36, 39, 182
Hyperreactivity, 26, 30, 113

I
Identical twins. See Monozygotic (MZ) twins
Impulsive behavior, 40, 113
In vitro fertilization, 205
Infant
    attachment. See Infant attachment
    behavior, 33, 35, 39
    disorganization, 42
    temperament, 42

Infant attachment
    adoption studies of, 20–21
    linkage studies of, 21–23
    twin studies of, 17–20
Informant, 62–63
Insensitive care, 13
Internalizing problems, 10, 13, 100, 115, 117,
    118, 120, 132, 152, 182, 191
Interpersonal relationships
    in adulthood, 6–7
    in childhood, 4–6
    gene function and, 8–9
    shared environmental influences on, 3, 5, 6
Intimate relationships, benefits of, 219–225
Intimidation, 157
Intrusion, 13

K
Kinship relationships, 216–217

L
Lifespan, 1–3, 7, 9, 157, 172, 173, 175, 179,
    194, 217
Lifespan development, 171, 206–207
Linkage studies
    of early parenting, 25–27, 35–36
    of infant attachment, 21–23
Louisville Twin Study Procedure (LTS), 17

M
Major depression (MD), 180, 222, 226
Maltreatment, 36, 37, 40
    physical, 30–32, 94
    sexual, 94
MAOA gene, 40, 248
Marital Adjustment Test, 179
Marital instability, 6
Marital quality, 1, 6, 179–183, 214–215
Marital relationships, 1, 139–141
Marital status, 176–177
Marriage, 175–176. See also Cohabitation;
        Pair-bonding
    and alcohol consumption, 223
    and cognition/cognitive change, 224–225
    and health, 220–222
    later-life, 212–215
    as legal/contractual relationship, 205
    and mental health, 222–223
    and other activities, 223–224
    remarriage, 6

same-sex, 212
self-repairing, 229
and smoking, 223
and socioeconomic status, 225
and wellbeing, 222–223
Mate diversification, 177, 211
Maternal depression, 16, 30, 35, 58, 59
Maternal mental health, 57
Maternal sensitivity, 16, 57, 253
Menopause, 215
Mental health, marriage and, 222–223
Middle adulthood, 171, 172, 206
Middle childhood, 5, 111, 144
Mineralocorticoid receptor (MR), 40–41
Minnesota Twin Registry, 176, 181
Molecular genetics, 246
Monozygotic (MZ) twins, 3, 17–18, 19, 28,
      29, 57, 59, 84, 85, 101, 104, 106, 108,
      119, 176
Multi-causal model of early parenting, 27
Multidimensional Personality Questionnaire,
      181, 216
Multidimensional Personality Questionnaire—
      Brief Form (MPQ-BF), 162

N
National Academy of Sciences-National
      Research Council (NAS-NRC), 176,
      182
National Longitudinal Study of Adolescent
      Health (ADDHealth), 105, 108
NEAD project, 191, 192
Negative affect, 152, 162, 225, 226, 229
Negative emotionality, 35, 39, 40, 64, 114,
      138, 157
Negative parenting, 26, 29, 32, 38, 39, 43,
      135, 185–186
Neonatal neurobehavioral organization, 41
Nonshared Environment and Adolescent
      Development (NEAD), 137
Novelty seeking, 21, 38
Nursing Child Assessment Teaching Scale
      (NCATS), 19

O
Older adults, interpersonal relationships in
      attachment theory, 207–209
      divorce, risk factors of, 216
      Erikson's theory of psychosocial-develop-
            ment, 206–207
      evolutionary and behavior genetic
            approaches to, 210–211

future directions of, 228–231
intimate relationships, benefits of, 219–225
kinship relationships, 216–217
loss of interpersonal relationships,
      225–226
marriage, 212–215
social convoy theory, 209–210
social support, 218–219
Overprotection, 13, 24
Over-responsiveness, 13
Oxytocin receptor (OXTR) gene, 22, 25–26,
      181

P
Pair-bonding, 176–177, 181, 207–208, 210,
      211, 214, 215, 217, 231. See also
      Marriage
Parent(s/ing)
      child effects on. See Child effects
      control, 26, 56
      demandingness, 56
      determinants of, 183–184
      dimensions, 24
      divorce, 85
      early. See Early parenting
      harshness, 27, 29, 35, 39
      hostility, 31
      maladaptive, 26, 27, 39
      negativity, 26, 29, 32, 38, 39, 43, 135,
            185–186
      over-reactive, 35
      positivity, 29, 36, 135, 185
      quality, 57
      responsiveness, 56
      self-reported, 24–25, 26, 40
      sensitive, 13, 36–38, 41
      stress, 36
      styles, 25, 185, 187
      theories of, 56–58
      warmth. See Warmth
      young children, gene-environment
            interplay in, 13–46. See also Young
            children, parenting
Parent-adolescent relationship, 133–136
Parent-based designs, 184–188
Parent-child relationships
      conservation of, 243–244
      environmental mediation of, 17
      genetic environmental contributions to,
            188–193
Parental Attitudes Toward Childrearing
      Questionnaire (PATC), 24

Parenting. *See* Early parenting; Young
    children, parenting
Passionate love, 231
Peer acceptance/rejection, 154
Peer affiliation
    deviant. *See* Deviant peer affiliation
    popular, genetic influences on, 155–156
Peer groups, 5, 93, 105, 107, 109, 110,
    112–116, 153, 155, 158–160, 165–167
Peer influence model, 99
Peer problems, 30, 31, 36
Peer relationships, problematic, 97–121
    developmental impact of, 100–101
    deviant peer affiliation. *See* Deviant peer
        affiliation
Peer victimization, 99–100
    gene × environment interactions, 115–118
    genotype-environment correlation,
        109–115
Peers
    antisocial peers, 99, 108
    deviant peers, 8, 98, 99, 100, 101, 104,
        105, 118, 120
    GE correlation, 134
    importance of, 97–98
    in middle childhood, 5
Perceived support, 218
Personality, 248, 249
    disorders, attachment and, 178
    traits, 132, 136, 172, 178, 250
Physical aggression, 88, 91, 106, 108, 109,
    112, 114, 157
Popularity, 151–167
    biological underpinnings of, 153–154
    current issues in, 154–155
    developmentally sensitive perspective on,
        151–152
    in females, 159–166
    genetic influences on peer affiliation in
        children, 155–156
    genetically-influenced traits/behaviors,
        types of, 157–159
    genetically informed perspective on,
        151–152
    overarching role of development, 153
Positive affect, 18, 162, 226
Positive parenting, 29, 36, 135, 185
Procreation, 206
Propensity to marry, 6
Prosocial behavior, 29, 66, 91
    parental negativity and, 29, 31–32

Q
Quantitative genetics, 246
Quebec Newborn Twin Study (QNTS), 106,
    108, 110, 111, 113, 114

R
Relational aggression, 108–109, 114
Relationship Scales Questionnaire (RSQ),
    177–178
Remarriage, 6, 228
Rivalry, 83, 88, 91, 138
Romantic attachment, 177, 178, 215
Rumor spreading, 99. *See also* Gossiping

S
Same-sex marriage, 212
Second hit, 244
Self-efficacy, 24
Self-repair relationships, 229
Sensation-seeking, temperamental, 36
Sensitive responsiveness, 13
Sexual intercourse, 180
Sexual intimacy, 206, 215
Shared environmental influences, 3, 57
    on adolescent family relationships,
        136–139, 141, 143, 156
    on adulthood, 6–7
    on attachment relationships, 178
    on childhood, 4–6
    on early parenting, 26, 28–32, 34, 43, 45
    on family stability, 250
    gene function and, 8–9
    on marital quality, 179
    on parenting in childhood, 58–68
    on problematic peer relationships, 104,
        106, 110, 119
    on sibling relationships, 83–93
Siblings, 5
    adolescent family relationships, 137–139
    future directions of, 92–93
    negativity in, 83–88
    as phenotype, 84–85
    positivity in, 88–91
    as shared environment source, 83–93
    socialization effects in development, 91–92
Single child policy, 137
Single nucleotide polymorphisms (SNPs), 144
Smoking, marriage and, 223
Social aggression, 106, 109

Social convoy theory, 209–210
Social exclusion, 99, 106, 108
Social homogamy model, 101
Social phobia, 13
Social Relations Model (SRM), 138, 139, 161
Social support, 218–219, 230
Socialization effects, 91–92
Socio-economic status (SES), 16, 30, 97
    marriage and, 225
Statistical power, 44
Step-family studies, of early parenting, 31–32
Strange Situation (SS), 15, 17, 18, 20–22, 41, 42
Stress hyperreactivity, 26, 30
Substance abuse disorders, 38
Suicide ideation, 100
Swedish Adoption Twin Study of Aging (SATSA), 212, 218

T
Telomere
    biology, 247
    length, 8–9
Temperament, 248, 249
Testosterone, 230–231
Twin studies, 3
    of early parenting, 24–25, 27–31, 33–34
    of infant attachment, 17–20
Twin Study of Behavioral and Emotional Development in Children (TBED-C), 155
Twin/Offspring Study (TOSS), 181, 187, 192
Twins, 3, 6
    co-twins, 176, 184, 186
    fraternal twins, 19, 28, 58, 105, 133. *See also* Dizygotic (DZ) twins
    identical twins, 58, 105, 135. *See also* Monozygotic (MZ) twins

U
Univariate genetic analysis, 67–68

V
Valence, 64–65
Vasopressin, 180–181
Verbal aggression, 88, 91, 112
Victimization, peer, 99–100
Vietnam Era Twin Registry (VETR), 177
Vietnam Era Twin Study of Aging (VETSA), 213, 214
Virginia Twin Registry, 180

W
Warmth
    adolescents, 135
    emotional, 135
    marital, 180
    parental, 29, 30, 134, 184, 185, 186, 189, 192, 243
    personal, 134
    sibling, 88, 92, 132, 138
Wellbeing, marriage and, 222–223
Work relationships, 193–194

Y
Young children, parenting, 13–15
    attachment theory, 15
    early parenting. *See* Early parenting
    environmental causation, 15–17
    infant attachment, 21–23
    parent-infant attachment, 17–20